Chemical Process: Design and Integration

Chemical Process: Design and Integration

Contributors

Shu-Kun Lin et al.

www.aurisreference.com

Chemical Process: Design and Integration

Contributors: Shu-Kun Lin et al.

Published by Auris Reference Limited

www.aurisreference.com

United Kingdom

Chemical Process: Design and Integration

ISBN: 978-1-78154-892-9

British Library Cataloguing in Publication Data
A CIP record for this book is available from the British Library

Printed in the United Kingdom

Exclusively distributed by CBS Publishers & Distributors Pvt. Ltd.

Sales & Distribution Rights only for India, Pakistan, Bangladesh, Sri Lanka, Nepal and Bhutan. This book is not to be sold outside these territories.

Contents

List of Abbreviations

AMV	Advanced Mean Value
APC	Advanced Process Control
BN	Bacterial Cellulose Nanocrystals
BS	Butadiene Sulfone
BTL	Biomass to Liquids
CFD	Computational Fluid Dynamics
CN	Cellulose Nanofibrils
CTL	Coal to Liquids
DES	Deep Eutectic Solvents
DMC	Dynamic Matrix Controller
DMSO	Dimethylsulfoxide
DP	Degree of Polymerization
EDM	Electrical Discharge Machine
GA	Genetic Algorithms
GER	Gross Energy Requirement
GTL	Gas to Liquids
HLB	Hydrophilic/Lipophilic Balance
HPLC	High Performance Liquid Chromatography
IPC	Individual Probabilistic Constraint
LUMO	Lowest Unoccupied Molecular Orbital
MA	Maleic Acid
MADS	Mesh Adaptive Direct Search
MAP	Modified Atmosphere Packaging
MCS	Monte Carlo Simulation
MPC	Model Predictive Controlled
MPP	Most Probable Point
MV	Mean Value
NIR	Near Infrared
OTR	Oxygen Transmission Rates
OTS	Operator Training Systems
PLA	Poly Lactic Acid
SA	Succinic Acid
SPC	Statistical Process Control
UDF	User Defined Function
VPO	Vanadium Phosphorus-Oxide
WGS	Water Gas-Shift
WVTR	Water Vapor Transmission Rate

List of Contributors

Shu-Kun Lin
Molecular Diversity Preservation International (MDPI), Sangergasse 25, Basel CH-4054 Switzerland.

Matthew J. Opgenorth
Department of Mechanical and Materials Engineering, University of Denver, Denver, USA

William E. McDermott
Applied Research and Technology Institute, University of Denver, Denver, USA

Peter Laz
Department of Mechanical and Materials Engineering, University of Denver, Denver, USA

Corinne S. Lengsfeld
Department of Mechanical and Materials Engineering, University of Denver, Denver, USA

Martin Zimmer
Department PV Production Technology and Quality Assurance, Fraunhofer Institute for Solar Energy Systems, Freiburg, Germany

Katrin Krieg
Department PV Production Technology and Quality Assurance, Fraunhofer Institute for Solar Energy Systems, Freiburg, Germany

Jochen Rentsch
Department PV Production Technology and Quality Assurance, Fraunhofer Institute for Solar Energy Systems, Freiburg, Germany

László Dobos
Department of Process Engineering , University of Pannonia, Egyetem Street 10, 8200 Veszprém, Hungary

András Király
Department of Process Engineering , University of Pannonia, Egyetem Street 10, 8200 Veszprém, Hungary

János Abonyi
Department of Process Engineering , University of Pannonia, Egyetem Street 10, 8200 Veszprém, Hungary

Mohammad Gias Uddin
Department of Textile Engineering, Ahsanullah University of Science and Technology, Dhaka 1208, Bangladesh

Robert Wojcieszak
CNRS UMR 8181, Unité de Catalyse et Chimie du Solide (UCCS), Université Lille 1 Sciences et Technologies, 59655 Villeneuve d'Ascq Cedex, France

Francesco Santarelli
CNRS UMR 8181, Unité de Catalyse et Chimie du Solide (UCCS), Université Lille 1 Sciences et Technologies, 59655 Villeneuve d'Ascq Cedex, France
Ecole Centrale de Lille, ECLille, Cité Scientifique, 59650 Villeneuve d'Ascq, France
Dipartimento di Chimica Industriale "Toso Montanari", Università di Bologna, Viale Risorgimento 4, 40136 Bologna, Italy

Sébastien Paul
CNRS UMR 8181, Unité de Catalyse et Chimie du Solide (UCCS), Université Lille 1 Sciences et Technologies, 59655 Villeneuve d'Ascq Cedex, France
Ecole Centrale de Lille, ECLille, Cité Scientifique, 59650 Villeneuve d'Ascq, France

Franck Dumeignil
CNRS UMR 8181, Unité de Catalyse et Chimie du Solide (UCCS), Université Lille 1 Sciences et Technologies, 59655 Villeneuve d'Ascq Cedex, France
Institut Universitaire de France, IUF, Maison des Universités, 103 Boulevard Saint-Michel, Paris 75005, France

Fabrizio Cavani
Dipartimento di Chimica Industriale "Toso Montanari", Università di Bologna, Viale Risorgimento 4, 40136 Bologna, Italy

Renato V Gonçalves
Institute of Chemistry, USP, Av. Professor Lineu Prestes, 748, São Paulo 05508-000, SP, Brazil.

Christoph R Müller
Institut für Technische und Makromolekulare Chemie (ITMC), RWTH Aachen University, Worringerweg 1, 52074 Aachen, Germany

Andreas Rosen
Institut für Technische und Makromolekulare Chemie (ITMC), RWTH Aachen University, Worringerweg 1, 52074 Aachen, Germany

Pablo Domínguez de María
Institut für Technische und Makromolekulare Chemie (ITMC), RWTH Aachen University, Worringerweg 1, 52074 Aachen, Germany
Sustainable Momentum, SL, Ap. Correos 3517, 35004 Las Palmas De Gran Canaria, Canary Islands, Spain

A Fachri
Chemical Engineering Department, University of Groningen, Nijenborgh 4, 9747 AG Groningen, The Netherlands
Faculty of Engineering, University of Jember, Kalimantan 37, Jember, Indonesia.

R M Abdilla
Chemical Engineering Department, University of Groningen, Nijenborgh 4, 9747 AG Groningen, The Netherlands

C B Rasrendra
Chemical Engineering Department, Faculty of Industrial Technology, Institut Teknologi Bandung, Ganesha 10, Bandung, Indonesia

H J Heeres
Chemical Engineering Department, University of Groningen, Nijenborgh 4, 9747 AG Groningen, The Netherlands

Joyeeta Mukherjee
Department of Chemistry, Indian Institute of Technology Delhi, Hauz Khas, New Delhi 110016, India

Munishwar Nath Gupta
Department of Biochemical Engineering and Biotechnology, Indian Institute of Technology Delhi, Hauz Khas, New Delhi 110016, India.

Yong Huang
School of Chemical and Biomolecular Engineering, Georgia Institute of Tech nology, Atlanta, GA 30332, USA
Specialty Separations Center, Georgia Institute of Technology, Atlanta, GA 30332, USA

Esteban E. Ureña Benavides
School of Chemical and Biomolecular Engineering, Georgia Institute of Tech nology, Atlanta, GA 30332, USA
Specialty Separations Center, Georgia Institute of Technology, Atlanta, GA 30332, USA

Afrah J. Boigny
School of Chemical and Biomolecular Engineering, Georgia Institute of Tech nology, Atlanta, GA 30332, USA

Zachary S. Campbell
School of Chemical and Biomolecular Engineering, Georgia Institute of Tech nology, Atlanta, GA 30332, USA

Fiaz S. Mohammed
School of Chemical and Biomolecular Engineering, Georgia Institute of Tech nology, Atlanta, GA 30332, USA
Specialty Separations Center, Georgia Institute of Technology, Atlanta, GA 30332, USA

Jason S. Fisk
The Dow Chemical Company, Midland, MI 48674, USA.

Bruce Holden
The Dow Chemical Company, Midland, MI 48674, USA.

Charles A. Eckert
School of Chemical and Biomolecular Engineering, Georgia Institute of Tech nology, Atlanta, GA 30332, USA
School of Chemistry and Biochemistry, Geor gia Institute of Technology, Atlanta, GA 30332, USA
Specialty Separations Center, Georgia Institute of Technology, Atlanta, GA 30332, USA

Pamela Pollet
School of Chemistry and Biochemistry, Geor gia Institute of Technology, Atlanta, GA 30332, USA
Specialty Separations Center, Georgia Institute of Technology, Atlanta, GA 30332, USA

Charles L. Liotta
School of Chemical and Biomolecular Engineering, Georgia Institute of Tech nology, Atlanta, GA 30332, USA
School of Chemistry and Biochemistry, Geor gia Institute of Technology, Atlanta, GA 30332, USA
Specialty Separations Center, Georgia Institute of Technology, Atlanta, GA 30332, USA

Felipe K Sutili
Biocatalysis and Organic Synthesis Group, Chemistry Institute, Federal University of Rio de Janeiro, Rio de Janeiro CEP 22941 909, Brazil

Halliny S Ruela
Biocatalysis and Organic Synthesis Group, Chemistry Institute, Federal University of Rio de Janeiro, Rio de Janeiro CEP 22941 909, Brazil

Daniel De O Nogueira
Biocatalysis and Organic Synthesis Group, Chemistry Institute, Federal University of Rio de Janeiro, Rio de Janeiro CEP 22941 909, Brazil
Faculdade de Farmácia, Federal University of Rio de Janeiro, Rio de Janeiro CEP22941909, Brazil.

Ivana CR Leal
Faculdade de Farmácia, Federal University of Rio de Janeiro, Rio de Janeiro CEP22941909, Brazil.

Leandro SM Miranda
Biocatalysis and Organic Synthesis Group, Chemistry Institute, Federal University of Rio de Janeiro, Rio de Janeiro CEP 22941 909, Brazil

Rodrigo OMA De Souza
Biocatalysis and Organic Synthesis Group, Chemistry Institute, Federal University of Rio de Janeiro, Rio de Janeiro CEP 22941 909, Brazil

Sandeep S Nair
School of Chemistry and Biochemistry, Georgia Institute of Technology, 500 10th Street, N.W, Atlanta, GA 30332, USA

JY Zhu
USDA Forest Service, Forest Products Laboratory, One Gifford Pinchot Drive, Madison, WI 53726, USA

Yulin Deng
School of Chemical and Biomolecular Engineering, Georgia Institute of Technology 500 10th Street, N.W, Atlanta, GA 30332, USA

Arthur J Ragauskas
Department of Chemical and Biomolecular Engineering, Department of Forestry, Wildlife, and Fisheries, Center for Renewable Carbon, University of Tennessee, Knoxville, TN 37996-2200, USA.

Juan Carlos Serrano-Ruiz
Departamento de Química Orgánica, Universidad de Córdoba, Campus de Excelencia Internacional Agroalimentario CeiA3, Edificio Marie Curie (C-3), Ctra Nnal IV-A, Km 396, Córdoba E-14014, Spain

Rafael Luque
Departamento de Química Orgánica, Universidad de Córdoba, Campus de Excelencia Internacional Agroalimentario CeiA3, Edificio Marie Curie (C-3), Ctra Nnal IV-A, Km 396, Córdoba E-14014, Spain

Juan Manual Campelo
Departamento de Química Orgánica, Universidad de Córdoba, Campus de Excelencia Internacional Agroalimentario CeiA3, Edificio Marie Curie (C-3), Ctra Nnal IV-A, Km 396, Córdoba E-14014, Spain

Antonio A. Romero
Departamento de Química Orgánica, Universidad de Córdoba, Campus de Excelencia Internacional Agroalimentario CeiA3, Edificio Marie Curie (C-3), Ctra Nnal IV-A, Km 396, Córdoba E-14014, Spain

Preface

In chemical engineering, process design is the design of processes for desired physical and/or chemical transformation of materials. Process design is central to chemical engineering, and it can be considered to be the summit of that field, bringing together all of the field's components. Process design can be the design of new facilities or it can be the modification or expansion of existing facilities. The design starts at a conceptual level and ultimately ends in the form of fabrication and construction plans. First chapter focuses on the nature of the chemical process. The objective of the second chapter is to develop and implement a design analysis of a fluid mixing nozzle using a coupled optimization and probabilistic approach. Third chapter gives an overview for a successful analytical method of the main components of an alkaline texturing bath by titration, HPLC, surface tension and NIR spectrometry. The aim of fourth chapter is to develop an optimization framework designed to determine optimal operating regimes of chemical processes by taking process constraints, desired maximum number (frequency) of constraint violations, and process uncertainties into consideration. The aim of fifth chapter is to evaluate the color levelness quality of fabric dyed with vegetable dyes. Sixth chapter presents the current state of the art on maleic acid synthesis from biomass-derived chemicals over homogeneous or heterogeneous catalysts. In seventh chapter, it is demonstrated that organocatalysts can be tuned to be used in different DES. Ninth chapter presents an overview of the use of biocatalysis as applied to the conversion of biomass into value added products. Biocatalysis is a more sustainable alternative to chemical catalysis. Eighth chapter reports an experimental study on the uncatalysed, thermal conversion of inulin to HMF in aqueous solutions in a batch set-up. In tenth chapter, comparisons are made with the solvent DMSO and an analogous sulfolene solvent—piperylene sulfone. In addition, recycling protocols for butadiene sulfone and piperylene sulfone are also presented. Eleventh chapter report the enzymatic esterification of steric hindered fructose derivative with free fatty acids derived from palm oil refining process (RePO) under continuous flow conditions at concentrations up to 0.5 M, increasing the productivity up to 100 mg. min^{-1}.g immob. enzyme^{-1}. Twelfth chapter aims to summarize the recent developments in various barrier films based on nanocellulose with special focus on oxygen and water vapor barrier properties. Thirteenth chapter is aimed to provide an overview of key continuous flow processes developed to date dealing with a series of transformations of platform chemicals including alcohols, furanics, organic acids and polyols using a wide range of heterogeneous catalysts based on supported metals, solid acids and bifunctional (metal + acidic) materials. Last chapter highlights on the microwave assisted chemical pretreatment of miscanthus under different temperature regimes.

Chapter 1

THE NATURE OF THE CHEMICAL PROCESS. 1. SYMMETRY EVOLUTION –REVISED INFORMATION THEORY SIMILARITY PRINCIPLE UGLY SYMMETRY

Shu-Kun Lin

Molecular Diversity Preservation International (MDPI), Sangergasse 25, Basel CH-4054 Switzerland.

ABSTRACT

Symmetry is a measure of indistinguishability. Similarity is a continuous measure of imperfect symmetry. Lewis' remark that "gain of entropy means loss of information" defines the relationship of entropy and information. Three laws of information theory have been proposed. Labeling by introducing nonsymmetry and formatting by introducing symmetry are defined. The function L ($L=\ln w$, w is the number of microstates, or the sum of entropy and information, $L=S+I$) of the universe is a constant (the first law of information theory). The entropy S of the universe tends toward a maximum (the second law law of information theory). For a perfect symmetric static structure, the information is zero and the static entropy is the maximum (the third law law of information theory). Based on the Gibbs inequality and the second law of the revised information theory we have proved the similarity principle (a continuous higher similarity–higher entropy relation after the rejection of the Gibbs paradox) and proved the Curie-Rosen symmetry principle (a higher symmetry–higher stability relation) as a special case of the similarity principle. The principles of information minimization and potential energy minimization are compared. Entropy is the degree of symmetry and information is the degree of nonsymmetry. There are two kinds of symmetries: dynamic and static symmetries. Any kind of symmetry will define an entropy and, corresponding to the dynamic and static symmetries, there are static entropy and dynamic entropy. Entropy in thermodynamics is a special kind of dynamic entropy. Any spontaneous process will evolve towards the highest possible symmetry, either dynamic or static or both. Therefore the revised information theory can be applied to characterizing all kinds of structural stability and process spontaneity. Some examples in chemical physics have been given. Spontaneous processes of all kinds of molecular interaction, phase separation and phase transition, including symmetry breaking and the densest molecular packing and crystallization, are all driven by information minimization or symmetry maximization. The evolution of the universe in general and evolution of life in particular can be quantitatively considered as a series of symmetry breaking processes. The two empirical rules − similarity rule and complementarity rule − have been given a theoretical foundation. All kinds of periodicity in space and time are symmetries and contribute to the stability. Symmetry is beautiful because it renders stability. However, symmetry is in principle ugly because it is associated with information loss.

INTRODUCTION

Symmetry has been mainly regarded as a mathematical attribute [1,2,3]. The Curie-Rosen symmetry principle [2]] is a higher symmetry−higher stability relation that has been seldom, if ever, accepted for consideration of structural stability and process spontaneity (or process irreversibility). Most people accept the higher symmetry−lower entropy relation because entropy is a degree of disorder and symmetry has been erroneously regarded as order [4]. To prove the symmetry principle, it is necessary to revise information theory where the second law of thermodynamics is a special case of the second law of information theory.

Many authors realized that, to investigate the processes involving molecular self-organization and molecular recognition in chemistry and molecular biology and to make a breakthrough in solving the outstanding problems in physics involving critical phenomena and spontaneity of symmetry breaking process [5], it is necessary to consider information and its conversion, in addition to material, energy and their conversions [6]. It is also of direct significance to substantially modify information theory, where three laws of information theory will be given and the similarity principle (entropy increases monotonically with the similarity of the concerned property among the components (figure 1) [7]) will be proved.

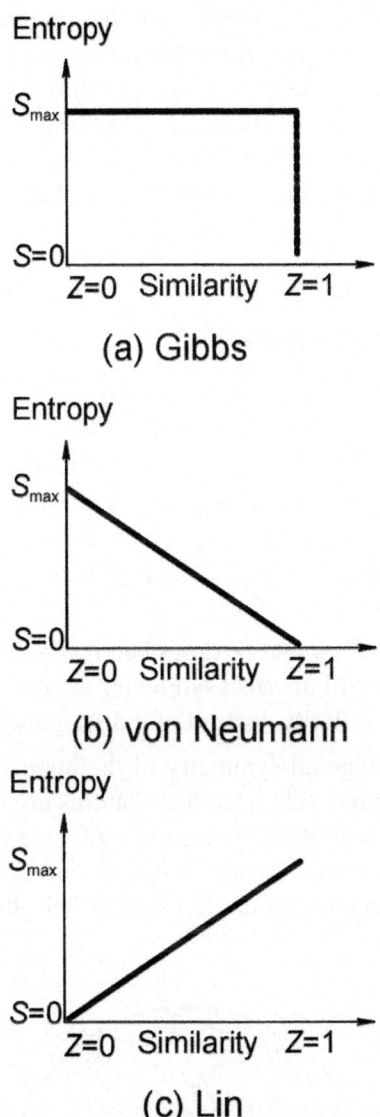

Figure 1. (a) Correlation of entropy (ordinate) of mixing with similarity (abscissa) according to conventional statistical physics, where entropy of mixing suddenly becomes zero if the components are indistinguishable according to the Gibbs paradox [21]. Entropy *decreases discontinuously*. figure 1a expresses Gibbs paradox statement of "same or not the same" relation. (b) von Neumann revised the Gibbs paradox statement and argued that the entropy of mixing *decreases continuously* with the increase in the property similarity of the individual components [21a,21b,21d,21j]. (c) Entropy *increases continuously* according to the present author [7] (not necessarily a straight line because similarity can be defined in different ways).

Thus, several concepts and their quantitative relation are set up: higher symmetry implies higher similarity, higher entropy, less information and less diversity, while they are all related to higher stability. Finally, we conclude that the definition of symmetry as order [4] or as "beauty" (see: p1 of ref. 8, also ref. 3) is misleading in science. Symmetry is in principle ugly. It may be related to the perception of beauty only because it contributes to stability.

DEFINITIONS

Symmetry and Nonsymmetry

Symmetry as a Greek word means *same measure* [1]. In other words, it is a measure of indistinguishability. A number w_s can be used to denote the measure of the indistinguishability and can be called the symmetry number [7d]. In some cases it is the number of invariant transformations. Clausius proposed to name the quantity S the entropy of the body, from the Greek word η τροπη, a *transformation* [9]. This may suggest that symmetry and entropy are closely related to each other [7d].

Only perfect symmetry can be described mathematically by group theory [2]. It is a special case of imperfect symmetry or continuous symmetry, which can be measured by similarity, instead of indistinguishability [7d].

Example 1. The bilateral symmetry of the human body is imperfect. We have our heart on the left side. Detailed features are only similar on the two sides. The two breasts on the left and the right side of women are normally somewhat different in size and shape. The beautiful woman Cindy (www.cindy.com) has a black birthmark on the left side of her otherwise very symmetric face.

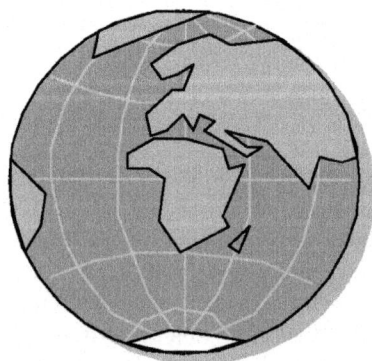

Figure 2. Our earth is not of perfect symmetry.

On the contrary, we define nonsymmetry as a measure of difference or distinguishability. Similarity can be defined as a continuous index of imperfect symmetry between the two limits: distinguishability (the lowest similarity) or nonsymmetry and indistinguishability (the highest similarity) or symmetry.

Entropy and Information

In statistical mechanics, entropy is a function of the distribution of the energy levels. The entropy concept in information theory is much more general. However, information theory, which has been used mainly in communication and computer science [10,11], is about the *process* of the communication channel. We will revise it to become a theory regarding the *structure* of a considered system. Process will be considered as a series of structures. Instead of defining *ensemble* (See:section 2.1, ref. [10]), we directly define that *the macroscopic structure* with regard to a certain kind of property X of a considered system as a triple (x, M_X, P_X) where the outcome x is a *microscopic structure* (a microstate), which takes on one of a set of possible microstates, $M_X = \{m_1, m_2, ..., m_i, ..., m_w\}$, having probabilities $P_X = \{p_1, p_2, ..., p_i, ..., p_w\}$ with probability $P(x=m_i)=p_i$, $p_i \geq 0$ and $\sum_{mi \in MX} P(x=m_i)=1$. The set M is mnemonic for microscopic structures (or microstates). For a dynamic system, the structure is a mixture among the w microstates (see examples illustrated in figure 3 and figure 4). In thermodynamics and statistical mechanics, the considered property is exclusively the energy level. In information theory, the property can be spin orientation itself (figure 3).

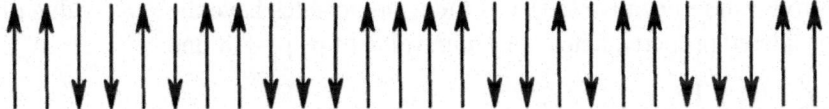

(a) A typical microstate as a freeze-frame of an array of spins undergoing up-and-down tumbling.

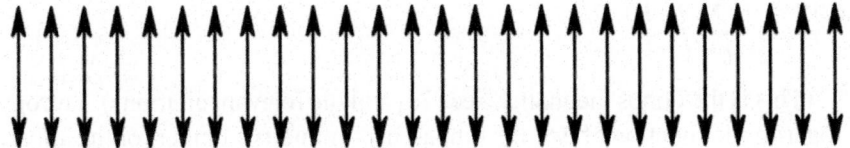

(b) The mixture of all w microstates.

Figure 3. Dynamic motion of spin-up and spin-down binary system. (a) A microstate. (b) The mixing of all microstates defines a symmetric macroscopic state.

ABACBC

(a) A microstate.

(b) The mixing of all 729 microstates.

Figure 4. One of 729 (3N=36=729) microstates. (b) Schematic representation of information loss due to dynamic motion. The pictures at the six positions are the same, hence they give a symmetric macroscopic state. All the three letters appear at a position with the same probability. All the microstates also appear with the same probability.

Entropy S of the structure regarding the property X is given by the following familiar Gibbs-Shannon expression [11] (See also section 2.4, ref. [10]).

$$S = -\sum_{i=1}^{w} p_i \ln p_i$$

(1)

with the understanding that $0\ln0=0$ and $1\ln1=0$. Because $1 \geq p_i \geq 0$ and $\ln p_i < 0$, entropy isnon-negative ($S \geq 0$). If the w microstates have the *same* value of the considered property, hence the *same* value of p_i, $p_i = 1/w$ and

$$-\sum_{i=1}^{w} \frac{1}{w} \ln \frac{1}{w} = \ln w = L$$

(2)

The maximal entropy is denoted as L, because

$$\ln w \geq -\sum_{i=1}^{w} p_i \ln p_i$$

(3)

(This is the Gibbs inequality, see [7d] and the relevant citations). Entropy is a logarithmic function of wS, the (apparent) symmetry number (or the apparent number of microstates of indistinguishable property, or the apparent number of the *equivalent* microstate which are of the *same* value of the considered property [7d]):

$$S = \ln w_S$$

(4)

Equation 4 is the familiar Boltzmann entropy expression. Combination of equations 1 and 4 leads to

$$w_S = \exp\left(-\sum_{i=1}^{w} p_i \ln p_i\right)$$

(5)

Obviously

$w \geq w_S$ (6)

Example 2. It can be illustrated by the entropy of a binary system (figure 4.1 in [10]) as a function of p_i. Coin tossing is a typical example of a binary system. Racemization of S and D enantiomers of N molecules at high temperature, a ferromagnetic system (spin-up and spin-down) of N spins at high temperature (figure 3) are examples in chemical physics. The number of microstates is $w=2^N$. The maximum entropy is $\ln 2^N$ when $p_1=p_2=0.5$ regarding the outcome of an interchanging enantiomer or a tumbling spin, or $p_1=p_2=...=p_w=1/2^N$ for all the w microstates. A microstate is a possible sequence of N times of tossing or a possible picture of N enantiomers or N spins (figure 3).

Let us recall Lewis' remark that "gain of entropy means loss of information, nothing more ......" [12] and define the relationship of entropy and information. Lewis' remark also gives the hint that information can be converted to entropy. First, a new logarithmic function L can be defined as the sum of entropy S and information I:

L=S+I (7)

L is mnemonic for Logarithmic function or the "Largest possible value" of either the logarithmic function S (equation 2) or I:

L=lnw (8)

Then, entropy is expressed as information loss [12]

S=L−I (9)

or in certain cases when the absolute values are unknown,

ΔS=ΔL−ΔI (10)

for a change between two structures.

Example 3. For a 1.44 MB floppy disk, L=1.44 MB whether the disk is empty or occupied with a file of the size of 1.44 MB. Let us use any available compression method to reduce the size of the original file to 0.40 MB. Then, I=0.40 MB, S=1.04 MB and L=1.44 MB.

From equations 4 and 8, a logarithmic expression of information

$I=\ln w_I$ (11)

can be given, and

$$w_I = w/w_S \tag{12}$$

where w_I is called the nonsymmetry number − the number of microstates of distinguishable property whereas w_S in equation 4 called symmetry number (vide infra, section 6).

The logarithmic functions S, I and L are all nonnegative in value, dimensionless and they are all macroscopic properties.

From equation 9 and the fact that in practice information can be recorded only in a static structure, we may define static entropy [7d]. A macroscopic static structure is described by a set of w microstates which are the w possible rearrangements or w possible transformations (recall the operations in group theory treatment of static structure which can be either an individual molecule or an assemblage) [2,8]).

In the entropy and information expressions, the unit is called nat (natural logarithmic unit). For a binary system, the unit is bit. Normally there is a positive constant in these expressions (e.g., the Boltzmann constant k_B in thermodynamics). Here we put the constant as 1. In thermodynamics, we may denote the traditionally defined thermodynamics entropy as

$$S_T = k_B S \tag{13}$$

and

$$E = k_B TS + F \tag{14}$$

where E is the total energy and F the Helmholtz potential.

Labeling and Formatting

In this paper similarity and its two limits (distinguishability and indistinguishability) will be frequently considered. Therefore, similarity and its significance should be clearly defined. Entropy or information (symmetry or nonsymmetry, vide infra) can be defined regarding a property X. Suppose there are n kinds of property X, Y, Z, ..., etc. which are independent. In order to enumerate the number of possible microstates regarding X, particularly in the cases of symmetry study, where we may encounter the real problem of indistinguishability, labeling with some of the other n-1 kinds of property Y, Z, ..., etc. is necessary. Microscopically the components in the considered system may appear as several kinds of property X, Y, Z, ..., etc. If only one kind of property is detected by the instrument (e.g., a chemical sensor to detect the existence of certain gas molecule) or by a machine (e.g., a heat engine which detects the average kinetic energy of the gas molecules [6c]), the others can be used for labeling purposes to distinguish the otherwise indistinguishable individual components. We have the following postulate (or definition):

Definition: The distinguishability of the components (microstates, individual molecules or phases) due to the *labeling* does not influence the similarity or indistinguishability of the individuals or the existing distinguishability of the components (microstates, individual molecules or phases) regarding the considered property.

In other words, if a difference in a certain kind of property Y does not influence the indistinguishability of the components regarding the considered X, the introduced or the existing difference in Y is defined as labeling.

Example 4. For a coin tossing example (example 2), the precondition for p1=p2=0.5 is that the relevant property of the two sides of the coin are the same (indistinguishable, or symmetric). The two different figures on the two sides of the coin can be regarded as two labels. Similarly, for enantiomers and spin orientations if the energy levels are the same, the chirality of the two different enantiomers D and L or the two different orientations can be used as two labels.

Example 5. Suppose information is recorded by a set of individual items, for example, the symbols 0 and 1 in a binary system used in computer science. The color or font size or the position index 1234567 does not influence the indistinguishability of a string of seven zeros: 0000000 or 0000000 or 1234567. Therefore these properties can be used for labeling.

Example 6. DNA is the genetic substance due to the distinguishability of the four organic bases (two pyrimidines − cytosine and thymine, and two purines − adenine and guanine). This kind of distinguishability does not influence the symmetric backbone structure (the periodicity of the sugar-phosphate backbone) along the axis that can be detected by X-ray diffraction [13]. Therefore the different organic bases can be used for labeling.

Example 7. The information retrieval problem. The distinguishable ID number or phone number belonging to an individual person does not influence the similarity of characters of the individuals or the indistinguishability of family names or given names.

Example 8. For an *ideal* gas model in thermodynamics, all the monatomic ideal gases are the same (e.g., He and Ar). All the diatomic ideal gases are also the same. The difference (e.g., nitrogen N_2 and oxygen O_2 gases, or the ordinary water H_2O and deuterated water D_2O) can be regarded solely as labeling which does not influence the indistinguishability of the gases the heat engine experiences [6c].

Because the ideal gas model is very important, we summarize the conclusion made in the preceding example in the following theorem:

Theorem: To a heat engine, all the ideal gases are indistinguishable.

All ideal gases have the following state equation as detected in a heat engine:

$$PV=k_B NT \tag{16}$$

where P is the pressure and T the temperature. The definition of labeling is very important for a final resolution of the Gibbs paradox of entropy of mixing (see section 8.1.2).

Definition: The indistinguishability of the components (microstates, individual molecules or phases, etc.) due to *formatting* does not influence the distinguishability of the individuals or the existing indistinguishability of the components (microstates, individual molecules or phases) regarding the considered property.

Normally formatting produces periodicity in a static structure. In DNA (example 6), the periodicity of the sugar-phosphate backbone can be regarded as formatting. In figure 3, the spin array is formatted as a string with an equal interval. In figure 4, the letters are formatted as the same font size and as a string with an equal interval. If the system is formatted with some of the other kinds of property Y, Z, ..., etc. the amount of information recorded by using the distinguishability of the property X will not change regardless of whether we create symmetry regarding many other kinds of property. It will be soon made clear that formatting increases the stability of the system at the sacrifice of the information recording capacity. For a memory system like a hard disk in a computer, formatting is the necessary preparation for information recording to provide certain stability of the system.

THE THREE LAWS AND THE STABILITY CRITERIA

Parallel to the first and the second laws of thermodynamics, we have:

The first law of information theory: the logarithmic function L ($L=\ln w$, or the sum of entropy and information, $L=S+I$) of an isolated system remains unchanged.

The second law of information theory: Information I of an isolated system decreases to a minimum at equilibrium.

We prefer to use information *minimization* as the second law of information theory because there is a stability criterion of potential energy *minimization* in physics (see section 6). Another form of the second law is the same in form as that of the second law of thermodynamics: *for an isolated system, entropy S increases to a maximum at equilibrium.* For other systems (closed system or open system), we define (see: p. 623 of ref. [14])

universe = system + surroundings   (17)

and treat the universe formally as an isolated system (Actually we have always done this in thermodynamics following Clausius [9]). Then, these two laws are expressed as the following: *The function L of the universe is a constant. The entropy S of the universe tends toward a maximum.* Therefore, the second law of information theory can be used as the criteria of structural stability and process spontaneity (or process irreversibility) in all cases, whether they are isolated systems or not. If the entropy of system + surroundings increases from structure A to structure B, B is more stable than A. The higher the value ΔS for the final structure, the more spontaneous (or more irreversible) the process will be. For an isolated system the surroundings remain unchanged.

At this point we may recall the definition of equilibrium. Equilibrium means the state of indistinguishability (a symmetry, the highest similarity) [7], e.g., as shown in figure 5, the thermal equilibrium between parts A and B means that their temperatures are the same (See also the so-called zeroth law of thermodynamics [14]).

We have two reasons to revise information theory. Firstly, the entropy concept has been confusing in information theory [15] which can be illustrated by von Neumann's private communication with Shannon regarding the terms entropy and information (Shannon told the following story behind the choice of the term "entropy" in his information theory: "My greatest concern was what to call it. I thought of calling it 'uncertainty'. When I discussed it with John von Neumann, he had a better idea: 'You should call it entropy, for two reasons. In the first place your uncertainty function has been used in statistical mechanics under that name, so it already has a name. In the second place, and more important, no one knows what entropy really is, so in a debate you will always have the advantage' " [15a]). Many authors use the two concepts entropy and information interchangeably (see also ref. 7d and citations therein). Therefore a meaningful discussion on the conversion between the logarithmic functions S and I is impossible according to the old information theory. Secondly, to the present author's knowledge, none of several versions of information theory has been applied to physics to characterize structural stability. For example, Jaynes' information theory [16] should have been readily applicable to chemical physics but only a parameter similar to temperature is discussed to deduce that at the highest possible value of that temperature-like parameter, entropy is the maximum. Jaynes' information theory or the so-called maximal entropy principle has been useful for statistics and data reduction (See the papers presented at the annual conference MaxEnt). Brillouin's negentropy concept [17] has also never been used to characterize structural stability.

To be complete, let us add the third law here also:

The third law of information theory: For a perfect crystal (at zero absolute thermodynamic temperature), the information is zero and the static entropy is at the maximum.

The third law of information theory is completely different from the third law of thermodynamics although the third law of thermodynamics is still valid regarding the dynamic entropy calculation (However, the third law of thermodynamics is useless for assessing the stabilities of different static structures). A more general form of the third law of information theory is "for a perfect symmetric *static* structure, the information is zero and the *static* entropy is the maximum". The third law of information theory defines the static entropy and summarizes the conclusion we made on the relation of static symmetry, static entropy and the stability of the static structure [7d] (The static entropy is independent of thermodynamic temperature. Because it increases with the decrease in the total energy, a negative temperature can be defined corresponding to the static entropy [7a]. The revised information theory suggests that a negative temperature also can be defined for a dynamic system, e.g., electronic motion in atoms and molecules [7a]).

The second law of thermodynamics might be regarded as a special case of the second law of information theory because thermodynamics treats only the energy levels as the considered property X and only the dynamic aspects. In thermodynamics, the symmetry is the energy degeneracy [18,19]. We can consider spin orientations and molecular orientations or chirality as relevant properties which can be used to define a static entropy. What kinds of property besides the energy level and their similarities are relevant to structural stability? Are they the so-called observables [18] in physics? Should these properties be related to energy (i.e., the entropy is related to energy by temperature-like intensive parameters as differential equations where Jaynes' maximal entropy principle might be useful) so that a temperature (either positive or negative [7a]) can be defined? These problems should be addressed very carefully in future studies.

Similar to the laws of thermodynamics, the validity of the three laws of information theory may only be supported by experimental findings. It is worth reminding that the thermodynamic laws are actually postulates because they cannot be mathematically proved.

SIMILARITY PRINCIPLE AND ITS PROOF

Traditionally symmetry is regarded as a discrete or a "yes-or -no" property. According to Gibbs, the properties are either the same (indistinguishability) or not the same (figure 1a). A continuous measure of static symmetry has been elegantly discussed by Avnir and coworkers [20]. Because the maximum similarity is indistinguishability (the sameness) and the minimum is distinguishability, corresponding to the symmetry and nonsymmetry, respectively, naturally similarity can be used as a continuous measure of symmetry (section 2.1). The similarity refers to the considered property X of the components (see the definition of labeling in section 2.3) which affects the similarity of the probability values of all the $w$microstates and eventually the value of entropy (equation 1).

Gibbs paradox statement [21] is a higher similarity–lower entropy relation (figure 1a), which has been accepted in almost all the standard textbooks of statistical mechanics and thermodynamics. The resolution of the Gibbs paradox of entropy–similarity relation has been very controversial. Some recent debates are listed in reference [21]. The von Neumann continuous relation of similarity and entropy (the higher the similarity among the components is, the lower value of entropy will be, according to his resolution of Gibbs paradox) is shown in figure 1b [21j]. Because neither Gibbs' discontinuous higher similarity–lower entropy relation (where symmetry is regarded as a discrete or a "yes-or -no" property) nor von Neumann's continuous relation has been proved, they can be at most regarded as postulates or assumptions. Based on all the observed experimental facts [7d], we must abandon their postulates and accept the following postulate as the most plausible one (figure 1c):

Similarity principle: The higher the similarity among the components is, the higher the value of entropy will be and the higher the stability will be.

The components can be individual molecules, molecular moieties or phases. The similarity among the components determines the property similarity of the microstates in equation 1. Intuitively we may understand that, the values of the probability p_i are related to each other and therefore depend solely on the similarity of the considered property X among the microstates. If the values of the component property are more similar, the values of p_i of the microstates are closer (equation 1 and 2). The similarity among the microstates is reflected in the similarity of the values of p_i we may observe. There might be many different methods of similarity definition and calculation. However, entropy should be always a monotonically increasing function of any kind of similarity of the relevant property if that similarity is properly defined.

Example 9. Suppose a coin is placed in a container and shaken violently. After opening the container you may find that it is either head-up or head-down (The portrait on the coin is the label). For a bent coin, however, the property of head-up and the probability of head-down will be different. The entropy will be smaller and the symmetry (shape of the coin) will be reduced. In combination with example 2, it is easy to understand that entropy and similarity increase together.

Now, let us perform the following proof: the Gibbs inequality,

$$\ln w \geq -\sum_{i=1}^{w} p_i \ln p_i$$

(18)

has been proved in geometry (see the citations in [7d]). As lnw represents the maximum similarity among the considered w microstates [18], the general expression of entropy $-\sum_{i=1}^{w} p_i \ln p_i$ must increase continuously with the increase in the property similarity among the w microstates. The maximum value of lnw in equation 2 corresponds to the highest similarity. Finally, based on the second law of the revised information theory regarding structural stability that says that *the entropy of an isolated system (or system + environment) either remains unchanged or increases*, the similarity principle has been proved.

Following the convention of defining similarities [7d,7e] as an index in the range of [0,1], we may simply define

$$Z = \frac{S}{L} = \frac{\ln w_S}{\ln w} = \frac{-\sum_{i=1}^{w} p_i \ln p_i}{\ln w}$$

(19)

as a similarity index (figure 1c). As mentioned above, Z can be properly defined in many different expressions and the relation that entropy increases continuously with the increase in the similarity will be still valid (However, the relation will not be necessarily a straight line if the similarity is defined in another way). For example, the N^2 similarities ρ_{ij} in a table

$$
\begin{matrix}
\rho_{11} & \rho_{12} & \cdots & \cdots & \rho_{1N} \\
\rho_{21} & \rho_{22} & \cdots & \cdots & \rho_{2N} \\
\cdots & \cdots & \cdots & \cdots & \cdots \\
\cdots & \cdots & \cdots & \cdots & \cdots \\
\rho_{N1} & \rho_{N2} & \cdots & \cdots & \rho_{NN}
\end{matrix}
$$

(20)

can be used to define a similarity value of a system of N kinds of molecule [7c,7e].

CURIE-ROSEN SYMMETRY PRINCIPLE AND ITS PROOF

It is straightforward to prove the higher symmetry−higher stability relation (the Curie-Rosen symmetry principle [2]) as a special case of the similarity principle. Because higher similarity is correlated with a higher degree of symmetry, the similarity principle also implies that entropy can be used to measure the degree of symmetry. We can conclude: *The higher the symmetry (indistinguishability) of the structure is, the higher the value of entropy will be.* From the second law of information theory, the higher symmetry−higher entropy relation is proved.

The proof of the higher symmetry−higher entropy relationship can be performed by contradiction also. Higher symmetry number−lower entropy value relation can be found in many textbooks (see citations in [7] and p. 596 of [19]) where the existence of a symmetry would result in a decrease in the entropy value:

$$\Delta S = -\ln\sigma \tag{21}$$

where σ denotes the symmetry number and $\sigma \geq 1$ Let the entropy change from S' to S is

$$\Delta S = S' - S = -\ln\sigma \tag{22}$$

Then the change in symmetry number would be a factor

$$\sigma = wS \rangle wS \tag{23}$$

and

$$S' - S = -\ln\frac{w_{S'}}{w_S} = -\ln w_{S'} - \left(-\ln w_S\right) \tag{24}$$

where wS and wS' are the two symmetry numbers. This leads to an entropy expression $S = -\ln w_S$. However, because any structure would have a symmetry number $w_S \geq 1$, entropy would be a negative value, which contradicts the definition that entropy is always positive. Therefore neither $\Delta S = -\ln\sigma$ nor $S = -\ln w_S$ are valid. The correct form should be equation 4 ($S = \ln w_S$, $w_S \geq 1$). In combination with the second law of information theory, the higher symmetry−higher stability relation (the symmetry principle) is also proved.

Rosen discussed several forms of symmetry principle [2]. Curie's causality form of the symmetry principle is that *the effects are more symmetric than the causes*. The higher symmetry−higher stability relation has been clearly expressed by Rosen [2]:

For an isolated system the degree of symmetry cannot decrease as the system evolves, but either remains constant or increases.

This form of symmetry principle is most relevant in form to the second law of thermodynamics. Therefore the second law of information theory might be expressed thus: an isolated system will evolve spontaneously (or irreversibly) towards the most stable structure, which has the highest symmetry. For closed and open systems, the isolated system is replaced by the universe or system + environment.

Because entropy defined in the revised information theory is more broad, symmetry can include both static (the third law of information theory) and dynamic symmetries. Therefore, we can predict the symmetry evolution from a fluid system to a static system when temperature is gradually reduced: the most symmetric static structure will be preferred (vide infra).

A COMPARISON: INFORMATION MINIMIZATION AND POTENTIAL ENERGY MINIMIZATION

A complete structural characterization should require the evaluation of both the degree of symmetry (or indistinguishability) and the degree of nonsymmetry (or distinguishability). Based on the above discussion we can define entropy as the logarithmic function of symmetry number w_S in equation 4. Similarly, the number w_I can be called nonsymmetry number in equation 11.

Other authors briefly discussed the higher symmetry–higher entropy relation previously (see the citation in [2]). Rosen suggested that a possible scheme for symmetry quantification is to take for the degree of symmetry of a system the order of its symmetry group (or its logarithm) (p.87, reference [2]). The degree of symmetry of a considered system can be considered in the following increasingly simplified manner: in the language of group theory, the group corresponding to L is a direct product of the two groups corresponding to the values of entropy S and information I:

$$G=G_S \times G_I \tag{25}$$

where GS is the group representing the observed symmetry of the system, GI the nonsymmetric (distinguishable) part that potentially can become symmetric according to the second law of information, and G is the group representing the maximum symmetry. In the language of the numbers of microstates

$$\omega=\omega_S \cdot \omega_I \tag{26}$$

where the three numbers are called the maximum symmetry number, the symmetry number and the nonsymmetry numbers, respectively. These numbers of microstate could be the orders of the three groups if they are finite order symmetry groups. However, there are many groups of infinite order such as those of rotational symmetry, which should be considered in detail in our further studies.

The behavior of the logarithmic functions $L=\ln w$, $S=\ln wS$ and their relation

$$L=S+I \tag{7}$$

can be compared with that of the total energy E, kinetic energy EK and potential energy EP which are the eigenvalues of H and K and P in the Hamiltonian expressed conventionally in either classical mechanics or quantum mechanics as two parts:

$$H = K + P \tag{27}$$

The energy conservation law and the energy minimization law regarding a spontaneous (or irreversible) process in physics (In thermodynamics they are the first and the second law of thermodynamics) says that

$$\Delta E = \Delta E_K + \Delta E_P \tag{28}$$

and $\Delta E = 0$ (e.g., for a linear harmonic oscillator, the sum of the potential energy and kinetic energy remains unchanged), $\Delta E_P \leq 0$ for an isolated system (or system +environment). In thermodynamics, Gibbs free energy G or Helmholtz potential F (equation 14) are such kinds of potential energy. It is well known that the minimization of the Helmholtz potential or Gibbs free energy is an alternative expression of the second law of thermodynamics. Similarly,

$$\Delta L = \Delta S + \Delta I \tag{29}$$

For a spontaneous process of an isolated system, $\Delta L = 0$ (the first law of information theory) and $\Delta I \leq 0$ (the second law of information theory or the minimization of the degree of nonsymmetry) or $\Delta S \geq 0$ (the maximization of the degree of symmetry). For an isolated system,

$$\Delta S = -\Delta I \tag{30}$$

For systems that cannot be isolated, spontaneous processes with both $\Delta S > 0$ or $\Delta S < 0$ for the systems are possible provided that

$$\Delta S > 0 \tag{31}$$

for the universe ($\Delta S = 0$ for a reversible process. $\Delta S < 0$ are impossible process). The maximum symmetry number, the symmetry number and the nonsymmetry number can be calculated as the exponential of the maximum entropy, entropy and information, respectively. These relations will be illustrated by some examples and will be studied in more detail in the future.

The revised information theory provides a new approach to understanding the nature of energy (or energy-matter) conservation and conversion. For example the available potential energy due to distinguishability or nonsymmetry can be calculated. This can be illustrated in a system undergoing spontaneous mass or heat transfer between two parts of a chamber (figure 5). The distinguishability or nonsymmetry is the cause of the phenomena. Many processes are irreversible because a process does not proceed from a symmetric structure to a nonsymmetric structure.

Figure 5. Mass (ideal gas) transfer and heat transfer between two parts of a chamber after removal of the barrier. The whole system is an isolated system.

INTERACTIONS

From the second law of information theory, any system with interactions among its components will evolve towards an equilibrium with the minimum information (or maximum symmetry). This general principle leads to the criteria of equilibrium for many kinds of system. For example, in a mechanical system the equilibrium is the balance of forces and the balance of torques.

We may consider symmetry evolution generally in two steps. Step one: bring several components (e.g., parts A and B infigure 5) to the vicinity as individually isolated parts. We may treat this step as if these parts have no interaction. Step two: let these parts interact by removing electromagnetic insulation, thermal insulation, or mechanical barrier, etc. (e.g., the right side of figure 5).

For step one, similarity analysis will be satisfactory. For the example shown in figure 5, we measure if the two parts A and B are of the same temperature (if they are, there will be no heat transfer), the same pressure (if they are, there will be no mass transfer) or the same substances (if they are, there will be no chemical reactions.

We may calculate the total logarithmic function L total, total entropy (S total) and total information of an isolated system of many parts for the first step. In the same way, we may calculate a system of complicated, hierarchical structures. Suppose there are M hierarchical levels (i=1,2,...,M, e.g., one level is a galaxy, the other levels are the solar system, a planet, a box of gas, a molecule, an atom, electronic motion and nuclear motion inside the atom, subatomic structures, etc.) and N parts (j=1,2,...,N, e.g., different cells in a crystals, individual molecules, atoms or electrons, or spatially different locations). It should be greater than the sum of the individual parts at all the hierachical levels,

$$S_{\text{total}} \geq \sum_{i,j} S_{i,j}$$

(32)

because any assembling and any interaction (coupling, etc.) among the parts or any interaction between the different hierachical levels will lead to information loss and entropy increase. Many terms of entropy due to interaction should be included in the entropy expression.

Example 10. The assembling of the same molecules to form a stable crystal structure is a process similar to adding many copies of the same book to the shelves of one library. Suppose there are 1GB information in the book. Even though there are 1,000,000 copies, the information is still the same and the information in this library is

$$I_{\text{library}} = \sum_{j=1}^{1000000} I_j = I_1 = 1 \text{ GB}$$

The entropy is 999999 GB.

Example 11. Interaction of the two complementary strands of polymer to form DNA. The combination leads to a more stable structure. The system of these two components (part 1 and part 2) have entropy greater than the sum of the individual parts $(S_{1+2} \geq 2S_1)$ because

$$I_{1+2} = \sum_{j=1}^{2} I_j = I_1 \leq 2I_1$$

However, operator theory or functional analysis might be applied for more vigorous mathematical treatment, which will be presented elsewhere. Application of the revised information theory to specific cases with quantitative calculations also will be topics of further investigations. We will only outline some applications of the theory in the following sections.

Finally, we claim that due to interactions, the universe evolves towards maximum symmetry or minimum information.

DYNAMIC AND STATIC SYMMETRIES

There are two types of symmetries: *dynamic* symmetry and *static* symmetry. Both are related to information loss as schematically illustrated in figure 6 and figure 7. Dynamic entropy is the logarithm of the number of the microstates of identical property (or identical energy level in thermodynamics).

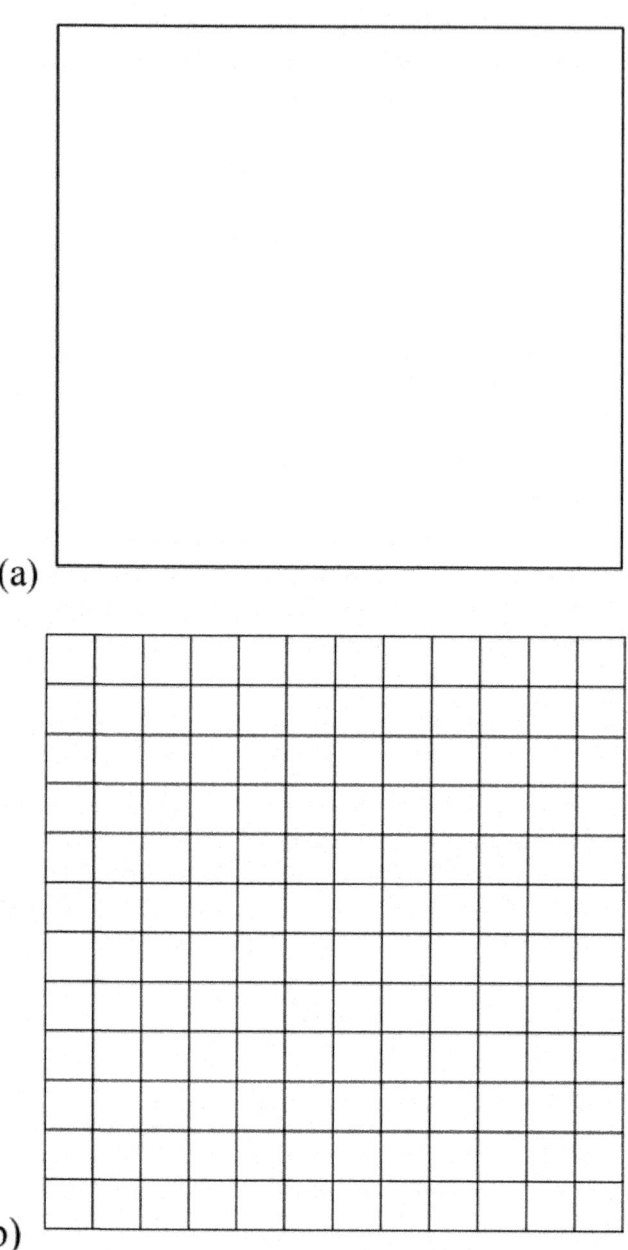

Figure 6. Schematic representation of a highly symmetric system of an ideal gas (dynamic symmetry, very homogenous and isotropic) (a) or a perfect crystal (static symmetry) (b) and the information loss. The highly symmetric paintings like (a) and (b) found in many famous modern art museums are the emperor's new clothes. These paintings are "reproduced" here without courtesy.

The Similarity Principle

If all the other conditions remain constant,
the higher the similarity among the components
is, the higher the value of entropy of mixing,
assembling or chemical bond formation
process will be.

The higher the similarity among the components
is, the more spontaneous the mixing, the
assembling or the chemical bond formation
process will be, and the more stable the mixture,
the assemblage or the chemical bond will be.

(a)

(b)

Figure 7. A static structure with certain text (a) and graphic (b) information recorded must have dramatically reduced symmetry.

The example of dynamic symmetry is the homogeneity of fluid systems of many particles, such as a gas of many molecules. The homogeneity of an ideal gas is due to the dynamic symmetry of the fluid phase. The static symmetry number can be easily estimated for a crystal [7].

Dynamic Symmetry

Fluid Systems

Because information is recorded in static structures, systems of molecular dynamic motion have zero value of information and, therefore, the logarithmic functions L and S have equal value at the equilibrium state. In order to accommodate the kinetic energy of the system, the structure of the system cannot stay in only one of many accessible microstates. The macroscopic properties – entropy and symmetry (homogeneity, isotropicity) of an idea gas used in a heat engine can be considered first because the ideal gas model is most important in thermodynamics.

Example 12. Ideal gas mass transfer and heat transfer between two parts. Entropy of an ideal gas in thermodynamics is a special kind of dynamic entropy. For a free expansion process of an ideal gas within an isolated system with two parts of different pressure (figure 5), $\Delta E=0$, $\Delta G<0$, $\Delta I<0$, $\Delta S>0$ and $\Delta L=0$ in thermodynamics and in our revised information theory treatment. Symmetry is higher at the final equilibrium structure: both sides are of the same pressure and temperature at the final structure.

Gibbs Paradox

In this context let us resolve Gibbs paradox of entropy of mixing [21]. It says that the entropy of mixing *decreases discontinuously* with an increase in similarity. It has a zero value for mixing of the indistinguishable subsystems (figure 1a). The isobaric and isothermal mixing of one mole of an ideal fluid A and one mole of a different ideal fluid B (figure 5 and figure 8) has the entropy increment

$$(\Delta S)_{distinguishable}=2R\ln2=11.53 \text{ J K}^{-1} \tag{33}$$

where R is the gas constant, while

$$(\Delta S)_{indistinguishable}=0 \tag{34}$$

for the mixing of indistinguishable fluids [21]. It is assumed that the two equations (33) and 33) are also applicable to the formation of solid mixtures and liquid mixtures (citations in [7]). Gibbs paradox statement of entropy of mixing has been regarded as the theoretical foundation of statistical mechanics [23], quantum theory [24] and biophysics [21]. It is certainly a most important problem in information theory if one intends to apply information theory to physics: e.g., Jaynes, the father of the maximal entropy principle [16], also considered this problem [21]. The resolutions of this paradox have been very controversial for over one hundred years. Many famous physicists confidently claim that they have resolved this paradox, in very diverse and different ways. A sustaining problem is that they do not agree with one another. For example, besides numerous other resolutions [21], von Neumann [21] provided a well-known quantum mechanical resolution of this paradox with which not many others are in full agreement [21d].

Figure 8. Ideal gas mixing in a rigid chamber.

Gibbs paradox is a problem regarding the relation of indistinguishability (symmetry) and entropy. It can be easily resolved if we recall the definition of labeling (section 2.3). For a rigid container, mixing two ideal gases is an identical process whether they are indistinguishable or distinguishable ideal gases, provided that the two gas chambers are parts of a rigid container (Because deformation changes the shape symmetry and entropy [7], we suppose that the container is rigid. The interesting topic of shape symmetry evolution or deformation [7] will be discussed in detail elsewhere): $\Delta E=0$, $\Delta G=0$, $\Delta(k_B TS)=0$, and $\Delta S=0$. As has been actually shown in experiment there is no change in the total energy, in the Gibbs free energy and there is no heat effect for the isobaric, isothermal mixing process. Therefore, the entropy change must be zero in both cases whether it is a mixing of the same ideal gases or of different ideal gases (equation 35):

$$(\Delta S)_{distinguishable}=0 \qquad (35)$$

Local Dynamic Motion in Solids (Crystals)

Another excellent example of local dynamic motion in a static structure is Pauling's assessment of residual entropy of ice [7,24]. The spin-up and spin-down binary system is shown in figure 3. The information loss and symmetry increase due to local dynamic motion can be further illustrated by typewriting different fonts at one location, as shown in figure 4. In many cases, the formation of certain static periodic structures can be regarded as formatting (section 2.3).

Static Symmetry

If the temperature is gradually reduced, a system of many molecules may become a static structure. There are many possible static structures. In principle, for a condensed phase, the static structure can be any of the w microstates accessible before the phase transition. Symmetric static structure (crystal) and nonsymmetric static structure must have different information (figure 6 and figure 7). Our theory predicts that the most symmetric microstate will be taken as the most stable static structure (the third law of information theory and the symmetry principle). Before the phase transition at higher temperature, the system should be of the highest possible dynamic symmetry (figure 9, for example). After the phase transition to form a solid phase, the system should evolve to the highest possible static symmetry (figure 10).

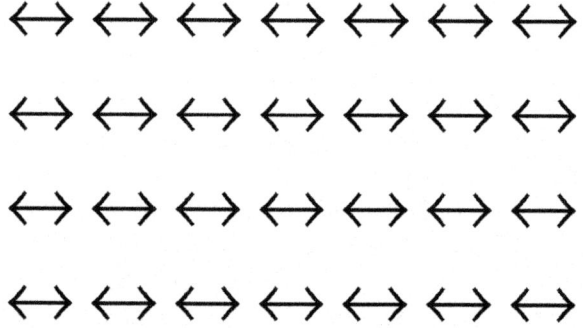

Figure 9. A two-dimensional static array with local dynamic symmetry. At a position, the two orientations have identical probability. The highest local dynamic symmetry leads to the least information. The information loss is equivalent to printing many pages of a book on one page which will lead to symmetry (every position will be the same hybrid of the two orientations) and information loss.

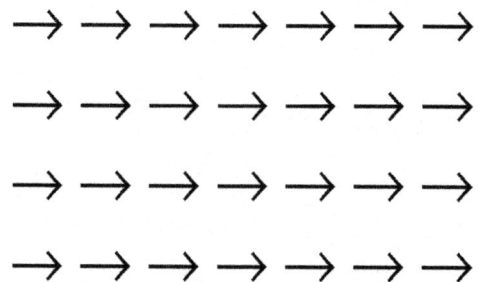

Figure 10. A schematic representation of a two-dimensional static system of high static symmetry with only one spin orientation.

Actually these examples are discussed in the famous Ising model (part III of the book [15]) and the related phase transition problem [6]. The structure can be easily characterized by the spin orientation indistinguishability (a symmetry).figure 3 is a one-dimensional array of spins. A two-dimensional system of spins at fixed lattices are given in figure 9(dynamic symmetry) and 10 (static symmetry).

The expression of the entropy of mixing has been applied in the same way to the mixing processes to form gaseous, fluid or solid mixtures (see any statistical mechanics textbook). Therefore, in the Ising model, the entropy of a noncrystal structure would have a higher entropy due to the entropy of mixing *different* species according to the traditional expression (figure 1a). However, the most symmetric static structure (figure 10, for instance) has the highest static entropy according to our third law. The present author believes that the traditional way of calculating the entropy of mixing is the largest flaw in solid state physics in particular and in physics in general. Prigogine's dissipative structure theory (which has been claimed to have solved such kind of symmetry breaking problems [4,25]), Wilson's method of renormalization group [5,26] and many other theories have been proposed. The symmetry breaking problem remains to be solved. According to our theory, the static symmetry (and the static entropy) should dominantly contribute to the stability of a ferromagnetic system in the spin-parallel static state below the Curie temperature. Due to the static symmetry, a perfect crystal has the highest static entropy (the third law of information theory). This conforms to perfectly the observed highest stability of crystal structure among all the possible static (solid) structures. The perfect crystal structure is equivalent to a newly formatted disk (hard disk or floppy disk) which has the highest symmetry and the least (or zero) information (figure 2b). Noncrystal solid structures are less stable (This prediction has already been confirmed by an abundance of experimental observations [7]; ask any experimental chemists or material scientists!).

PHASE SEPARATION AND PHASE TRANSITION (SYMMETRY BREAKING)

When the thermodynamic temperature decreases, there will be a phase separation where different substances separate as a result of the spontaneous assembling of the indistinguishable substances.

Similarity and Temperature

Let us illustrate the relation of similarity in thermodynamics and the thermodynamic temperature T with the simplest example. The energy level similarity between two energy levels E_a and E_b is calculated from their

Boltzmann factors $e^{-\frac{E_a}{kT}}$ and $e^{-\frac{E_b}{kT}}$. The similarity will approach the maximum if temperature increases and the minimum if the temperature approaches zero:

$$\lim_{T\to\infty} e^{-\left|\frac{\Delta E}{kT}\right|} = \lim_{T\to\infty} e^{-\left|\frac{E_a-E_b}{kT}\right|} = 1 \tag{36}$$

$$\lim_{T\to 0} e^{-\left|\frac{\Delta E}{kT}\right|} = \lim_{T\to 0} e^{-\left|\frac{E_a-E_b}{kT}\right|} = 0 \tag{37}$$

Jaynes' maximal entropy principle [15,16] may be applied to discuss the relation of similarity and a temperature-like parameter. Generally speaking, similarity increases with the increase in the absolute value of temperature $|T|$ (For a system of negative temperature where $T \leq 0$, similarity increases with the increase in $|T|$; see ref. 7a).

Phase Separation, Condensation, and the Densest Packing

The similarity among components will decrease at a reduced temperature (equation 37). Because a heterogeneous structure with components of very different properties has high information and is unstable, phase separation for a system of multiple components will follow the similarity principle: *different components spontaneously separate. The components of the same (or very similar) properties mix or merge to form a homogeneous phase. The different components separate as a consequence of the assembling of components of the most similar (or the same) properties.* Spontaneous phase separation and its information loss (or symmetry increase) as well as the opposite process can be illustrated in figure 11 [7]. Therefore, if phase separation is desirable, we decrease the temperature (equation 37); if the mixing of components is required, we increase the temperature (equation 36).

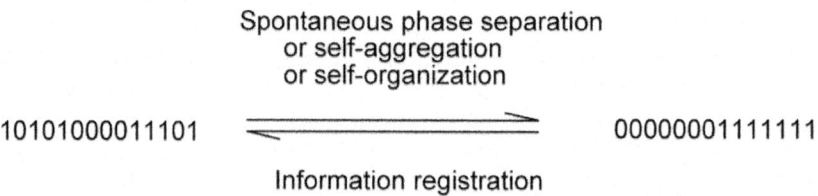

Figure 11. Informational registration as a binary string and the information loss due to a possible spontaneous process.

Spontaneous phase separation has been the main chemical process to separate different substances and to purify substances. For example, even enantiomers can be separated in this way [7d].

Condensation from a gaseous phase and many examples of the densest molecular packing in the formation of crystal structures and other solid state structure can be explained by our revised information theory as a phase separation of a binary system of two "species": substance and vacuum. In this approximation, all molecules (not necessarily the same kind of molecule or one kind of molecule) can be taken as the species "1" and all the parts of free space are taken as species "0". At a reduced thermodynamic temperature, substance and vacuum both occupy the space, but otherwise are so different that they will separate as a consequence of assembling the same species together as two "phases", one is the condensed substance phase, the other is a bulky "phase" of vacuum (figure 11).

Phase transition and Evolution of the Universe and Evolution of Life

The evolution of the Universe is a series of phase separations and phase transition. At extremely high temperature, even matter and antimatter can apparently coexist. At high temperature, the similarity is high (equation 36). When the temperature is reduced, the increase of the entropy can be achieved either by phase separation or phase transition as spontaneous processes. At reduced temperature, matter aggregates with matter. Then a mixture of matter and antimatter are not stable because their similarity is extremely low. The biosphere has only L-amino acids and D-sugars [27]. Our theory can shed light on the solution of the symmetry breaking phenomena during the universe evolution in general and the molecular evolution of life in particular. It is clear that at every stage of symmetry breaking, the critical phenomenon is characterized by the system's tendency towards the highest possible dynamic symmetry at higher temperature and the highest possible static symmetry at lower temperature.

SIMILARITY RULE AND COMPLEMENTARITY RULE

Generally speaking, *all* intermolecular processes (molecular recognition and molecular assembling or the formation of any kinds of chemical bond) and intramolecular processes (protein folding [28], etc.) between molecular moieties are governed either by the similarity rule or by the complementarity rule or both.

Similarity rule (a component in a molecular recognition process loves others of like properties, such as hydrophobic interaction, π-stacking in DNA molecules, similarity in softness of the well-known hard-soft-acid-base rules) predicts the affinity of individuals of *similar* properties. On the contrary, complementarity rule predicts the affinity of individuals of certain *different* properties. Both types of rule still remain strictly empirical. The similarity rule can be given a theoretical foundation by the similarity principle (figure 1c) [7] after rejection of Gibbs' (figure 1a) and revised (figure 1b) relations of entropy–similarity.

All kinds of donor–acceptor interaction, such as enzyme and substrate combination, which may involve hydrogen bond, electrostatic interaction and stereochemical key-and-lock docking [30] (e.g., template and the imprinted molecular cavity [6]), follow the complementarity rule. For the significance of the complementarity concept in chemistry, see the chapter on Pauling in the book [1].

Definition: Suppose there are n kinds of property X, Y, Z, ..., etc. (See the definition of entropy and labeling in section 2) and n=l+m. For a binary system, if the two individuals contrast in l kinds of property (negative charge-positive charge or convex and concave, etc.) and exactly the same for the rest m kinds of property, the relation of these two components is complementary. An example is given in figure 12.

Figure 12. The print and the imprint [6b] are complementary.

Firstly, for stereochemical interaction (key-and-lock docking) following a complementarity rule it can be treated as a special kind of phase separation where the substance and the vacuum (as a different species) separate (cf. section 8). The final structure is more "complete", more integral, more "solid" and more symmetric. More generally speaking, the components in the structure of the final state become more similar due to the property *offset* of the components and the structure is more symmetric. The calculation of symmetry number, entropy and information changes during molecular interaction are numerous topics for further studies.

Complementarity principle: The final structure is more symmetric due to the property offset of the components. It will be more stable.

In the recent book by Hargittai and Hargittai [1] many observations showed that in our daily life symmetry created by combination of parts means beauty. We may speculate that, in all these cases, the stability of the interaction of the symmetric static images (crystal, etc.) and symmetric dynamic processes (periodicity in time, etc.) may play an important role in our perception. Perception of visual beauty might be our visual organ's interaction with the images or a sequences of images and a tight interaction might result if the similarity rule and the complementarity rule are satisfied.

Normally we consider complementarity of a binary system (two partners). However, it can be an interaction among many components. The component motifs are *distinguishable* (for a binary system they are *contrast*) in the considered property or properties.

Chemical bond formation and all kinds of other interaction following the similarity and complementarity rules will lead to more stable *static* structures. Actually there are also similarity and complementarity rules for the formation of a stable *dynamic* system. The resonance theory is an empirical rule [29]. Von Neumann tried very hard to use the argument of entropy of mixing but his final conclusion regarding entropy–similarity relation (figure 1b) was wrong and cannot be employed to explain the stability of quantum systems [21]). Pauling's resonance theory can be supported by the two principles – similarity principle and complementarity principle for a dynamic system regarding electron motion and electronic configuration [7]. The final structure as a hybrid or a mixture of a "complete" set (having all the microstates or all the available canonical structures of very similar energy levels, similar configuration, etc. see 1 and 2 in figure 13a) should be more symmetric. The most complete set of microstates will lead to the highest dynamic entropy and the highest stability. Other structures (e.g., 3 in figure 13a) cannot contribute significantly to the structure because they are very different from the structure of 1 or 2.

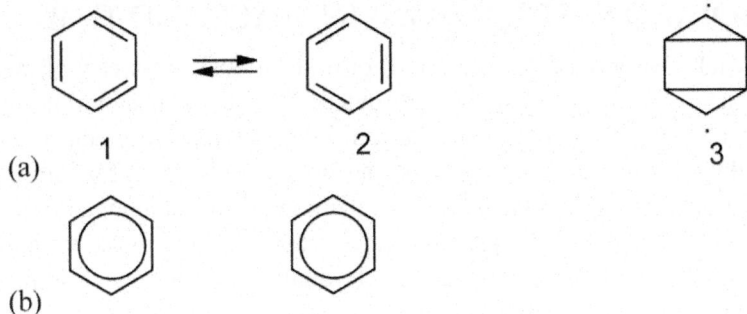

(a)   1   2   3

(b)

Figure 13. Information loss due to dynamic electronic motion. The different orienta-tions of the valence bond benzene structures 1 and 2 in (a) can be used for recording information. However, the oscillation makes all the individual benzene molecules the same as shown in (b).

To explain the stability of the systems other than molecular systems, one may replace molecules by individuals or components under consideration and then calculate in a similar way as for molecular systems. At least the relative structural stability may be assessed for different structures where the similarity principle is applicable.

Finally we should emphasize that the similarity rule is always more significant than the complementarity rule, because most properties of the complementary components are the same or very similar ($m>l$). Examples in chemistry are the HSAB (hard-soft-acid-base) rule where the two components (acid and base) should have similar softness or similar hardness (see any modern texts in inorganic chemistry). Another example is the complementary pair of LUMO (lowest unoccupied molecular orbital) and HOMO (highest occupied MO) where the energy levels of the MO are very close (see a modern textbook of organic chemistry). The formation of a chemical bond can be illustrated by a stable marriage (example 13).

Example 13. Successful marriage might be a consequence of the highest possible similarity and complementarity between the couple. The stability of the relation is directly determined by how identical the partners are: both man and woman have interests in common, have the same goals, common beliefs and principles, and share in wholesome activities. Love and respect between the two are the complementarity at work which is the more interesting aspect of the marriage due to their contrasting attributes: One is submissive, the other not. A beautiful marriage is determined by how adaptable both of them are regarding differences, rather than by how identical they are. God created man and woman to make one as a complement of the other (Genesis 2:18).

PERIODICITY (REPETITION) IN SPACE AND TIME

The beauty of periodicity (repetition) or translational symmetry of molecular packing in crystals must be attributed to the corresponding stability. Formatting of a floppy disk or a hard disk will create symmetry and erasure of all the information. Formatting is a necessary preparation for stable storage of information where certain kinds of symmetry are maintained: e.g., books must be "formatted" by pages and lines where periodicity is maintained as background. The DNA molecules are "formatted" by the sugar and phosphate backbone as periodic units.

Steady-state is a special case of a dynamic system. Its stability depends on the symmetry (periodicity in time). The long period of chemical oscillation

[1,4] happens in a steady-state system. Cars on a highway must run in the same direction at the same velocity. Otherwise, if one car goes much slower, there will be a traffic accident (Curie said "nonsymmetry leads to phenomena" [2]), and the unfortunate phenomenon is the collision. Many kinds of cycle (heart-beating cycle, sleep-and-wake-up cycles, etc.) are the periodicity in time, which contributes stability, very much the same as the Carnot cycle of an ideal heat engine. However, exact repetition of everyday life must be very boring, even though it makes life simple or easy and stable (vida infra).

FURTHER DISCUSSIONS: BEAUTIFUL SYMMETRY AND UGLY SYMMETRY [31]

To stress the important points of the present theory and to apply it in a relaxed manner to much more general situations, let us follow the style of discussing symmetry concept in a recent book [1] and clarify the relation of symmetry with other concepts. Whether symmetry is beauty is of particular interest as the relationship of symmetry and beauty has been the subject of serious scientific research [32].

Order and Disorder

According to common intuition, the standard of beauty is that beauty correlates with less chaos or more order which in turn correlates with more information and less entropy. Our discussions of the following will be based on this general correlation of beauty–orderliness–information. Therefore, a consideration of this problem may be reduced to a proper definition of order (Table 1) and disorder.

Table 1. Two definitions of order

Definition	Order = periodicity or symmetry	Order = nonsymmetry and difference
Formation	Generated spontaneously	Not spontaneously
Examples	Chemical oscillation (symmetry and periodicity in time) or crystals (symmetry and periodicity in space)	Gas A stays in the left chamber and B in the right chamber to follow the order as shown in figure 5
Reference	"Order Out of Chaos" [4]	The present work
Comment	Challenges the validity of the second law	Conforms to the second law

However, unfortunately there are two totally different definitions of order or orderliness. Because they are opposite in meaning, it is only possible for us to conduct a meaningful discussion in chemistry and physics if only the suitable one of the two definitions is taken and consistently used. The first one regards "ordered" pattern as of the "order out of chaos" through "self-organization" [4]. Because the system is allowed to evolve spontaneously [4] the organized structure must be closer to an equilibrium state. It is well known that a system closer to equilibrium has a higher value of entropy of the universe or of an isolated system. It follows that a more "ordered" system has more entropy [4]. This kind of misconception is due to the correlation of higher entropy with lower symmetry because symmetry has been regarded as order. In this definition of order [4], order is the periodicity or symmetry generated spontaneously or through a "self"-action. This conclusion obviously violates the second law of thermodynamics [4], as commented by the present author [33].

In the other and the proper definition of order, order is generated and maintained by the surroundings (including the experimenter or information register and their equipment, etc.) [7]. This order is achieved by applying confinements or constraints to the concerned system such as containers to restrict molecules spatially, or applying forces to produce distinguishability (and information). One example is shown in figure 5 where the two parts are confined and they are distinguishable. The social order is generated and maintained by applying discipline to its members to create distinguishability. Without discipline, if everyone in a country claims to be as powerful as the president and behaves in the same manner, the permutation symmetry is high and the society is in chaos. If two or more book locations are changed and the order of the library is unchanged, there is permutation symmetry in this library. This library either has many (to say 10000) copies of the same book, or is in a totally chaotic situation. Some years ago, when the present author did experiments using a refrigerated room where thousands of compounds prepared by the professor's former students were stored, he decided that symmetry was by no means order: If one brings any symmetry to the storeroom and accordingly reshapes the professor's nonsymmetrical arrangement of the precious samples (most of them are themselves asymmetric molecules) to "make order", the professor would react angrily. Now the present author is supervising a sample archive center [34] and has an even stronger picture that the archive order of the samples in a stock room is generated and maintained by confining samples to distinguishable containers and locations.

The statement that "a static structure (not necessarily a crystal), which is a frozen dynamic structure, is more orderly than a system of dynamic motion" is correct. However, when two *static* structures are directly compared, the saying that "a symmetric static structure has more order or is more orderly than a nonsymmetrical static structure" will be completely incorrect and misleading if disorder has been defined as entropy or information loss.

As mentioned in section 8.2 and section 9, for stability reasons, a system applied for storing information always has symmetries for certain kinds of property (e.g., the periodicity of the sugar-phosphate backbone for DNA in example 6). In most cases information is recorded by using only one of several properties. The symmetric structure regarding other properties can be taken as formatting (section 2.3). The properly defined order or the genetic information in a DNA is recorded by the nonsymmetric structural feature (the 4 bases), not at all by the sugar or phosphate which are responsible for the symmetry. Strictly speaking, symmetry is not order; it is chaos.

Symmetry and Diversity

Intuitively, it can be easily understood that entropy and symmetry increase together if we simply use some common sense: it is indistinguishable (which is symmetry, see figure 6); and there is no information (or there is a large amount of entropy)! Whereas a similarity (Z) of a system can be defined as

$$Z = \frac{S}{L} = \frac{\ln w_S}{\ln w}$$

(19)

the diversity of the system (D) can be defined as

$$D = \frac{I}{L} = \frac{\ln w_I}{\ln w}$$

(38)

For more details, see ref. 7c-7e. Why are diversities [7], such as molecular diversity, biodiversity, cultural diversity, etc. appreciated by many people, and why is diversity beautiful? The new theory gives a clear answer. If a collection of 10000 stamps has extremely high similarity, which means that all of them are of the same figure and size, even though this kind of stamp might be by itself very interesting, the permutation symmetry is obvious: If the positions of any two stamps are exchanged, the album of stamps remains the same. This collection is not beautiful and not precious because it lacks diversity. A library of 10000 copies of the same book is not nice either. Based on this idea, as a chemist, the present author initiated MDPI [34], an international project for collecting diverse chemical samples to build up molecular diversity. He has not collected 10000 samples of the same compound, even though it might be very interesting, e.g., the famous C_{60} molecule. The molecule C_{60} is itself beautiful not because of its symmetry, but because of its distinct structure and property compared to many other molecules and its contribution to the diversity of molecules we have studied.

For similar reason, it is true that we may feel a symmetric object beautiful if we have many other objects of less symmetry in our collection or in our experience. This symmetric object contributes to the diversity.

Diversity is beautiful. Symmetry is not. Coffee with sugar and cream is an interesting drink because of its diversity in taste (sweet and bitter) and color (white and dark). Diversity in a mixture makes the so-called high throughput screening of bioactivity testing possible [7]. A country (e.g., USA or Switzerland) is considered a nice place because of its tolerance to all kinds of diversity (racial diversity, cultural diversity, religious diversity, etc.). Without such appreciation of diversity, this country would become much less colorful and less beautiful. If everyone behaves the same as you, looks the same as you, and there is a lot of symmetry, the world would be truly ugly. Democracy might be regarded as a sort of social or political diversity.

Unfortunately a system with high diversity is less stable. Synthetic organic chemists know that samples with impurities are less stable than highly purified samples. A mixture of thousands of compounds from a combinatorial synthesis is stable because the properties of the compounds are very similar and they normally belong to one type of molecule with the same parent structure. A mixture of acid and base is not stable. Storing a bottle of HCl and a bottle of NH_3 together is not a good idea.

A useful or beautiful system (a collection) should have a compromise between stability (due to indistinguishability or permutation symmetry and entropy) and diversity (due to distinguishability, nonsymmetry, diversity and information). As we have discussed before on complementarity (section 9 and the example), there can be many components or individuals that are complementary. A system satisfying this kind of complementarity is the most beautiful one with the most stability due to indistinguishability (symmetry) and the information due to distinguishability (nonsymmetry). A stable diversity is beautiful and practically useful for all kinds of diversity preservation [7].

Ugly Symmetry

Assembling

Symmetry has been defined as beauty in many books and research papers [3,8,31]. According to the *Concise Oxford Dictionary*, symmetry is beauty resulting from the right proportion between the parts of the body or any whole, balance, congruity, harmony, keeping (page 1 of ref. 8).

Almost every modern arts museum collects a few paintings which are very symmetric, sometimes completely indistinguishable between the parts with the same pure white color (or gray or other color) overall (figure 6a). Symmetry is beautiful? OK, that's it! Why not? It is ugly because the emperor is topless and bottomless, even though many great intellectuals can cheerfully enjoy his most beautiful clothes (figure 6a). The most symmetric and most fussy paintings of those post-impressionists are the emperor's new clothes.

Crystals might be more beautiful than fluids because a solid structure has obviously lost all the dynamic symmetry such as isotropicity (see figure 6a). In a crystal, isotropicity does not exist: the properties along various directions become different (see figure 6b). This might be the reason why we like chocolate or ice cream in a certain form and may not like melted chocolate or runny ice cream soup.

Individual Items

High speed motion around a fixed point and a fixed axis will create a figure of spherical and cylindrical symmetry, respectively. The electronic motion in a hydrogen atom around the proton creates a spherical shape for the hydrogen atom. These are many examples showing that the dynamic motion of a system results in an increased symmetry. These systems are stable but not necessarily beautiful. For example, a photo showing the face and shape of a girl figure skater with no symmetry is more beautiful than a photo of a mixture of the images facing many directions, or a mixture of structures at different angular displacements.

The famous molecule buckminsterfullerene C_{60} is symmetric [1] and very stable. We can predict that any modified C_{60} mother structure with reduced symmetry will be less stable than the symmetric C_{60}.

Symmetric static structure is stable but not necessarily beautiful. A beautiful model, whether she stands or sits, never poses in a symmetric style before the crowd. A fit and beautiful body differs from the more symmetric, spherical shape of an overweight body. A guard in front of a palace stands in a symmetric way to show more stability (and strength). A Chinese empress sits in a symmetry style because stability is more interesting to her. The pyramid in Egypt is stable due to its symmetry. However, it is by no means the most beautiful construction.

Information will be lost if the ink is extremely faint or the same color as the background. A crystal is more stable than a noncrystal solid because the former has high static symmetry, a perfect symmetry without any information (figure 6b). Children understand it: on the walls of children's classrooms and bedrooms a visitor may find a lot of paintings, drawings and even scrawls done by the kids. If you do not put some colorful drawings there, children will create them themselves (e.g., figure 7b). The innocent children like to get rid of any ugly symmetry surrounding them. If your kids destroy symmetry on the walls, they are doing well.

Perfect symmetry is boring [32,1], isn't it? The very symmetric female face of oriental beauty [32] can be much more attractive if it is made less symmetric by a nonsymmetric hairstyle or by a nonsymmetric smile. The combination of certain symmetry (or symmetries) contributing stability according to our theory and certain nonsymmetry (or distinguishability, contributing certain interesting information) is the ideal beauty. The earth (figure 2) is beautiful because of its combination of symmetry and nonsymmetry. If it were of perfect symmetry and we could not distinguish North America, Europe or Asia, the world would be a deadly boring planet.

"Symmetry Is Beauty" Has Been Misleading in Science

Beauty and its relation to symmetry has been a topic of serious scientific investigations [32]. This is fine. However, it has been widely believed that if the beholder is a physicist or a chemist, beauty also means symmetry. This may have already practically misled scientific research funding, publication and recognition. The situation in chemistry may be mentioned. Experimental chemists doing organic or inorganic synthesis may have more difficulty in synthesizing a specific structure of less symmetry than that of a highly symmetric one. However, a large number of highly symmetric structures can be much more easily published in prestigious journals (*Angewandt Chemie*, for example), because these molecular structures have been taken as more beautiful. Consequently other synthetic chemists have been regarded as intellectually lower achievers than those preparing highly symmetric structures. It has been shown by the history of chemical science and demonstrated by the modern arts of chemistry, particularly organic synthesis, that chemists endeavor to seek for asymmetry related to both space and time (nonsymmetry is the cause of phenomena [2]) not at all for symmetry [31]. The symmetric buckminsterfullerene C_{60} is beautiful. However, many derivatives of C_{60} have been synthesized by organic chemists. These derivatives are less symmetric, more difficult to produce and might be more significant. None of the drugs (pharmaceuticals) discovered so far are very symmetric. Very few bioactive compounds are symmetric. Because all of the most important molecules of life, such as amino acids, sugars, and nucleic acids, are asymmetric, we can also attribute beauty to those objects that are practically more difficult to create and yet practically more significant. The highest symmetry means equilibrium in science and death in life [31].

Sometimes, even graphic representation and illustration can be biased by the authors to create false symmetry. One example is shown in figure 14.

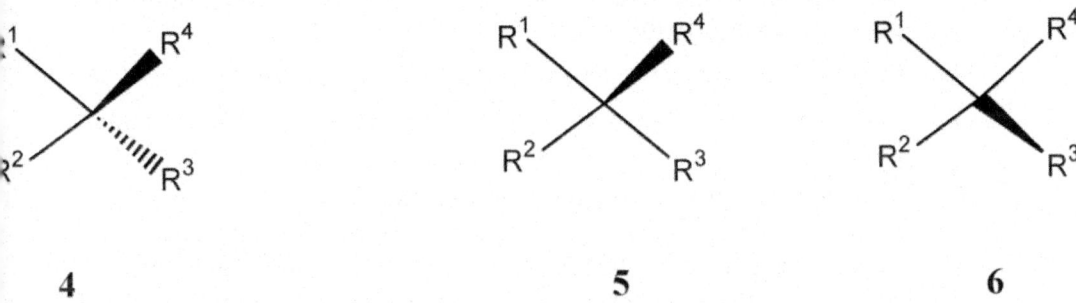

4   **5**   **6**

Figure 14. Stereochemical representation with false mirror symmetry in structure **4**: Both wedges have the thick ends at R^3 and R^4 placed identically away from the center [35]. The one-wedge representation (**5** or **6**) offers unambiguity and esthetic appeal [35a].

CONCLUSION

The structural stability criteria of symmetry maximization can be applied to predict all kinds of symmetry evolution. We have clarified the relation of symmetry to several other concepts, namely higher symmetry, higher similarity, higher entropy, less information and less diversity and they are all related to higher stability. This lays the very necessary foundation for further studies in understanding the nature of the chemical process.

ACKNOWLEDGEMENTS

The author is very grateful to Dr. Kurt E. Geckeler, Dr. Peter Ramberg, Dr. S. Anand Kumar, Dr. Alexey V. Eliseev and Dr. Jerome Karle and numerous other colleagues for their kind invitations to give lectures on the topics of ugly symmetry and beautiful diversity. Dr. Istvan Hargittai and Dr. J. Rosen kindly reviewed this paper.

REFERENCES AND NOTES

1. (a)Hargittai, I. *Symmetry, A Unifying Concept*; Shelter Publications, Inc.: Bolinas, California, 1994. (b)Hargittai, I.; Hargittai, M. *Our Own Image, Personal Symmetry in Discovery*; Kluwer: New York, 2000.

2. (a)Rosen, J. *Symmetry in Science*; Springer: New York, 1995. (b)Rosen, J. *A Symmetry Primer for Scientists*; Wiley: New York, 1983. A book review: Lin, S.-K. *Entropy*, 1999, 1, pp. 53–54. http://www.mdpi.org/entropy/htm/e1030053.htm.

3.   Weyl, H. *Symmetry*; Paperback Reprint edition, Princeton University Press: New York, 1989.

4.   Prigogine, I.; Stengers, I. *Order out of Chaos*; Heinemann: London, 1984.

5.   Ma, S. K. *Modern Theory of Critical Phenomena*; W. A. Benjamin: London, 1976.

6.   (a)Chemistry can be comprehended as an *information science*, see: Lehn, J. M. Some Reflections on Chemisry-Molecular, Supramolecular and Beyond. In *Chemistry for the 21st Century*; Keinan, E., Schechter, I., Eds.; Wiley-VCH: Weinheim, 2001; pp. 1–7.(b)Günter, W. Templated Synthesis of Polymers – Molecularly Imprinted Materials for Recognition and Catalysis. In *Templated Organic Synthesis*; Diederich, F., Stang, P. J., Eds.; Wiley-VCH: Weinheim, 2000; pp. 39–73. A book review: Lin, S.-K. *Molecules*, 2000, 5, pp. 195–197. http://www.mdpi.org/molecules/html/50200195.htm.

7.   Lin, S.-K. Understanding structural stability and process spontaneity based on the rejection of the Gibbs paradox of entropy of mixing. *J. Mol. Struct. Theorochem*, 1997, 398, pp. 145–153. http://www.mdpi. org/lin/lin-rpu.htm.   Lin, S.-K. Gibbs paradox of entropy of mixing: Experimental facts, its rejection, and the theoretical consequences. *EJ. Theor. Chem.* 1996, 1, pp. 135–150. http://www.mdpi.org/lin/lin-rpu.htm. Lin, S.-K. Molecular Diversity Assessment: Logarithmic Relations of Information and Species Diversity and Logarithmic Relations of Entropy and Indistinguishability After Rejection of Gibbs Paradox of Entropy of Mixing.*Molecules*, 1996, 1, pp. 57–67. http://www.mdpi.org/lin/lin-rpu. htm. Lin, S.-K. Correlation of Entropy with Similarity and Symmetry. *J. Chem. Inf. Comp. Sci.* **1996**, *36*, 367–376. http://www.mdpi.org/lin/lin-rpu.htm.   Lin, S.-K. *Diversity Assessment Based on a Higher Similarity-Higher Entropy Relation after Rejection of Gibbs Paradox*; Preprint, physics/9910032. 1999. http://xxx.lanl.gov/abs/physics/9910032.

8.   Elliott, J. P.; Dawber, P. G. *Symmetry in Physics*; Macmilan: London, 1979.

9.   Clausius, R. *Annalen der Physik und Chemie 1865, 125, 353-400. For its English translation, see: Magie, W. F. A Source Book in Physics*; McGraw-Hill: New York, 1935. http://webserver.lemoyne.edu/ faculty/giunta/clausius.html.

10. MacKay's e-book on information theory is very readable, see: MacKay, D. J. C. *Information Theory, Inference, and Learning Algorithms; Draft 2.2.0*, 2000. http://wol.ra.phy.cam.ac.uk/mackay/MyBooks.html.

11. Shannon, C. E. *Bell Sys. Tech. J.* **1948**, *27*, 379-423; 623-656.Shannon, C.E. *Mathematical Theory of Communication*, http://cm.bell-labs.com/cm/ms/what/shannonday/paper.html.

12. Lewis, G. N. *Science* **1930**, *71*, 569.

13. Watson, J. D. *The Double Helix*; The New American Library: New York, 1968.

14. Halliday, D.; Resnick, R.; Walker, J. *Fundamentals of Physics*, 4th Ed. ed; J. Wiley & Sons, Ltd.: New York, 1993.

15. (a)Chowdhury, D.; Stauffer, D. *Principles of Equilibrium Statistical Mechanics*; Wiley-VCH: Weinheim, 2000; pp. 158–159. Lambert, F. L. *J. Chem. Educ.* **1999**, *76*, 1385. (c)Denbigh, K. G.; Denbigh, J. S. *Entropy in Relation to Incomplete Knowledge*; Cambridge University Press: London, 1985.

16. Jaynes, E. T. *Phys. Rev.* **1957**, *106*, 620–630.

17. Brillouin, L. *Science and Information Theory*, 2nd Ed. ed; Academic Press: New York, 1962.

18. Wehrl, A. *Rev. Mod. Phys.* **1978**, *50*, 221–260.

19. Atkin, P. W. *Physical Chemistry*, 4th Ed. ed; Oxford University Press: London, 1990; p. 324.

20. Zabrodsky, H.; Peleg, S.; Avnir, D. *J. Am. Chem. Soc.* **1992**, *114*, 7843–7851.Zabrodsky, H.; Peleg, S.; Avnir, D. *J. Am. Chem. Soc.* **1993**, *115*, 8278–8289.Zabrodsky, H.; Avnir, D. *J. Am. Chem. Soc.* **1995**, *117*, 462–473.

21. Lesk, A. M. On the Gibbs paradox: what does indistinguishability really mean? *J. Phys. A: Math. Gen.* **1980**, *13*, L111–L114. (b)van Kampen, N. G. *The Gibbs Paradox, In Essays in Theoretical Physics*; Parry, W. E., Ed.; Pergamon: Oxford, 1984; pp. 303–312. Kemp, H. R. Gibbs' paradox: Two views on the correction term. *J. Chem. Educ.* **1986**, *63*, 735–736. Dieks, D.; van Dijk, V. Another look at the quantum mechanical entropy of mixing. *Am. J. Phys.* **1988**, *56*, 430–434. Richardson, I. W. The Gibbs paradox and unidirectional fluxes. *Eur. Biophys. J.* **1989**, *17*, 281–286. Lin, S.-K. Gibbs paradox and its resolutions. *Ziran Zazhi* **1989**, *12*, 376–379. Wantke, K.-D. A remark on the Gibbs-paradox of the entropy of mixing of ideal gases. *Ber. Bunsen-Ges. Phys. Chem.* **1991**, *94*, 537. (h) Jaynes, E. T. The Gibbs paradox. In *Maximum Entropy and Bayesian Methods*; Smith, C. R., Erickson, G. J., Neudorfer, P. O., Eds.; Kluwer Academic: Dordrecht, 1992; pp. 1–22. Blumenfeld, L. A.; Grosberg, A. Y. Gibbs paradox and the notion of construction in thermodynamics and statistic physics. *Biofizika* **1995**, *40*, 660–667. (j)von Neumann, J.

Mathematical Foundations of Quantum Mechanics; Princeton University Press: Princeton, 1955; Ch. 5. Lin, S.-K. *Gibbs paradox of Entropy of Mixing Website*, http://www.mdpi.org/entropy/entropyweb/gibbs-paradox.htm.

22. Schrödinger, E. *Statistical Mechanics*; University Press: London, 1952; p. 58.

23. Pauli, W. *Pauli Lectures on Physics: Vol.3,Thermodynamics and the Kinetic Theory of Gases*; MIT Press: Cambridge, MA, 1973; p. 48.

24. Pauling, L. The Structure and Entropy of Ice and of Other Crystals with Some Randomness of Atomic Arrangement. *J. Am. Chem. Soc.* **1935**, *57*, 2680–2684.

25. (a)Kondepudi, D.; Prigogine, I. *Modern Thermodynamics: From Heat Engines to Dissipative Structures*; Wiley: Chichester, UK, 1998; chapters 19-20. A book review: Lin, S. -K. *Entropy*, 1999, 1, pp. 148–149. http://www.mdpi.org/entropy/htm/e1040148.htm.

26. Wilson, K. *Rev. Mod. Phys.* **1983**, *55*, 583–600.

27. Frank, P.; Bonner, W. A.; Zare, R. N. On One Hand But Not the Other: The Challenge of the Origin and Survival of Homochirality in Prebiotic Chemistry. In *Chemistry for the 21st Century*; Keinan, E., Schechter, I., Eds.; Wiley-VCH: Weinheim, 2001; pp. 173–208.

28. Duan, Y.; Harvey, S. C.; Kollman, P. A. Protein Folding and Beyond. In *Chemistry for the 21st Century*; Keinan, E., Schechter, I., Eds.; Wiley-VCH: Weinheim, 2001; pp. 89–101.

29. (a)*The resonance theory is an empirical rule because the chemical bond stability has never been explained. See: Pauling, L. The Nature of the Chemical Bond*, 3rd ed.; Cornell University: Ithaca, 1966.(b) It should be recalled that resonance is a term used in physics, where the oscillation is the amplitude of the induced oscillation is greatest if the angular frequency of the driving force is the same as the natural frequency of an oscillator [14].

30. Fischer, E. *Ber. Dtsch. Chem. Ges.* **1894**, *27*, 2985–1993.

31. Adapted from the lectures of the present author, e.g, see: Lin, S.-K. Ugly Symmetry; Lecture at the 218th ACS national meeting in New Orleans, Louisiana. 22-26 August 1999. http://www.mdpi.org/lin/uglysym1.htm).

32. E.g., see: (a) Enguist, M.; Arak, A.

33. Prigogine indeed questioned the validity of the second law of thermodynamics, see also an editorial: Lin, S.-K.*Entropy*, 1999, 1, pp. 1–3. http://www.mdpi.org/entropy/htm/e1010001.htm).

34. (a) See the http://www.mdpi.org website.An editorial: Lin, S.-K. *Molecules*, 1997, 2, pp. 1–2. http://www.mdpi.org/molecules/edito197. htm.

35. Lin, S.-K.; Patiny, L.; Yerin, A.; Wisniewski, J. L.; Testa, B. One-wedge convention for stereochemical representations.*Enantiomer*, 2000, 5, pp. 571–583. http://www.mdpi.org/molecules/wedge.htm. Juaristi, E.; Welch, C. J. *Enantiomer*. [PubMed]Lin, S.-K. A proposal for the representation of the stereochemistry of quatrivalent centres.*Chirality* **1992**, *4*, 274– 278. Gal, J. Rootworm pheromones – The Root of a stereochemical mixup. *Chirality* **1992**, *4*, 68–70. Testa, B. On flying wedges, crashing wedges, and perspective-blind stereochemists. *Chirality* **1991**, *3*, 159– 160. Simonyi, M.; Gal, J.; Testa, J. *Trends Pharmacol. Sci.* **1989**, *10*, 349–354.

Chapter 2

A COMBINED PROBABILISTIC AND OPTIMIZATION APPROACH FOR IMPROVED CHEMICAL MIXING SYSTEMS DESIGN

Matthew J. Opgenorth[1], William E. McDermott[2], Peter Laz[1], Corinne S. Lengsfeld[1]

[1]Department of Mechanical and Materials Engineering, University of Denver, Denver, USA

[2]Applied Research and Technology Institute, University of Denver, Denver, USA

ABSTRACT

A design analysis of a mixing nozzle was performed using a combination of probabilistic and optimization techniques. A novel approach was utilized where probabilistic analysis was used to reduce the number of geometric constraints based on sensitivity factors. An optimization algorithm used only the most significant parameters to maximize mixing. A second probabilistic analysis was performed after optimization was complete in order to quantitatively predict the effects of manufacturing tolerances on mixing performance. This process for automated design is attractive over full parameter optimization techniques due to the computational efficiency resulting from an intelligent reduction in evaluated variables.

INTRODUCTION

Computational fluid dynamics (CFD) analyses, with software packages like Fluent™ (ANSYS, Inc.), have become accepted techniques for solving complex fluid flows using numerical implementation of fluid mechanics principles. CFD-driven optimization has been explored by several researchers [1-3]. Peigin et al. has proposed a design tool that implements multi-constrained optimization of shape design driven by Genetic Algorithms (GA) coupled with CFD. The benefit to GA is that they can handle a large number (20+) of design variables; however, with a large number of constraints, GA are computationally expensive. Peigin et al. reports 15-18 hours for a single point optimization and on average 8-12 optimization steps [1]. A similar multiconstrained optimization strategy has been implemented on a ship's hull. Again the downside is the time constraints, where they report approximately 10 days for 20 shape generations on a PC-based cluster [2]. Many researchers are using these GA's in order to optimize a system with numerous variables.

GA's have also been used by Carroll (1996) for the modeling of complex chemical mixing systems. The GA was coupled with the Blaze II code [4]. The Blaze II code can address up to 500 chemical reactions and 40 species. It also contains 1-D fluid dynamic equations, with mixing terms derived from 2-D equations, that can be used to model axis symmetric and 2-D flow fields [4]. The GA used 5 parameters with either 32 or 16 discrete possibilities resulting in approximately 2 million permutations. Even with a 2-D model and only 5 variable parameters the GA required 8 days of continuous runtime [3].

The objective of the current project is to develop and implement a design analysis of a fluid mixing nozzle using a coupled optimization and probabilistic approach. Due to the potentially large computational times to run GA optimization, our approach is to reduce the number of constraints though probabilistic analysis in order to use single objective optimization techniques. This technique was applied to mixing nozzles for a generic chemical mixing system. Based on the insight into the features that most affect performance provided by the probabilistic analysis, optimization has the potential to improve design performance while also reducing the weight of the system. Additionally, a probabilistic analysis of the optimized design can confirm the original optimization parameters and provide insight into the effects of manufacturing tolerances of specific geometric variables. This will in turn, reduce manufacturing costs by determining which dimensions are important to the mixing performance of the nozzle and which geometric tolerances can be loosened. A clear benefit to this design approach is that it allows for fast optimization by reducing the number of constraints through probabilistic analysis, as well as, assessing the impact of manufacturing tolerances.

METHODOLOGY

Currently, there is no commercially available tool in industry to perform fluid mechanics analyses, optimization and probabilistic analysis in the proposed integrated manner. There are however, software packages that can execute individual components of the process. Our approach is to use commercially available CFD, probabilistic and optimization software and interface them together with custom scripting. The probabilistic software, or optimization routine, manages the CFD code by importing any number of variables within ranges and/or distributions set by the user. **Figure 1** and 2 show a diagram linking the processes mentioned above.

CFD Model

The CFD software used for the simulations was Fluent™ Version 6.3.26 [5]. It is a widely used computational software package for modeling fluid flow and heat transfer in complex geometries. Easy mesh generation and ability to refine or coarsen the mesh autonomously based on the flow solution are just some of the features that make this CFD package extremely versatile and ideal for automation. Gambit ® was the pre-processor used for the solid modeling and mesh generation.

The algorithm employed was the pressure-based Navier-Stokes solution algorithm. Typically, this algorithm is used for low velocity incompressible flows, which fits this case where the flow velocity is about 83 m/s. The momentum equation provides the velocity field and the density is calculated from the equation of state. Other governing equations include energy and species conservation. The turbulence model chosen was the standard k-epsilon turbulence model. This model is robust and suitable for initial iterations, initial alternative design screenings, and parametric studies. The k-epsilon model will be ideal for the automated analysis where many different shapes will be analyzed. A no-slip boundary condition was placed at the walls. The primary inlet utilized a user-defined function (UDF) that modeled a fully developed fluid flow.

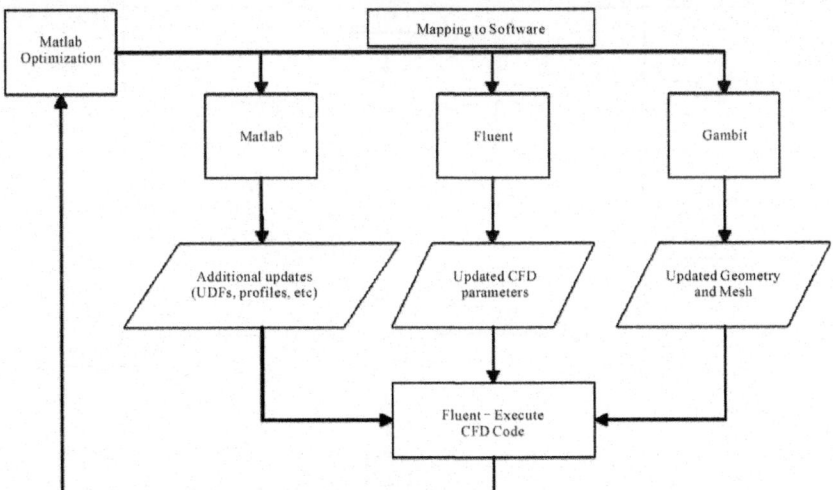

Figure 1. Diagram of the probabilistic/optimization and CFD interface. Each program is linked by custom scripts.

Definition of the boundary conditions is critical to both accuracy of the model and computational efficiency, the importance of which should not be under estimated when venturing into optimization and probabilistic analysis. Considerable attention was paid to identifying the best algorithmic parameters in order to achieve convergence in the minimum number of iterations, reliably, and with physically correct solutions regardless of geometry. It is also noteworthy to mention that this process was carried out in parallel on 8 processors in order to reduce computational time.

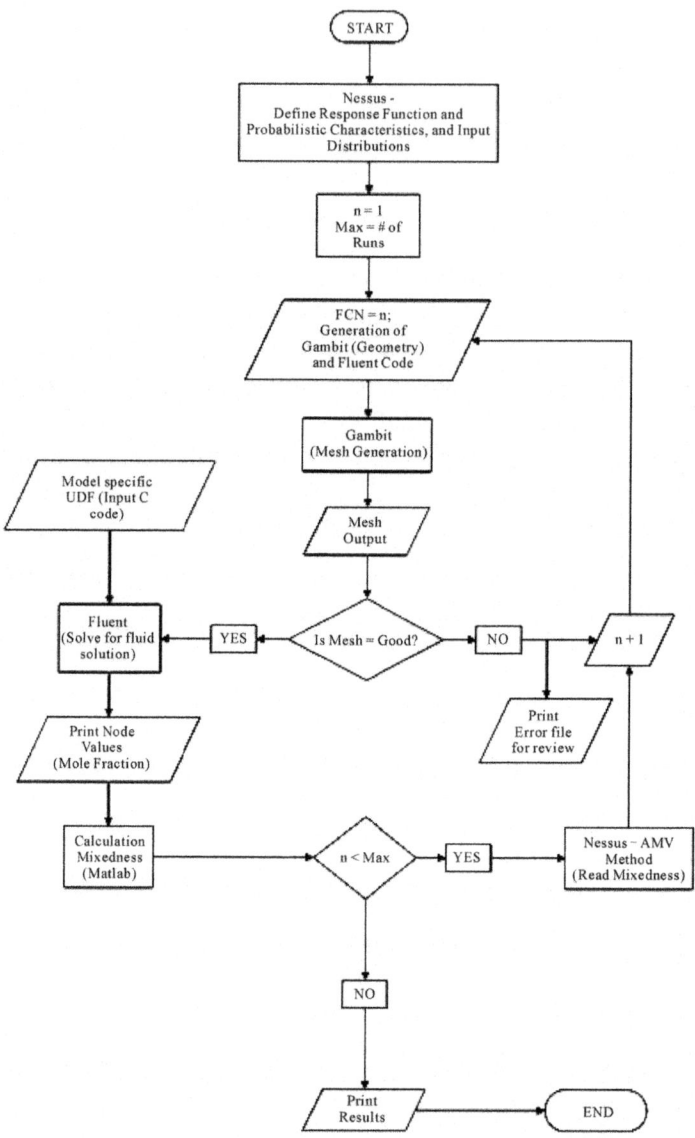

Figure 2. Example flow chart of the probabilistic interface using the AMV method.

Probabilistic Model

As uncertainty is inherent in physical systems, a probabilistic analysis model input variables as distributions and then predicts a distribution of performance. Based on the distribution of performance, sensitivity and importance factors are used to identify critical parameters. The probabilistic software, Nessus®, implements a variety of probabilistic methods that vary in efficiency and accuracy of the solution. The most commonly used probabilistic method is Monte Carlo [6]. The Monte Carlo method generates random values for each variable according to its distribution and then predicts the distribution of performance through repeated trials. As the accuracy of the prediction is dependent on the number of trials performed, the Monte Carlo method is computationally expensive. The mean-value (MV) family of methods are approximate, but considerably more efficient than the Monte Carlo method. They create a mean-based response function and compute the most probable point (MPP), which is the shortest distance from the origin to the limit state surface and represents the combination of stochastic variables resulting in a specific level of performance [6]. In this study, the Advanced Mean Value (AMV) method was applied, which uses a higher order approximation to determine the MPP [7]. The number of trials used for the AMV method is 1 + the number of variables + the number of probability levels desired [7]. While the AMV method is a discrete and approximate solution, it has shown excellent agreement with Monte Carlo analysis for monotonic system. This has been shown in other applications which utilize FE analysis with geometric perturbations and realistic loading conditions [8,9]. The major advantage of the AMV is that it requires a small number of trials, which saves significant computational time when the CFD analysis requires multiple iterations to reach convergence. Both of these probabilistic methods provide sensitivity factors identifying and quantifying the contributions of each variable on performance of the system.

The AMV method was utilized early in the design process to identify the variables that contributed significantly to the behavior of the system reducing the number of parameters required for performance optimization. By reducing the number of variables in the optimization, the computation time and required resources are dramatically reduced. Instead of using a sensitivity study perturbing a single variable at a time, a probabilistic approach was used to allow for the interaction affects between the various input parameters. During the early stages of the design process, the emphasis was on identifying the important and unimportant variables. Following optimization, AMV methods were employed to evaluate the effects of manufacturing tolerances on system performance.

Optimization Model

Numerical optimization techniques are designed to minimize an objective function subject to constraints, with many algorithms developed over the past several decades [10]. In general, the algorithms require a starting point, x_0, and then iterate or step until there is no more progression, or the approximate solution falls within a user-defined tolerance. Typically, algorithms follow one of two types of strategies, line search or trust region. This study implemented a trust region [11] strategy in order to account for geometric changes that may result in the fluid domain acting non-linearly. A common problem in line searches is the fixed step size can causes them to miss a local minimum, whereas the step size in the trust region search is not fixed and, therefore, has a better opportunity to find a minimum that is close to the current point.

The success and efficiency of an optimization is contingent on selection of an appropriate algorithm and an accuracy characterization of the problem. It was not known whether the variables would behave linear, nonlinear, or convex, only that they could hold any value between the bounds. The optimization algorithm had to be suitable for a continuous objective function with variables that are constrained by simple bounds and can solve for linear, nonlinear and convex variables.

A trust-region algorithm was implemented, which utilizes an active-set algorithm, for the optimiz-ation analysis. An active-set algorithm will employ linear techniques to estimate the active-set at each iteration and then solve an equality constrained quadratic program to generate a step [12]. This method was used because it tends to yield more exact solutions and is less sensitive to the initial starting point than interior point methods. Another benefit, in our case, of the active-set algorithm is that it uses a gradient projection method when only bounds are applied to the constraints [13]. The gradient projection method attempts to speed up the solution process within the active-set, but is only utilized when the variables are bounded. It consists of two different stages. First, the search direction will be along the path of steepest decent from the current point. The second stage investigates the face of the feasible region using the active-set constraints [12]. This can significantly reduce the optimization time.

Interfacing Model

To facilitate communication between the all of the software packages, custom interfacing was developed to build CFD models with perturbed parameters and calculate performance parameters from the analysis outputs. Interfacing was performed with components written in Matlab®, Dos and C. In addition, checks were performed to ensure mesh quality to prevent analyses that would fail or highly skewed elements, which may lead to convergence issues. This is noteworthy because the automated process can potentially take days and even weeks to run and computational efficiency will be a driving factor, especially as more complex flows are examined. An imported UDF specifies the physical boundary conditions for the specific system. Once the CFD simulation converges to a solution for the flow field; it calculates the values of a user defined fluid property (i.e. mole fraction of interested species) via a UDF and prints the results to a file. A script is also utilized to calculate the performance parameter and print the results for analysis by either the probabilistic or optimization routines. The performance parameter used is Mixedness, which is defined by:

$$Mixedness = 1 - \frac{\sum \left| M_f - M_{f\_Homogeneous} \right|}{n \cdot M_{f\_Homogeneous}}$$

$$\text{(1)}$$

The degree of mixing is measured by the ratio of the integral value for species mole fraction (M_f) across an exit plane divided by the homogeneous mole fraction ($M_{f\_Homogeneous}$), where n is the number of nodes within the exit plane. Increased mixing of species in chemical systems should result in greater chemical efficiencies and better performance. This interfacing routine is continued until all of the probabilistic or optimization trials are completed.

PROBLEM DESCRIPTION

CFD Model

The combined probabilistic and optimization approach is demonstrated for a low flow, subsonic Hydrogen-Iodide Chlorine (HICl) Laser. As opposed to other chemical lasers the HICl Laser has yet to reach its expected performance potentially due in part to a lack of homogeneous distribution of its excited chemical species.

Additionally:

1. the subsonic HICl Laser has simple mixing geometry, employing a repetitive nozzle array that allows the computational domain to be reduced based on simple lines of symmetry;

2. the cross-flow injection geometry would be easy to perturb within the code;

3. validation data exists for injection cross-flow at speeds similar to the HICl.

The geometry for the subsonic HICl Laser consists of a rectangular flow channel (**Figure 3**(a)). There are four rows of secondary flow injection nozzles on the top and bottom plates of the cavity (**Figure 3**(b)). The primary inlet flow for the subsonic HICl Laser is a mixture of helium (He) and hydrogen (H_2). The secondary flow through the nozzle plates has two different mixtures that are being injected into the primary flow. The first three rows inject a mixture of helium and hydrogen-iodide (HI), where the fourth injects a mixture of helium and nitrogen-trichloride (NCl_3). **Table 1** shows the boundary conditions applied during this design optimization run. A constant cross-sectional area was maintained on each inlet, but the aspect ratio was allowed to vary.

It would be computationally expensive to model the entire flow channel for the CFD analysis; consequently, planes of symmetry are used again to reduce the size of the modeled fluid domain (**Figure 4**). The fluid domain was modeled and meshed using text commands within a journal file. This allowed for easy modification during the automated process.

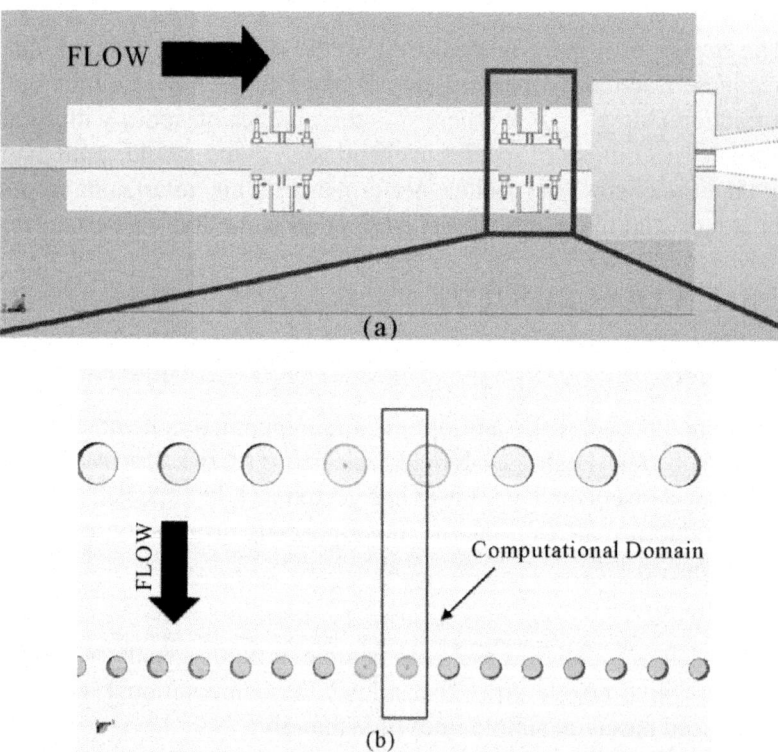

Figure 3. (a) A section view of the laser cavity showing the secondary inlet nozzle plates, and (b) detailed view of one of the nozzle plates. Note: the system has two identical sets of nozzle plates.

Table 1. Flow conditions for subsonic HICl laser

Initial Conditions and Constants		
Pressure	50	torr
Outlet Velocity	83	m/s
Temperature	300	K
Gas Constant	62400	Torr-cm^3/K-mole
Avogadro's Number	6.02E+23	molecules/mole

The algorithms, boundary conditions, and meshing strategies for the CFD simulation were validated using experimental data from subsonic (yet compressible flows) jets in cross flow [14]. Downstream velocity and temperature profiles obtained from the CFD simulations showed good comparison (**Figure 5**) to the experimental data of Dizene et al. (2000). This effort highlighted the need to accurately describe the velocity profile at the entrance boundaries and confirmed that the dynamic adaptive grid technique and algorithms selected worked well in these types of fluid flow conditions.

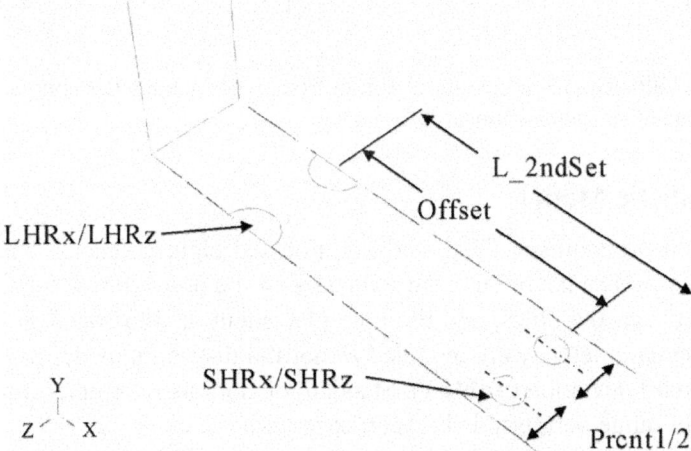

Figure 4. Schematic of the variables and geometry for the first probabilistic analysis.

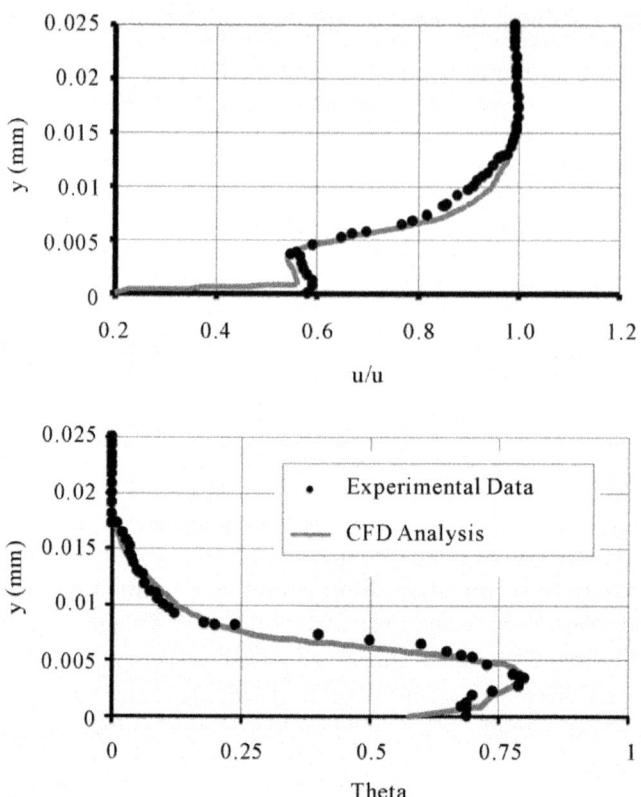

Figure 5. Velocity and Temperature profiles from experimental data compared to CFD simulations of jet in crossflow at X/D = 4.

Probabilistic Model

The initial probabilistic CFD analysis modeled eight geometric variables as distributions. The parameters were the size of the holes in two directions, the offset between the holes, and the hole placement, as illustrated in **Figure 4**. All of the parameters were modeled as normal distributions defined by mean and standard deviation (**Table 2**). Standard deviations were determined by the maximum range permissible by geometric limits.

Optimization Model

Optimization was then applied to determine the values of the variables within the constraints that maximized Mixedness. The four important variables, LHRx, LHRz, SHRx, and LHRz can be combined into two by keeping the cross sectional areas of the orifices constant. Therefore, only two variables,

LHRz and SHRz, need to be perturbed for the HICl Laser optimization. The objective function and constraints for the optimization is stated below:

Maximize: M = f(LHRz, SHRz)   (2)

Subject to: 0.0055 cm $\leq$ LHRz ≤ 0.023 cm

0.010 cm $\leq$ SHRz ≤ 0.042 cm   (3)

The optimization routine was used to determine the values of LHRz and SHRz in each analysis. The initial guess in the optimization algorithm were the mean values from the probabilistic analysis. The automated process updated the geometry, generated the mesh, performed the CFD analysis and extracted the results to compute the Mixedness parameter. Since the maximum Mixedness is desired, the minimization of the ratio of HI concentrations to the maximum concentration (the second term in Equation (1) was calculated. The optimization process continues until the Mixedness value has converged to within a 0.01% tolerance. In addition, initial guesses near the lower and upper bounds of the design space were also evaluated to check for local minimums that may be on either side of the mean orifice dimensions.

Table 2. Probabilistic variables with respective mean, standard deviation, and distribution

Name	Mean (mm)	StDev (mm)	Distribution
LHRx	0.1125	0.067	Normal
LHRz	0.1125	0.067	Normal
SHRx	0.1	0.07	Normal
SHRz	0.1	0.07	Normal
OffsetH	0.2025	0.05	Normal
L_2ndSet	1.5	0.33	Normal
Prcnt1	0.0	0.3	Normal
Prcnt2	0.0	0.3	Normal

RESULTS AND DISCUSSION

Identification of Important Parameters

Figure 6 shows the importance levels associated with each variable that was perturbed in the modeled geometry. The importance factors represent the design vector or value of each parameter that defines the MPP, which is proportionate to the output measure at the specified probability. As they are reported in the standard normal space, importance factors are relative measures and the sum of the squares for each measure will equal 1. The importance levels identify the variables that contribute the most to the reliability of the design; therefore, it is deduced from **Figure 5** that the radii of the inlet orifices contribute the most to either increase or decrease of the response variable, Mixedness. The other important observation is that PRCNT1, PRCNT2, OFFSETH and L_2NDSET had little to no effect on the Mixedness of the system. Recall, they had the largest standard deviation of all the variables. Now, the elimination of four of the eight variables within the geometry allows for the focus on only the radii of the orifices.

Design Optimization

All of the optimization analyses converged to the same value with the ratio of LHRz to SHRz equal to 0.6. This lower bound is both a computational and manufacturing limit due to the small size of the orifices. **Figure 7** shows the contours of all the data taken from the three different optimization runs. The green area shows the larger Mixedness parameters and the path that the optimization algorithm followed. The optimization required between 18 and 29 analyses and from 5 to 9 optimization iterations.

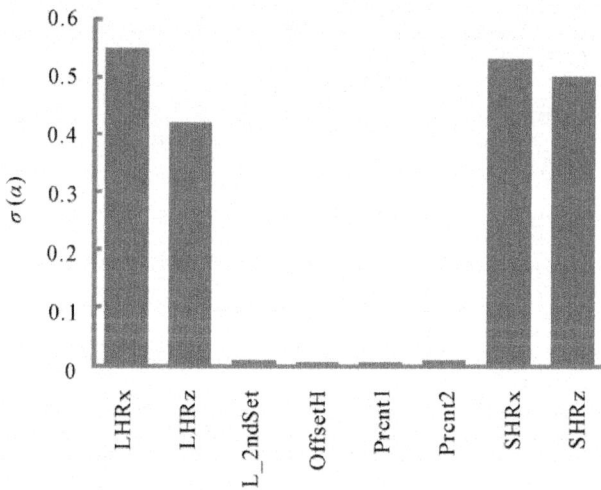

Figure 6. Histogram of the Importance Levels due to the first probabilistic analysis. It can be seen that the orifices affect the Mixedness the most.

The contour plot of the optimization analysis shows the optimized solution and the various starting points (Figure 8). Since, the only information acquired was along the optimization path, little reliable information can be extracted from the other areas of the plot. This does, however; show the optimal orifice aspect ratio for LHRz and SHRz is approximately equal to 4 and 6, respectively, and resulted in an increased Mixedness of nearly 10%. The major diameters of both orifices are parallel to the fluid flow.

In order to ensure that the global optimum was found, a Monte Carlo simulation was performed using the probabilistic CFD model to characterize the design space used in the optimization analysis. The response surface (Figure 9) with 200 Monte Carlo trials and the optimization results confirmed the findings of the optimization. Notably, the horizontal bands indicate that the LHRZ dimension has little effect on the fluid system. For example, starting at the optimized location (SHRz = 0.010 cm, LHRz = 0.006 cm), changes in LHRz within the design space did not affect the Mixedness value by more than 1%. Knowledge of these relationships is helpful in the design and manufacturing processes. The LHRz orifice can be a simple, circular geometry (which will be easier to create) and the mixing of the system will stay approximately the same compared to if it were elliptical. The response surface is non-monotonic over the optimization design space. The Monte Carlo results also reaffirm that the optimal configuration of the orifices, where the small orifice is located upstream to large orifice.

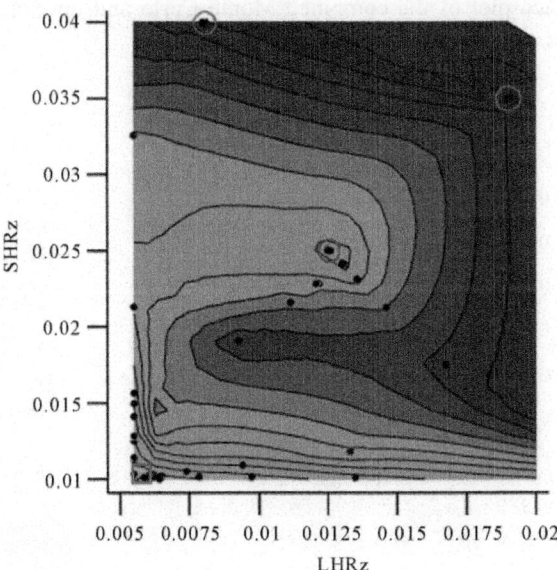

Figure 7. Contour plot of the optimization analysis with data point of each iteration. The optimized geometry converged near the lower bounds, where both orifices are elliptical and the major diameter is parallel to the fluid flow. The red circles locate the three different starting points and the red box indicates the optimum.

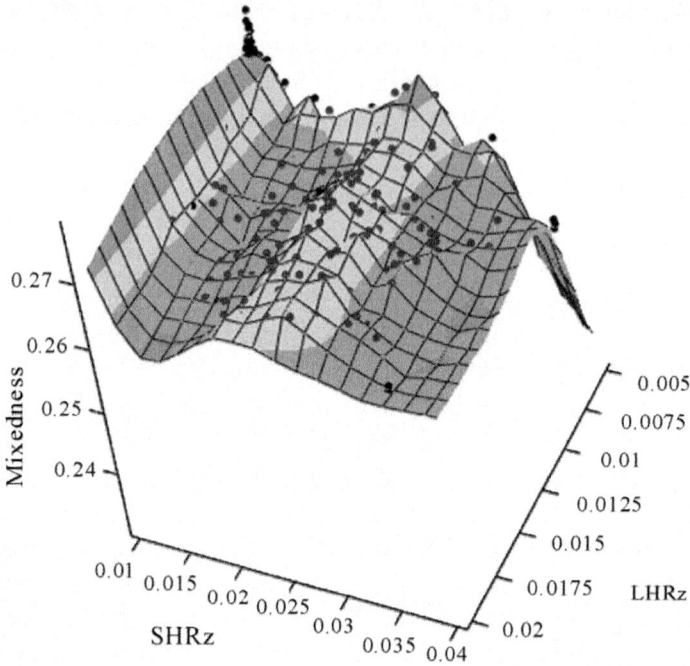

Figure 8. Surface plot of the combined Monte Carlo and optimization simulations with data points.

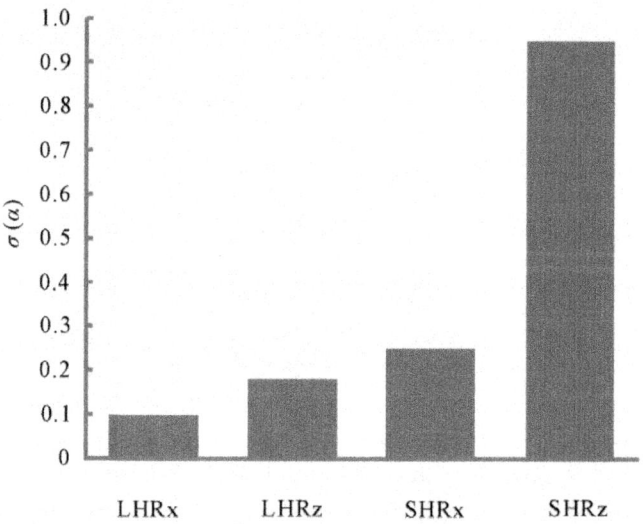

Figure 9. Histograms of the importance levels from the manufacturing sensitivity simulation.

Manufacturing Sensitivities

Based on the optimized design, the probabilistic platform was used to investigate the variability in Mixedness due to the manufacturing tolerances. The most economical way to create these elliptical geometries at the small dimensions required is through a wire electrical discharge machine (EDM). In this analysis, the geometry from the optimization simulation with mean values for LHRz and SHRz equal to 0.006 and 0.010 cm, respectively, will be used. The tolerances on these dimensions are +/− 0.001 cm, a tolerance that can typically be achieved by a wire EDM in a cost effective manner. The AMV method incorporated seven CFD simulations. The values for the AMV simulation are given below in **Table 3**.

The results for the importance of the four parameters, LHRz, LHRx, SHRz, and SHRz, are show in**Figure 1**. These results illustrate the same trends as the whole system. First, the LHRx, LHRz, and SHRx dimensions have virtually no importance on the Mixedness of the system. For the manufacturing design, the relaxation of tolerances on the LHRz dimensions would reduce the cost of fabrication and still maintaining a high Mixedness number. The importance of SHRz dimension does affect the system and the resulting Mixedness number. This result is liberal because the orifices' aspect ratios did change slightly. Since the X and Z dimensions were not dependent on each other, the X or Z radii could get larger or smaller while the other could do the same. Therefore, the results showed an increase in Mixedness, but if the aspect ratios were the same the Mixedness should not change. However, the Mixedness number reduces by ~3% if the SHRz dimension is decreased by 0.001 cm and increases by ~3% if it is decreased by 0.001 cm.

CONCLUSIONS

The purpose of this research was to create and implement an efficient computational design tool to combine optimization and probabilistic modeling to provide insight into how to improve chemical mixing systems performance. This has provided insight into the sensitivities of several different parameters that affect chemical mixing systems. Baseline CFD models were created for subsonic and mixing nozzles. The interfacing tools designed worked with minimal interaction from the user. The automated process for the probabilistic analysis requires input variables and perturbations from the user (assuming the geometry is created). Once all the inputs are specified, the interface carries out numerous evaluations of a fluid system. The optimization interface operated in much the same way. It required a starting point, as well as, constraints on the system. These actions were implemented on a subsonic mixing nozzle and results were obtained. Finally, the parallel processing technique enabled complex flows to be optimized in a timeframe consistent with real-time design processes (i.e. less than one week). For the AMV probabilistic analysis with 8 variables, the clock time was approximately 5 hours. The optimization process had a clock time on the order of 3-4 days. All the computations took place on a HP xw8400 Workstation with 2 - 2.33 GHz Xeon Quad Core Processors and 4 GB RAM.

Table 3. Probabilistic variables for manufacturing sensitivities on the optimized orifice geometry

Name	Mean	StDev	Distribution
LHRZ	0.0060	0.00034	Normal
LHRX	0.0265	0.00034	Normal
SHRz	0.0636	0.00034	Normal
SHRx	0.0100	0.00034	Normal

It was discovered, using the above computational tools, that in a HICl laser mixing nozzle, elliptical orifices with the major diameter parallel to the flow direction increased the mixing within the system by roughly 10%. Haven and Kurosaka (1997) similarly found through experimentation that the injection port geometry had a powerful influence on penetration in the near field [15]. For this case the optimum aspect ratio of the larger orifice to be approximately 6 and showed that the small injection orifice should be placed in front of the larger in a staggered alignment pattern.

The second evaluation phase of this work explored the impact of manufacturing tolerances on Mixedness. It was shown that the tolerance on SHRz plays the largest role on mixing quality, and that the shape of the small diameter secondary injection orifice does not have a great effect on the Mixedness. In fact, the Mixedness only decreases by 0.7%.

NOMENCLATURE

L_2ndSet = length to second set of orifices

LHRx = first orifice x-direction's radius

LHRz = first orifice z-direction's radius

M_f = mole fraction

$M_{f\_Homogeneous}$ = homogeneous mole fraction across downstream plane

N = number of nodes

OffsetH = Distance between first and second orifices within each set

Prcnt1,2 = percent offset between centers of first and second orifices

SHRx = second orifice x-direction's radius

SHRz = second orifice z-direction's radius

ACKNOWLEDGEMENTS

Financial support for this project was funded by the DOD Joint Technology Office; "Hybrid Iodine Laser Testing," Contract No. FA8632-05-C-2461.

REFERENCES

1.   S. Peigin and B. Epstein, "Multiconstrained Aerodynamic Design of Business Jet by Cfd Driven Optimization Tool," Aerospace Science and Technology, Vol. 12, No. 2, 2008, pp. 125-134. doi:10.1016/j.ast.2007.03.001

2.   Y. Tahara, D. Peri, E. Campana and F. Stern, "Computational Fluid Dynamics-Based Multiobjective Optimization of a Surface Combatant Using a Global Optimization Method," Journal of Marine Science and Technology, Vol. 13, No. 2, 2008, pp. 95-116.doi:10.1007/s00773-007-0264-7

3.   D. L. Carroll, "Chemical Laser Modeling with Genetic Algorithms," AIAA Journal, vol. 34, no. 2, 1996, pp. 338-346.

4.   L. H. Sentman, M. Subbiah and S. W. Zelazny, "Blaze II: A Chemical Laser Simulation Computer Program," Bell Aerospace Textron, TR H-CR-77-8, 1977.

5.   FLUENT_Inc., FLUENT Documentation, 2006.

6.   A. Haldar and S. Mahadevan, "Probability, Reliability and Statistical Methods in Engineering Design," John Wiley & Sons, Inc., New York, 2000.

7.   Y.-T. Wu, H. R. Millwater and T. A. Cruse, "Advanced Probabilistic Structural Analysis Method for Implicit Performance Functions," AIAA Journal, vol. 28, no. 9, 1990, pp. 1663-1669.

8.   S. K. Easley, et al., "Finite Element-Based Probabilistic Analysis Tool for Orthopaedic Applications," Computer Methods and Programs in Biomedicine, Vol. 85, No. 1, 2007, pp. 32-40. doi:10.1016/j.cmpb.2006.09.013

9.   P. J. Laz, J. Q. Stowe, M. A. Baldwin, A. J. Petrella and P. J. Rullkoetter, "Incorporating Uncertainty in Mechanical Properties for Finite Element-

Based Evaluation of Bone Mechanics," Journal of Biomechanics, Vol. 40, No. 13, 2007, pp. 2831-2836.doi:10.1016/j.jbiomech.2007.03.013

10. J. Nocedal and S. J. Wright, "Numerical Optimization. Springer Series in Operations Research," Springer-Verlag, New York, 1999.

11. G. N. Vanderplaats, "Numerical Optimization Techniques for Engineers," Third Edition, Vanderplaats Research and Development, Inc., 2001.

12. R. H. Byrd and R. A. Waltz, "An Active-Set Algorith for Nonlinear Programming Using Parametric Linear Programming," Department of Industrial and Systems Engineering, University of Southern California, Los Angeles, 2007.

13. MATLAB, The MathWorks Inc., 1994-2008

14. R. Dizene, J. M. Charbonnier, E. Dorignac and R. Leblanc, "Experimental Study of Inclined Jets Cross Flow Interaction in Compressible Regime. I. Effect of Compressibility in Subsonic Regime on Velocity and Temperature Fields," International Journal of Thermal Sciences, Vol. 39, No. 3, 2000, pp. 390-403. doi:10.1016/S1290-0729(00)00219-2

15. B. A. Haven and M. Kurosaka, "Kidney and Anti-Kidney Vortices in Crossflow Jets," Journal of Fluid Mechanics, Vol. 352, 1997, pp. 27-64. doi:10.1017/S0022112097007271

Chapter 3

ONLINE PROCESS CONTROL OF ALKALINE TEXTURING BATHS: DETERMINATION OF THE CHEMICAL CONCENTRATIONS

Martin Zimmer, Katrin Krieg, Jochen Rentsch

Department PV Production Technology and Quality Assurance, Fraunhofer Institute for Solar Energy Systems, Freiburg, Germany

ABSTRACT

Almost every monocrystalline silicon solar cell design includes a wet chemical process step for the alkaline texturing of the wafer surface in order to reduce the reflection of the front side. The alkaline texturing solution contains hydroxide, an organic additive usually 2-propanol and as a reaction product silicate. The hydroxide is consumed due to the reaction whereas 2-propanol evaporates during the process. Therefore, the correct replenishment for both components is required in order to achieve constant processing conditions. This may be simplified by using analytical methods for controlling the main components of the alkaline bath. This study gives an overview for a successful analytical method of the main components of an alkaline texturing bath by titration, HPLC, surface tension and NIR spectrometry.

INTRODUCTION

The wet chemical alkaline texturing is still an important process step during the fabrication of monocrystalline silicon solar cells [1-3] . The texturing process takes place in an aqueous solution of potassium or sodium hydroxide and the organic compound 2-propanol (IPA) at a certain temperature chosen between 70°C and 80°C. The hydroxide is consumed due to the reaction with silicon whereas 2-propanol is not consumed, but evaporates during the process due to its low boiling point (82°C). This changes the bath's composition. This texturing process results in a reduction of the reflection of the silicon wafer surface due to a complete coverage of the surface with randomly distributed micro pyramids. The mechanism of the pyramid formation is not completely understood yet, but the influence of the chemical concentrations [4,5] as well as that of the accumulating etching products [6], was subject of many studies and is known as critical. Hence, a complete analysis and control of the chemical concentrations in alkaline texturing baths is the prerequisite for a continuous processing at a high quality level. In this paper, we present different approaches for the analysis of the components of alkaline texturisation baths by titration, high performance liquid chromatography, surface tension and near-infrared spectroscopy. In order to evaluate the analytical methods, the component concentrations in the texturisation bath were measured during several processes at a semi-industrial batch process plant.

EXPERIMENTAL

Reference Sample Preparation

The validation of the titration was achieved using a set of reference samples. Samples were prepared from sodium hydroxide pellets (NaOH p.a., Merck, Darmstadt, Germany), sodium metasilicate powder (Na$_2$SiO$_3$·5H$_2$O, Carl Roth GmbH, Karlsruhe, Germany), sodium carbonate powder (Na$_2$CO$_3$, Carl Roth GmbH, Karlsruhe, Germany) and 2-propanol (IPA Puranal, Honeywell, Seelze, Germany). The silicon concentration ranged between 0 and 50 g/l and sodium carbonate between 0 g/l and 9 g/l, each sample in 25 g/l NaOH and 50 g/l IPA (Table 1). The validation of the high performance liquid chromatography (HPLC) was achieved using reference samples with 40 g/l IPA in different amounts of KOH and dissolved silicon in KOH in the molar ratio of Si:KOH of 1:1.83 (Table 2).

Titration

All titrations were performed with a Titrando (Deutsche Metrohm, Filderstadt, Germany) using a 20 ml burette and a pH-electrode (Mettler-Toledo, Greifensee, Switzerland). Titration curves, equivalence point recognition and concentration calculation were evaluated with the titration software tiamo 1.1 assuming K$_2$SiO$_3$. A volume of 5 to 10 ml of the alkaline solution was titrated against 0.5 M hydrochloric acid (Merck, Darmstadt, Germany).

The titer was determined with certified sodium carbonate (Merck, Darmstadt, Germany). If the silicon to potassium hydroxide molar ratio exceeded a factor of two, a previous dosing of 1 M sodium hydroxide took place.

High Performance Liquid Chromatography

The determination of 2-propanol (IPA) by high performance liquid chromatography (HPLC) was achieved by a system consisting of the isocratic HPLC pump P680 (Dionex, Sunnyvale, USA) and an Acclaim Organic Acid column (4.0 × 150 mm) with a following refraction index detector at 35°C (Shodex RI-101, Showa Denko, Tokyo, Japan). The mobile phase of the chromatography system was a 30 mM methane sulfonic acid. The injection loop contained a volume of about 1 µl for an automated two position valve (EV700-100-S2, Rheodyne, USA). The samples were not diluted for measurements. The chromatograms were evaluated with the chromatography software Chromeleon 6.70. Peak area was used for the quantitative analysis. The external calibration of 2-propanol was done using 2-propanol in water with the automated injection valve and an auto sampler.

Table 1. Measured NaOH, Si and Na_2CO_3 concentrations and recovery rates for reference samples, prepared from pure chemicals. The initial target concentration was determined by weighting the used substances. Mean ± standard deviation, n = 3 titrations

Sample	Target			Titrated NaOH		Titrated Si		Titrated Na_2CO_3	
	NaOH	Si	Na_2CO_3	NaOH	Recovery	Si	Recovery	Na_2CO_3	Recovery
	(g/l)	(g/l)	(g/l)	(g/l)	(%)	(g/l)	(%)	(g/l)	(%)
Ref A1	26.0	0.0	0.0	25.7 ± 0.3	99	No EP			
Ref A2	25.6	14.9	0.0	24.6 ± 1.5	96	14.3 ± 1.2	96		
Ref A3	25.7	30.0	0.0	26.7 ± 1.9	104	27.9 ± 0.8	93		
Ref A4	25.4	50.0	0.0	26.4 ± 1.5	104	49.4 ± 0.5	99		
Ref A5	26.6	30.0	3.0	24.9 ± 0.1	94	30.4 ± 0.2	101	No EP	(0)
Ref A6	24.9	30.0	6.0	26.3 ± 0.8	105	29.4 ± 0.9	98	6.1 ± 0.4	102
Ref A7	25.1	30.0	9.1	25.7 ± 1.2	102	29.3 ± 0.6	98	8.6 ± 0.1	94

Table 2. Alkaline samples for HPLC consisting of 2-propanol and as matrix potassium hydroxide and silicate (dissolved silicon in KOH). Given are target concentrations as well as characteristic peak data. Negative KOH values appear since the silicon to KOH solution ratio is smaller than 1:2 and silicon is calculated as K_2SiO_3. Mean ± standard deviation, n = 3 titrations and 4 HPLC analyses

Sample	Target concentration			Titrated inorganic		HPLC inorganic		HPLC IPA		
	KOH	Si	IPA	KOH	Si	peak area	peak height	peak area	peak height	IPA
	(g/l)	(g/l)	(g/l)	(g/l)	(g/l)	(μRIU)	(μRIU min)	(μRIU)	(μRIU min)	(g/l)
Ref C1	0	10	40.1	−5.1 ± 0.1	11.0 ± 0.02	27.5 ± 0.6	142 ± 2	33.0 ± 0.3	131 ± 1	39.7 ± 0 4
Ref C2	0	20	40.0	−10.3 ± 0.3	21.1 ± 0.06	55.6 ± 1.0	223 ± 2	32.5 ± 0.5	130 ± 2	39.1 ± 0.6
Ref C3	20	0	40.1	20.9 ± 0.1	No EP	5.1 ± 0.0	48 ± 0	32.6 ± 0.2	125 ± 1	39.3 ± 0.3
Ref C4	15	5	40.1	13.4 ± 0.2	5.1 ± 0.04	27.5 ± 0.6	101 ± 1	33.2 ± 0.7	131 ± 2	40.0 ± 0.9
Ref C5	0	0	41.1	-	-	-	-	33.0 ± 0.6	118 ± 2	39.8 ± 0.7

Surface Tension Measurement

The surface tension of the alkaline solution was measured with a SITA Science Line t60 (SITA Messtechnik, Dresden, Germany), using the bubble pressure method. A direct measurement with the capillary in the process bath did not produce the required reproducibility; hence the measurement took place in a customized overflow measurement cell (Figure 1, Zitt-Thoma, Freiburg, Germany).

Near-Infrared Spectroscopy

Near infrared (NIR) spectra were collected continuously with a FT-NIR process spectrometer FTPA 2000-200 (ABB, Quebec, Canada) directly through the 1/2 in PFA tube of the circulation line which is commonly used in industrial etching plants. In this study, 128 scans from 8000 to 11,000 cm^{-1} were averaged for one spectrum at a resolution of 32 cm^{-1}; the measurement cycle was 30 s. The calibration set up and partial least square calibration model (PLS) as described in Zimmer et al. [4], was used to calculate the KOH, silicon and IPA concentration from the spectra.

Online Measurements at the Etching Plant

All texturisation and analytical experiments were done in a fully automated, semi-industrial scale batch process plant (Stangl Semiconductor, Fürstenfeldbruck, Germany). The alkaline process bath had a volume of 105 l and was divided into an inner process bath of 54 l, where the texturisation took place, and an outer buffer tank. The medium was pumped from the buffer tank to the process unit and circulated via overflow. All samples for the titration were taken from the process bath, the online measurement sites for the near-infrared spectroscopy, surface tension measurement and liquid chromatography are installed on the pressure side of the circulation pump (Figure 1).

RESULTS AND DISCUSSION

Titration of Alkaline Hydroxide, Dissolved Silicon and Alkaline Carbonate

A first method for the determination of alkaline hydroxides and silicates with one single titration was given by Grosvenor [7]. This method uses sodium citrate to improve the detection of the first equivalence point e.g. silicate of the titration curve. Due to efforts in technology in the last ten years, all equivalence points of hydroxide, silicate and carbonate were found without addition of sodium citrate in this study.

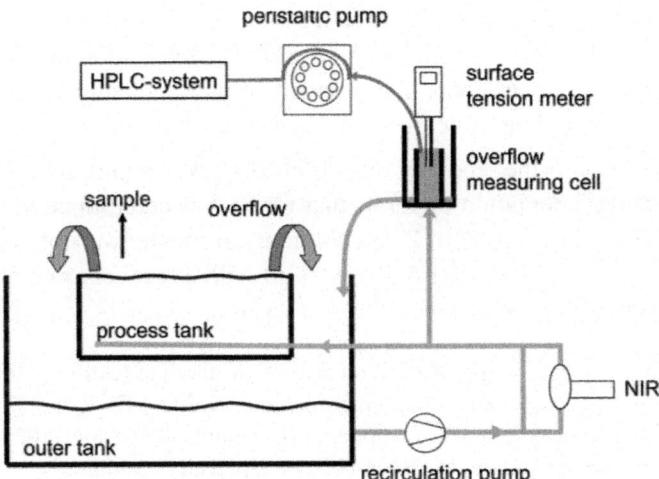

Figure 1. Process bath of the semi-industrial etching plant and installed analytical devices, surface tension meter and HPLC at the recirculation line and NIR at a bypass of the recirculation line.

An acid base titration of an alkaline texturing solution which contains 25.7 g/l NaOH, 29.3 g/l dissolved Si, 8.6 g/l Na_2CO_3 and 50.1 g/l 2-propanol shows three equivalence points (Figure 2).

According to Grosvenor [7] and the pKa values of the components, the volume of the first equivalence point (V1) represents the neutralization of NaOH and the first neutralization of the dibasic sodium silicate. The volume between first and the second equivalence point (V2) is assigned to the complete neutralization of sodium silicate and the first protonation of sodium carbonate. The complete neutralization of the bicarbonate ions occurs between the second and the third equivalence point (V3). All equivalence points were assigned to certain pH windows which enable the distinction between the different species in the absence of one component (Table 3).

The presence of silicate and hydroxide delivers two equivalence points (EP1 and EP2, Table 3), the same pattern as a pure silicate solution. It might be that the EP2 of a pure silicate solution is assigned wrongly as EP3; it can be treated as EP2. Carbonate with and without hydroxide gives EP2 and EP3. The presence of hydroxide can be distinguished by the volumes between the equivalence points. In case of a pure silicate solution, the volume of EP2-EP1 is equal to that of EP1.

The lower titration limit for silicon can be estimated from Figure 3, where samples with a constant NaOH concentration and increasing silicon amount were titrated. A silicon concentration of 1 g/l shows a peak shoulder to the left of the main peak (EP2). At silicon concentrations higher than 2 g/l, a separate peak representing EP1 is formed. A mixture of different amounts of sodium silicate and sodium carbonate in a constant amount of sodium hydroxide were titrated in order to determine the recovery rate (Table 1).

The average of the recovery rates for NaOH was found to be 101% with a relative standard deviation of 3.7%, that for Si was determined to 97% with a relative standard deviation of 2.9%. Sodium carbonate was not detected for a concentration of 3 g/l, but for 6 g/l and 9 g/l with a recovery rate of 102% and 94% respectively.

Table 3. Equivalence points in titration curves of alkaline texturing solutions with different missing components assuming silicon as K_2SiO_3. The cross (X) shows the existing equivalence point for its component. The equivalence point EP3 only appears as carbonate is in the system. Potassium can be replaced by sodium

Component	EP 1 pH 9-11	EP 2 pH 6-9	EP 3 pH 2-6	Volume between EP
KOH		X		
KOH + Si	X	X		EP2 − EP1 > EP1
Si	X	X	(X)[a]	EP2 − EP1 = EP1
KOH + K_2CO_3		X	X	EP3 − EP2 > EP2
K_2CO_3		X	X	EP3 − EP2 = EP2
KOH + Si + K_2CO_3	X	X	X	
Si + K_2CO_3	X	X	X	

a.   The second equivalence point of the titration appears in the pH range of 5 - 6 which refers to EP3. It can be treated as EP2.

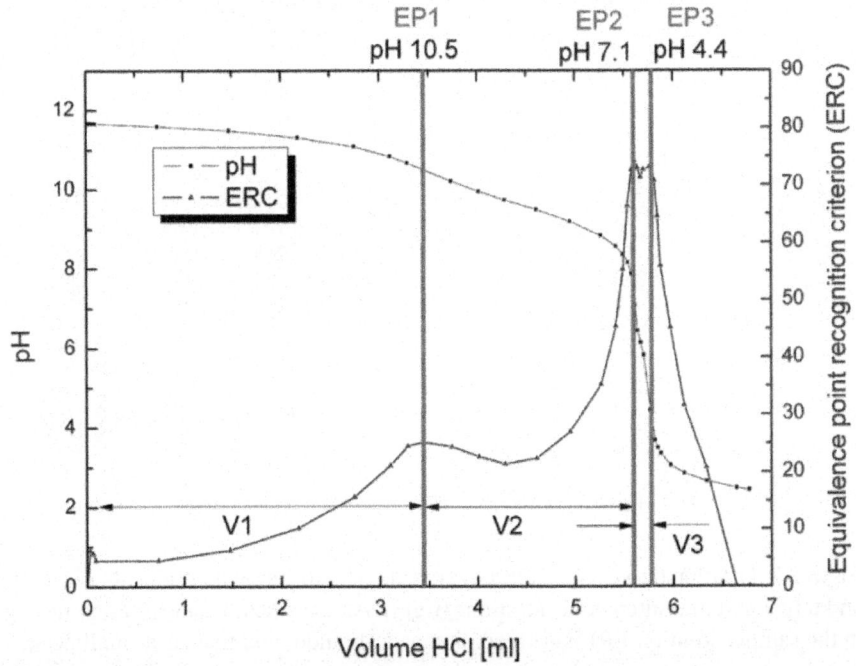

Figure 2. Titration curve for a mixture of hydroxide, silicate and carbonate. The equivalence points EP1, EP2 and EP3 of the components appear at characteristic pH values. From the equivalence points the volumes V1 to V3 can be calculated. V1 refers to the concentration of hydroxide and silicate, V2 refers to bisilicate and bicarbonate and V3 to carbonate.

Due to the alkalinity of the silicate, even a pure silicate solution without free sodium hydroxide is able to etch silicon. Therefore, silicate solutions with a stoichiometry of hydroxide to silicon smaller than two are possible resulting in a (formal) negative KOH value when K_2SiO_3 is assumed for concentration calculation (Table 2). This is indicated in a titration curve by the shift of EP1 towards lower volumes as EP2 stays at the same volume. As the ratio of hydroxide to silicon reaches one, EP1 completely disappears, but can be uncovered by a preceding dosing step of hydroxide.

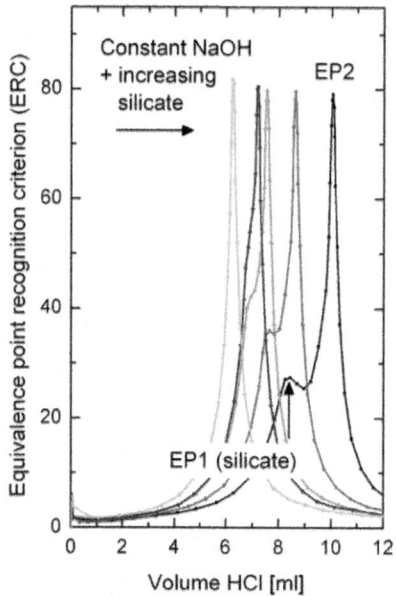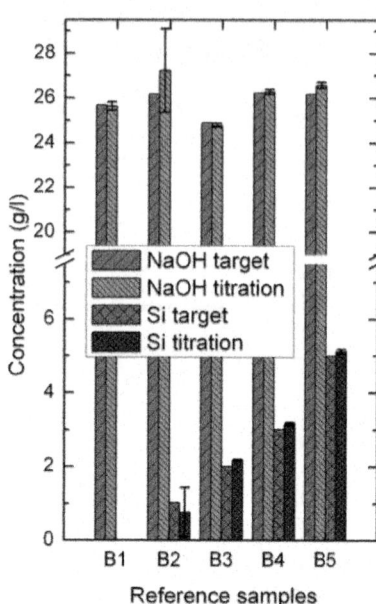

Figure 3. Titration curves of sodium hydroxide with different amounts of silicate (left) and target and measured concentrations (right). As the silicate concentration increases in the sample solution EP1 is formed, first as a shoulder and then as a small peak.

A fix volume of hydroxide is therefore dosed into the titration solution. The detection limit for dissolved silicon and sodium carbonate using an initial volume of 5 ml is in the range of 1 g/l. The detection limit increases with decreasing titration volume.

Determination of 2-Propanol Using an HPLC

The most important parameter to control in alkaline texturing baths is the concentration of the additive, in this case 2-propanol. The 2-propanol concentration in the texturing solution changes very quickly during the process, since the process temperature of 70°C to 80°C is near 2-propanol's boiling point (82°C). An undiluted solution consisting of potassium hydroxide, dissolved silicon as potassium silicate and 2-propanol shows three peaks in the HPLC chromatogram (Figure 4).

The inorganic components elute at 1.5 min, the negative injection peak at 1.9 min, and 2-propanol elutes at 2.1 min. The clear separation of the 2-propanol peak from the inorganic peak depends on the amount of the inorganic components e.g. potassium hydroxide and silicate. At high inorganic charges, the injection peak disappears that the inorganic peak and the 2-propanol peak merge. KOH makes the inorganic peak broader (Figure 4), high amounts of silicate increase the inorganic peak height (from 48 µRIU min for 0 g/l Si to 223 µRIU min for 20 g/l Si, Table 2).

The K^+ load can be correlated to the quotient of the peak area and the peak height. As the K^+concentration is too high the inorganic and 2-propanol merge depending on the separation quality of the column and the eluent concentration. If the peaks merge, the 2-propanol peak height is increased due to the tailing of the inorganic peak and has to be evaluated as a rider. If other additives than 2-propanol are used for texturisation which elute later than 2-propanol, this problem exists only for very high K^+ concentrations. The complete analysis of IPA requires four minutes and is therefore suitable for an insitu control of alkaline texturing baths.

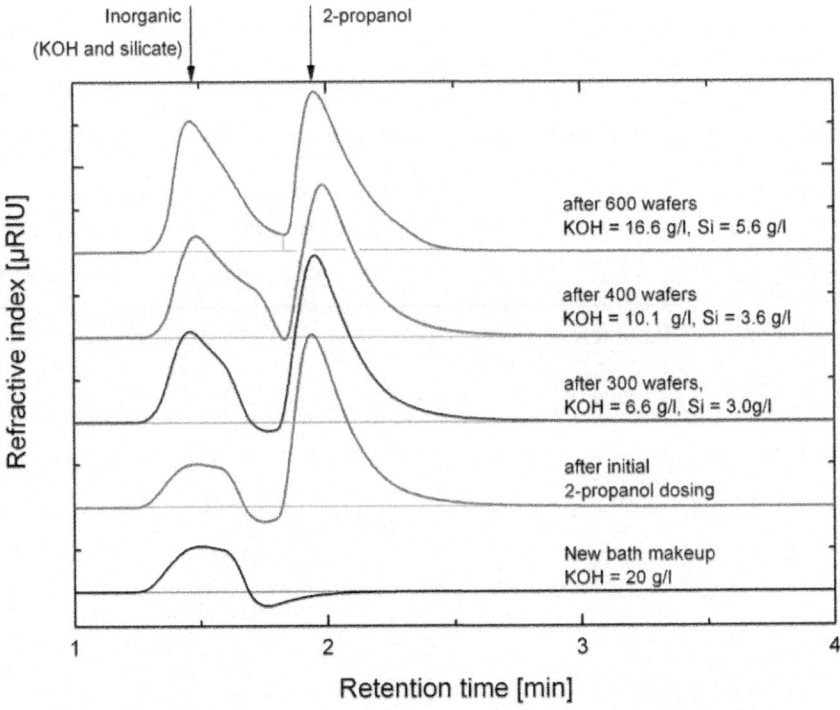

Figure 4. HPLC chromatograms of potassium hydroxide, potassium silicate and 2-propanol during the texturisation of six batches with 100 wafers. The inorganic components elute prior to the injection peak and prior to 2-propanol. A high amount of inorganic components (measurement after 600 wafers) inhibits a separating of the peaks.

Determination of 2-Propanol with Surface Tension Measurement

2-propanol is a surface active alcohol and lowers the surface tension of an aqueous solution. The surface tension decreases with increasing 2-propanol content and temperature of the solution. The dependence of the surface tension s on the 2-propanol concentration can be described by the Szyszkowski equation, which was expanded by three linear terms in Equation (1) [8], which describes the temperature dependence as well as the minor influences of KOH and $K_2Si(OH_2)O_2$.

$$\sigma_{solution} = \sigma_{max} - A \cdot \ln\left(1 + B \cdot c_{isopropanol}\right)$$
$$+ k_{KOH} c_{KOH} + k_{Si} c_{Si} + k_{Temp} T \tag{1}$$

The constants A, B, k_{KOH}, K_{Si} and k_{Temp} were determined empirically (Table 4), fitting measured surface tension to the known concentrations of a reference dataset. The measurement cycle for the surface tension measurement is 8 s, but depends on a running circulation pump and a reference analytical method.

Table 4. Fitted parameters s_{max} as the surface tension of pure water at 0°C and constants for Equation (1) for the determination of the 2-propanol concentration from the surface tension

Parameter	Value
σ_{max}	75.5831
A	15.2159
B	0.0928
k_{KOH}	−0.0683
k_{Si}	−0.1828
k_{Temp}	−0.1675

Determination of KOH and 2-Propanol with NIR Spectroscopy

In near infrared (NIR) spectra 2-propanol has its own vibration mode compared to water and can relatively easily be differentiated to aqueous solutions. This 3v (C-H) vibration mode shows a very sensitive signal at 8440 cm⁻¹ that can be used for its calibration. The NIR calibration model for 2-propanol, KOH and silicon is described in Zimmer et al. [4]. The NIR method is as well as the surface tension temperature dependent, but since the temperature was included in the calibration model, it can be calculated from the spectra. The fast measuring cycle of NIR spectroscopes allows redosing in dependence of the calculated concentrations from NIR.

Online Control of Concentrations

The shown analytical methods were applied for several alkaline texturing processes at a batch etching plant. The analytical data was recorded at the batch plant as shown in Figure 1. The online measuring is suitable for the online analysis for the concentrations of KOH, silicate and the additive and the control of the dosing (Figures 5 and 6).

Online measurements have the advantage that the response returns in a short time (seconds for surface tension and NIR). The online control of the 2-propanol control gives a good overlay of the HPLC and NIR data. The 2-propanol dosing only can be seen in the NIR data, the delay of the HPLC is too long for a quick dosing control, but the increase and decrease of the IPA concentration before and during the process can be seen. The HPLC might be suitable for an additive that does not evaporate and changes its concentration slowly. The KOH and Si concentration can also be measured by NIR and show the increase of silicon and a decrease of the KOH concentration. The KOH dosing before a texturing process results as a spike in the NIR values. The KOH concentration decreased in spite of KOH replenishment. Probably the replenishment was too little.

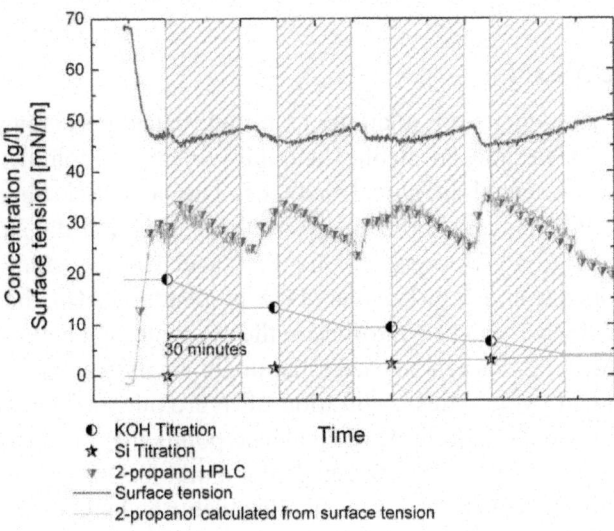

Figure 5. Concentration development in the alkaline texturisation bath during the texturing of four batches with 100 wafers each. Reference measurements were done with

titration (KOH, Si) and HPLC (2-propanol). Additionally the 2-propanol concentration calculated from online surface tension measurement is shown.

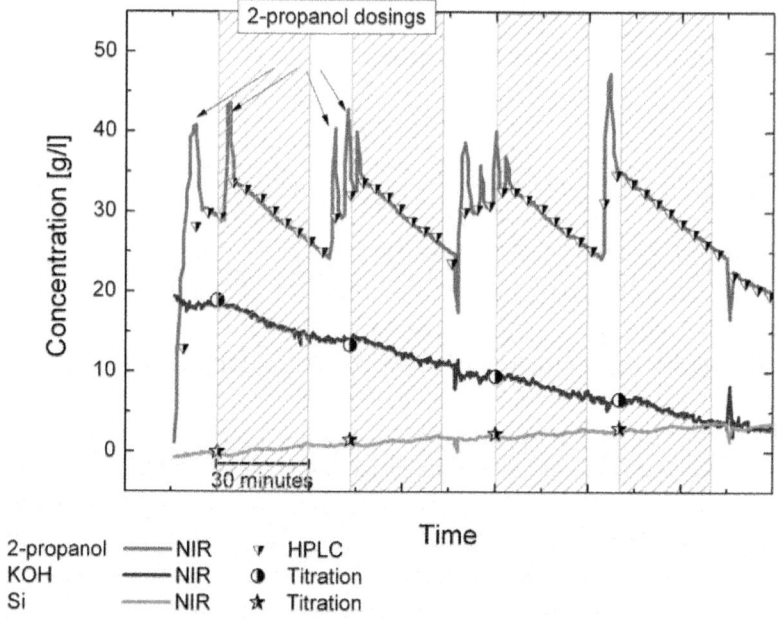

Figure 6. Concentration development as described in **Figure 5** with additionally plotted online-measurements with nearinfrared spectroscopy.

CONCLUSION

The main compounds of alkaline texturing baths consisting of alkaline hydroxide, silicate and an alcohol as additive can be completely analyzed by using the presented analytical methods. The acid base titration analyzes hydroxides, silicates and carbonates, and the HPLC and the surface tension analyze the additive 2-propanol. The NIR spectroscopy allows measuring hydroxide, silicate and 2- propanol and the temperature. Also, the dosing of the chemicals and water can be controlled in the concentration curves of the online data. All these methods can be used as an in-situ analytical method that differs in measurement cycle times. A titration of hydroxide, silicate and carbonate takes about 10 min due to three equivalence points that need to be found. An HPLC chromatogram has been finished after 4 min for 2-propanol. The surface tension and the NIR spectrum of an alkaline texturing bath are recorded in less than 20 s. The NIR method and the surface tension need a reference analytical method in order to calibrate them for the used chemicals. The advantage of both methods is that they do not consume any eluents or titration solutions.

REFERENCES

1. D. Biro, S. Mack, A. Wolf, A. Lemke, U. Belledin, D. Erath, B. Holzinger, E.-A. Wotke, M. Hofmann, L. Gautero, S. Nold, J. Rentsch and R. Preu, "Thermal Oxidation as a Key Technology for High Efficiency Screen Printed Industrial Silicon Solar Cells," Proceedings of the 34th IEEE Photovoltaic Specialists Conference, Philadelphia, 7-12 June 2009, pp. 1594-1599.

2. A. Goetzberger, J. Knobloch and B. Voss, "Crystalline Silicon Solar Cells," John Wiley & Sons Ltd., Chichester, 1998.

3. J. Rentsch, J. Ackermann, K. Birmann, H. Furtwängler, J. Haunschild, G. Kästner, R. Neubauer, J. Nievendick, A. Oltersdorf, S. Rein, A. Schütte, M. Zimmer and R. Preu, "Wet Chemical Processing for C-Si Solar Cells-Status and Perspectives," Proceedings of the 24th European Photovoltaic Solar Energy Conference, Hamburg, 21-25 September 2009, pp. 1113-1117.

4. P. K. Singh, R. Kumar, M. Lal, S. N. Singh and B. K. Das, "Effectiveness of Anisotropic Etching of Silicon in Aqueous Alkaline Solutions," Solar Energy Materials and Solar Cells, Vol. 70, No. 1, 2001, pp. 103-113. http://dx.doi.org/10.1016/S0927-0248(00)00414-1

5. I. Zubel and M. Kramkowska, "Etch Rates and Morphology of Silicon (h k l) Surfaces Etched in KOH and KOH Saturated with Isopropanol Solutions," Sensors and Actuators A: Physical, Vol. 115, No. 2-3, 2004, pp. 549-556.http://dx.doi.org/10.1016/j.sna.2003.11.010

6. K. Fisher and J. Cotter, "Investigation of Low Molarity Alkaline Texturing Solutions," Proceedings of the 15th International Photovoltaic Science & Engineering Conference, 2005, pp. 622-623.

7. V. L. Grosvenor, "Aqueous Potentiometric Titration of Silicate and Hydroxide in Alkali Silicate Solutions," American Chemical Society, Vol. 54, No. 4, 1982, pp. 837-838.

8. W. Adamson and A. P. Gast, "Phsical Chemistry of Surfaces," 6th Edition, Wiley, New York, 1997.

Chapter 4

ECONOMIC-ORIENTED STOCHASTIC OPTIMIZATION IN ADVANCED PROCESS CONTROL OF CHEMICAL PROCESSES

László Dobos, András Király, and János Abonyi

Department of Process Engineering , University of Pannonia, Egyetem Street 10, 8200 Veszprém, Hungary

ABSTRACT

Finding the optimal operating region of chemical processes is an inevitable step toward improving economic performance. Usually the optimal operating region is situated close to process constraints related to product quality or process safety requirements. Higher profit can be realized only by assuring a relatively low frequency of violation of these constraints. A multilevel stochastic optimization framework is proposed to determine the optimal setpoint values of control loops with respect to predetermined risk levels, uncertainties, and costs of violation of process constraints. The proposed framework is realized as direct search-type optimization of Monte-Carlo simulation of the controlled process. The concept is illustrated throughout by a well-known benchmark problem related to the control of a linear dynamical system and the model predictive control of a more complex nonlinear polymerization process.

INTRODUCTION

Due to the dynamic and significant changes of the economic environment performance assessment of process control is highlighted area of chemical engineering [1]. The aim of this paper is to develop an optimization framework designed to determine optimal operating regimes of chemical processes by taking process constraints, desired maximum number (frequency) of constraint violations, and process uncertainties into consideration.

Variance in the closed control loop caused by unmeasured disturbances and badly designed controllers might cause variations in the product quality. In case of increasing variance of process variables the probability and frequency of violation of quality requirements are increasing that might lead to the increase of the amount of less valuable offset products. Typical examples when reduced number of violations of the predetermined process constraints are acceptable can be found in the field of statistical process control (SPC [2]). In Statistical Process Control statistical tools are applied to monitor the performance of the production process and detect significant deviations that may later result in offset products.

In this paper a more sophisticated model-based approach is followed. Modern process analysis, monitoring, control, and optimization tools are mainly based on some kind of process model. It is obvious to utilize these process models also in the economic assessment and optimization. Usually the output of cost-benefit analysis is cost reduction or profit increment expressed by a cost function. These functions incorporate the costs of the operation, raw materials, current prices of products [3], and risks of malfunctions. In our economic-oriented optimization strategy the aim is to find steady state operation points (controller set points) where profit might be realized. This task is fulfilled at the supervisory control level [4].

The general approach for economic performance evaluation comprises the following steps: reduce the variance of the controlled variable and shift the set points (process mean) closer to the operation limits [5] without increasing the frequency of the violation. This operation is referred to as the improved control [6]. The variance reduction might mean to retune the existing controllers, or, in more radical cases, change the whole control strategy. The model-based predictive controllers (MPCs [7]) is highly applicable for variance-reduction purposes. Application of MPCs in the operative control level results in a multilayer optimization problem, since an MPC also minimizes its objective function. In this approach the upper layer is the supervisory control level which is responsible for economically optimal operation, and the lower layer is for variance reduction.

To handle uncertainty and effects of measurement noise in this paper a novel Monte-Carlo simulation-based approach is proposed. Monte-Carlo simulation is frequently applied in various areas [8]. This tool has also proven its efficiency in risk-related optimization of chemical processes; for example, it is applied in optimizing maintenance strategies of operating processes [9]. There is a common characteristic in these solutions: the stochastic nature of the studied system has to be modeled. In the applied methodology this simulation is related to the modeling of the unmeasured disturbances of the control loops. To handle this random effect, Monte-Carlo simulation is applied with the characterized noise. An economic cost function is calculated in every case to measure the economic efficiency of the process. Integrating this benefit analysis tool into the mesh adaptive direct search optimization algorithm—

where the task is to find the most beneficial steady state operation point—resulted in the proposed economic-oriented optimization framework. In the proposed multilayer optimization framework, the application of gradient-based methodologies for maximizing the economic throughput is not possible, thanks to the stochastic characteristics caused by the closed-loop variance. That is why the utilization of direct search methods is necessary. Mesh Adaptive Direct Search (MADS) [10] class of algorithms is a relatively new set of direct search methods for nonlinear optimization; that is, these algorithms are capable of calculate the extremums of a nonsmooth functions, like our economic objective function. Since the steady state operation points are mainly determined by the variance of the controlled variables, incorporating this effect into the model is inevitable. The created optimization framework functions as an industrial Advanced Process Control system, [3, 11].

The paper is organized as follows: in Section 2 the economic cost function-based multilayer optimization framework is introduced. In Section 3 the applied methodology is explained in detail. In Section 4, the efficiency of the proposed methodology is illustrated throughout a linear benchmark control problem and a Model Predictive Controlled (MPC) highly non-linear technology. In both cases an economic performance measure has been formalized as a basis for optimizing the set point signal. As base case of the benchmark example the process is controlled with a PI controller. To reduce the closed-loop variance caused by unmeasured disturbance a linear MPC is installed to replace the PI controller. With the reduction of the variance the set point of the controller can be moved closer to the process constraints which yields higher economic performance. Such economic-oriented optimization is carried out at two different risk levels. As a second example a non-linear process controlled by a linear MPC is considered, since this combination is widely applied in chemical process industry. In this case study the process variance is caused by an unmeasured disturbance, model mismatch, and noise added to the controlled variable. In this example the effect unmeasured disturbance with different amplitude is examined in detail. These examples show the realistic benefits of the proposed methodology.

ECONOMIC COST FUNCTION-BASED MULTILAYER OPTIMIZATION

The proposed framework is rather similar to Advanced Process Control (APC) systems applied in the chemical process industry, [3]. The scheme of this multilayer optimization problem is depicted in Figure 1. The main aspects and tasks that have to be taken into consideration in the different optimization levels will be introduced in the following subsections.

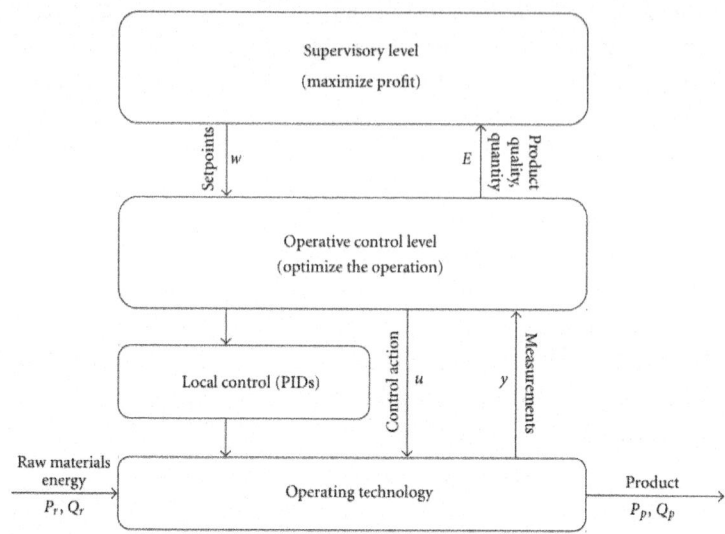

Figure 1: The layers of an economic optimization of an operating technology.

The Supervisory Control Level

The main task in the supervisor level is to maximize the economic throughput with varying the steady state set point signal. In general the economically optimal set point is close to the operation limits of process. That is why the reduction of the closed-loop variance is necessary. Thanks to process variance—caused by disturbances, noise, and so forth—there is a risk of process constraint violation which has to be taken when the new set point is determined. The essence of this economic optimization approach is depicted in Figure 2.

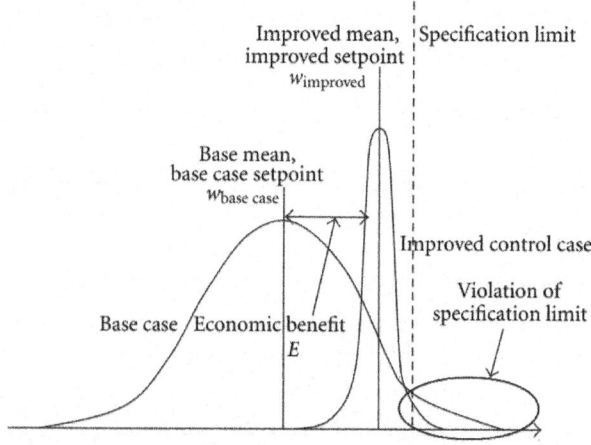

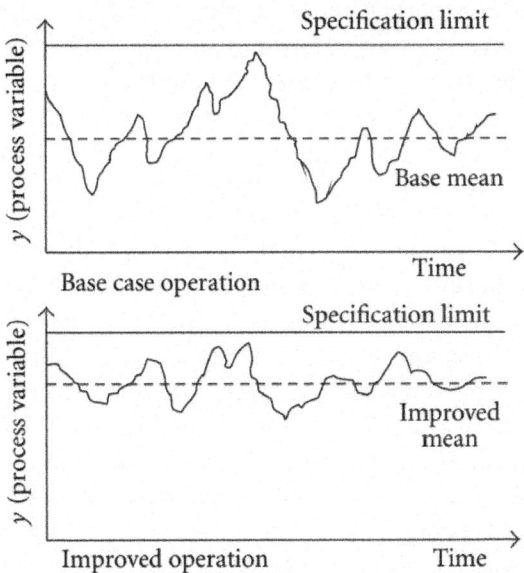

Figure 2: Approach to economic benefit estimation with variance reduction.

The aim of the economic-oriented process optimization framework can be formulated as maximizing a cost function. Such cost function mostly includes the cost of the actual operation, raw materials, and the value of the product as the following:

$$\max_{w} E = \sum_{i=1}^{N} P_i \cdot Q_i - \sum_{r=1}^{M} P_r \cdot Q_r, \tag{1}$$

where P_p, P_r and Q_p, Q_r are the prices and the quantities of the products and raw materials, respectively. In the optimization problem $w = [w_1, \dots, w_p]^T$ represents the setpoints of local controllers of the operative control level; p is the number of the controlled variables, denoted with y_i. The task of the operative control level can be summarized as y_i should be as close to w_i as possible.

Continuous economic improvement of process control is about to find the setpoint values of control loops with the possible highest economic performance. In steady state operation the value of (1) is often increasing by shifting the steady state operation point closer to the constraints of the process. To reach this goal, the reduction in variance of the key process variables is necessary with, for example, retuning the controller or even redesigning the existing control strategy. As Figure 2 illustrates, when the variance of the key process variables is reduced extra profit can be realized as the difference of the economic potential in the old and revised steady state operation points.

The cost function must be optimized with respect to the process constraints to ensure the required process safety and product quality. Constraints defined on the process variables can be expressed as follows:

$$y_{i,\min} \leq y_i, \leq y_{i,\max}, \quad i = 1, \ldots, p,$$

$$u_{j,\min} \leq u_j, \leq u_{j,\max}, \quad j = 1, \ldots, m, \tag{2}$$

where $y_{i,\min}$ and $y_{i,\max}$ are the lower and upper bounds of output variables, $u_{j,\min}$ and $u_{j,\max}$ are the lower and upper bounds of input variables, and p and m are the number of output and input variables, respectively. Thanks to uncertainties, such as process variation and disturbance, the probability of violating the predetermined process constraints is increasing by getting closer to them. A reasonable approach to handle the uncertainties in the constraints is to cast the problem in term of the probability of constraint is to cast the problem in term of the probability of constraint violation, which is the approach to be implemented in this paper.

The probability constraints can be expressed as

$$\Pr\{y_{i,\min} \leq y_i \leq y_{i,\max} \ i = 1, \ldots, p\} \geq 1 - \alpha, \tag{3}$$

or

$$\Pr\{y_{i,\min} \leq y_i \leq y_{i,\max}\} \geq 1 - \alpha_i \quad i = 1, \ldots, p, \tag{4}$$

where $\Pr\{\cdot\}$ is the operator of probability and α is the specified probabilistic violation level (demonstrated in Figure 2). The formulation of probability constraints means that satisfying process constraint of y_i is not required by 100% probability but a certain confidence level, $1 - \alpha$. Inequality (3) represents a so-called Joint Probabilistic Constraint (JPC) problem, which means that all process variables must be kept in the defined operation regime with maximum probability of violation of α. Inequality (4) is called is Individual Probabilistic Constraint problem (IPC) ([6]), where each process variable has a specified confidence level, $1 - \alpha_i$, to be satisfied. In this paper the second approach has been adopted.

The final goal in multilayer optimization is to maximize the accessible profit, determined by the cost function mentioned before in (1), by finding the optimal steady state operation point with respect to the process constraints, (2). Thanks to process variation a reasonable risk level has to be taken by defining the probability of process constrain violation (formalized as α). $1 - \alpha$ means a confidence level, which is a non-linear constraint in the economic optimization. This optimization problem represents the supervisory control layer.

The Operative Control Level

In optimization and control of complex production processes, the role of Model Predictive Controllers (MPCs) is increasing. The more and more widespread application is reasonable, thanks to the good variance reduction ability.

MPC is a model-based control algorithm where models are used to predict the behavior of process outputs of a dynamical system with respect to changes in the process inputs. The MPC uses the models and current plant measurements to calculate future moves in the manipulated variables, which will result in operation that honors all input and output variables' constraints (see (2)).

Predictive control uses the receding horizon principle. This means that after the computation of the optimal control sequence, only the first control action will be implemented; subsequently, the horizon shifted one sample and the optimization is restarted with new information about the measurements. That is the reason why the MPCs do not optimize the operation on the time horizon of the whole steady state operation, but consider just the horizon, implemented in the controller, and solve the optimization problem iteratively. With the help of Figure 3 the essence of the model predictive control is easily understandable.

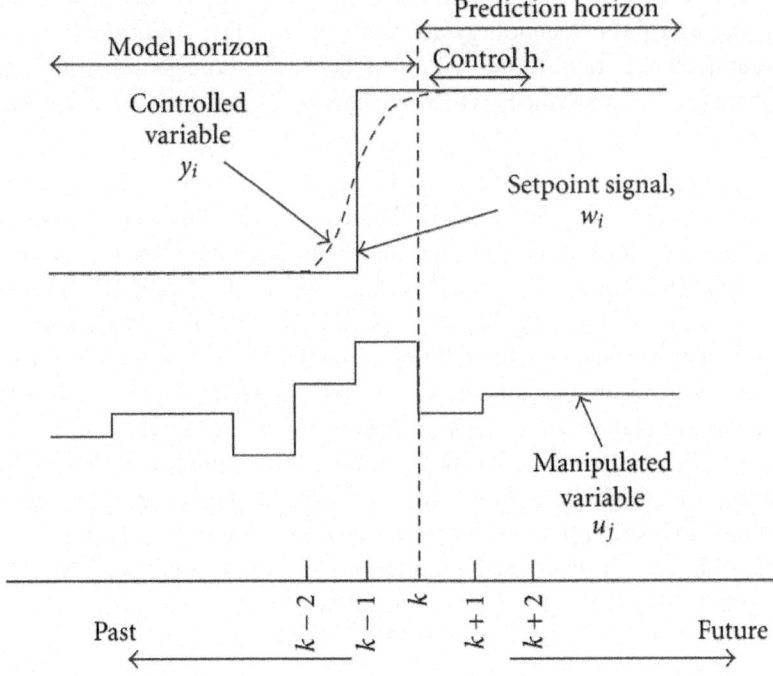

Figure 3: Illustration of essence of MPC in control of y_i by manipulating u_j.

Utilizing this control strategy in the operative control level the previously proposed multilayer optimization framework can be resulted (depicted in Figure 1), thanks to the control rule of MPCs, expressed as:

$$\min_{\Delta \mathbf{u}_{(k+j)}} \sum_{i=1}^{p} \sum_{j=1}^{H_p} \left(w_{i,k+j} - y_{i,k+j} \right)^2 + \sum_{i=1}^{m} \lambda_i \sum_{j=1}^{H_c} \Delta u_{i,k+j-1}^2, \tag{5}$$

where p and m are the numbers of the controlled variables and manipulated variables, respectively. In MPC control strategy the different number of controlled and manipulated variables is acceptable, since the interconnection between the different manipulated and controlled is considered in the process model, which is applied in the MPC. The tuning parameters of the controller are as follows: H_p and H_c are the length of prediction and control horizo, λ is the a factor for punishing the change of the control signal. Δu is the variation of the manipulated variable at a given time, which is calculated during the optimization method on control horizon.

MPC formulates an objective function which is used to find the optimal input sequence to eliminate the difference of the controlled variable and the set point in the prediction horizon. Since this objective function, (5), is designed to ensure smooth and stable operation it does not directly reflect the economic performance of the technology (formalized in (1)). Additionally this cost function does not count to the risk of violating the process constraints caused by unmeasured disturbances which appears as closed-loop variance (see (3)-(4)).

Constraint violations have also to be taken into account during economic performance optimization. This is the reason why this paper suggests the application of Monte Carlo simulation of the augmented process model and the model of the operative control level (see Figure 1). The result of the Monte Carlo simulation is an aggregated economic performance (e.g., mean of the economic performance of the individual runs). Due to the stochastic nature of the optimized system the gradient of this aggregated economic cost function is difficult to calculate. Hence the optimization algorithm should be gradient free-yet computationally very effective. To meet this requirement the application of the advanced Mesh Adaptive Direct Search methodology is proposed. In the following section the multilayer optimization framework and its two main building blocks—Monte Carlo simulation and Mesh Adaptive Direct Search methodology—are going to be introduced in detail.

STOCHASTIC MODELING ECONOMIC BENEFIT MAXIMIZATION WITH DIRECT SEARCH METHODOLOGY

Taking process variance into account the previously proposed economic-oriented objective function, (1), becomes a stochastic characteristic. To handle uncertainties Monte Carlo simulation is applied. The Monte Carlo method is applied frequently in solution of stochastic optimization problems, for example, in stochastic linear programming [12, 13]. Kjellstrom [14] was the first to use Monte Carlo estimators for the iterative improvement of convergence behavior in nonlinear stochastic optimization.

Due to the stochastic characteristics caused by the closed-loop variance the application of gradient-based methodologies for maximizing the economic throughput is not efficient. Integrating the simulation-based economic performance assessment methodology into a direct search optimization algorithm an effective optimization framework is obtained. Mesh Adaptive Direct Search (MADS) [10] class of algorithms is a relatively new set of direct search methods for nonlinear optimization; that is, these algorithms are capable of calculating the extremums a nonsmooth objective functions, like our economic objective function.

Our methodology is stated as follows:

1. economic performance assessment of the considered steady state operation point. It means applying a set point (w) and calculating the value of the economic cost function, (1), with respect to the process constraints, (2), and the value of the probability of constraint violation (see (3)-(4)). Because of considering the process variance as random phenomena, Monte Carlo simulation with multiple runs of augmented process simulator is applied to aggregate the effect of the random variances in a final economic cost function;

2. integrate the economic performance evaluation tool into the MADS optimization algorithm to find the economically optimal steady state operation point. The previously applied economic cost function, (2), has to be maximized with respect to the proposed constraints with varying setpoint signal (w). This algorithm can handle constraint limits of process variables, the certain confidence levels to violate these limits.

Using the methodology discussed above the optimization process is capable to isolate and handle all the disturbances technology has, whose nature is constant in time; thus it can be characterized statistically. These uncertainties are time homogeneous and static-time disturbances, like measurement noise or model error. In the following section the application way of Monte Carlo simulation and MADS optimization algorithm is introduced briefly.

Monte Carlo Simulation

Monte Carlo Simulation (MCS) methods are highly applied in the mathematical modeling problems where some kind of stochastic phenomena must be handled. In the proposed multilayer optimization framework process variance caused by unmeasured disturbances is considered. The Monte Carlo simulation consists of the following steps.

1. Define the domain of possible inputs.

2. Generate inputs from this domain randomly using a specified probability distribution.

3. Execute deterministic computation using the inputs.

4. Aggregate the results of the computations into the final result.

In engineering practice normal distribution is considered as an adequate assumption for characterizing uncertainties. At the modeling of the considered process the following steps are followed: at first the mathematical model of the process is created. Then noise and unmeasured disturbances of the control loops are characterized and random signals related to the real process variance are added to the corresponding input and output variables. The value of the economic objective function, (1), is calculated by aggregating the results of the individual Monte Carlo runs into a statistical economic performance. Since complex production processes are mostly characterized by non-linear process models the economic assessment and optimization needs an optimization algorithm which is able to handle the non-linear cost functions and constraints, (2), (3), and (4).

The Mesh Adaptive Direct Search Methodology

Since the calculation of the gradient of the economic objective function with respect to the steady state operation points is highly computational demanding and due to the Monte Carlo simulation the economic cost function is nonsmooth the application of gradient free optimization method is needed. Mesh Adaptive Direct Search (MADS) [10] is a relatively new set of direct search methods for nonlinear optimization. This algorithm is capable of minimizing a nonsmooth function, like our economic cost function (1) under the proposed constraints in (2) and (4). According to [10, 15], MADS can be interpreted as a generalization of Generalized Pattern Search (GPS) [16] algorithms, with the restriction to finitely many pool direction removed.

MADS is an iterative algorithm, where at each iteration a finite number of test points are generated. At the beginning of an iteration, the infeasible test points are filtered (discarded); that is, infinite objective value is assigned to it ($f(x) = +\infty$). Thereafter the feasible test points are evaluated by the objective function and compared with the current best objective function value found so far. Each of these test points lies on the current mesh, which is constructed from a finite set of n_D directions $D \in R^n$ and scaled by the mesh size parameter $\Delta^m_k \in R^n$. If we find a point with lower objective value than the current best one, this test point is a so-called improved mesh point and the iteration is a successful iteration.

Each iteration consist of two steps, the so-called SEARCH step and POLL step. SEARCH step can return any point of the underlying mesh; it is trying to find an unfiltered point. If it fails to generate an improved mesh point, then the second step, the POLL is invoked. POLL step consists of a local exploration around the current best solution, and the test points are generated in some directions scaled by the mesh size parameter. MADSs are novel in the number of usable directions, since In GPS, POLL directions belong to a finite set, while POLL direction in MADS belongs to a much larger set; in fact if the iteration number k goes to infinity, the union of the normalized POLL directions over all k becomes dense in the unit sphere. According to [10], this algorithmic construction allows stronger convergence. Another important difference between MADS and GPS is the so-called poll size parameter, Δ^p_k. This parameter determines the size of the frame where the POLL step can operate. In case of GPS, mesh size and poll size are equal ($\Delta^m_k = \Delta^p_k$), while in MADS these two parameters can differ. This difference is depicted in Figure 4. Additional pieces of information like convergence analysis or practical implementations can be found in [15].

$$\Delta^m_k = \Delta^p_k = 1$$

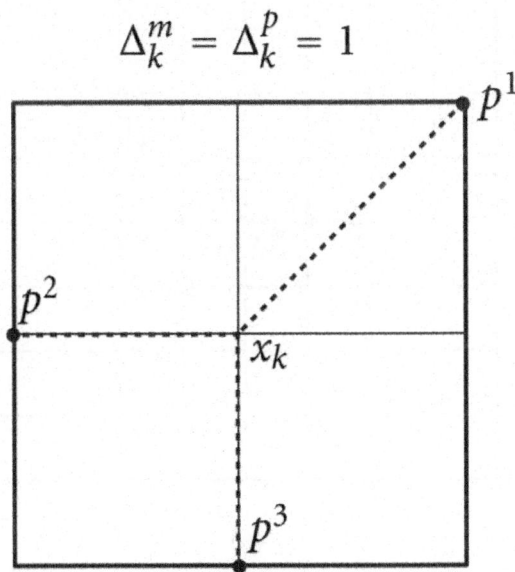

$$\Delta_k^m = \Delta_k^p = \frac{1}{2}$$

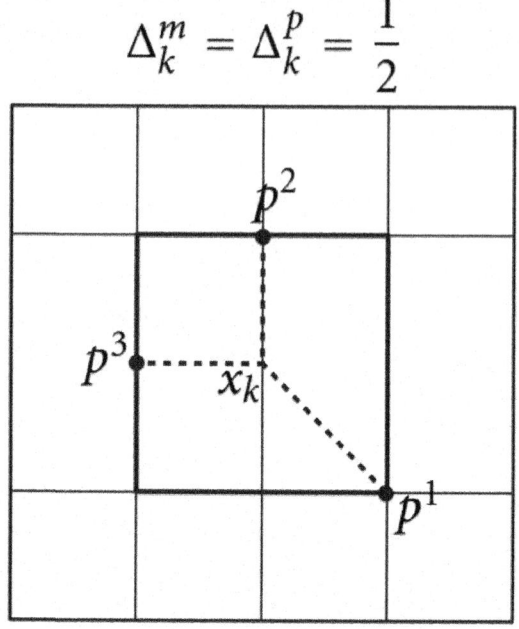

$$\Delta_k^m = \Delta_k^p = \frac{1}{4}$$

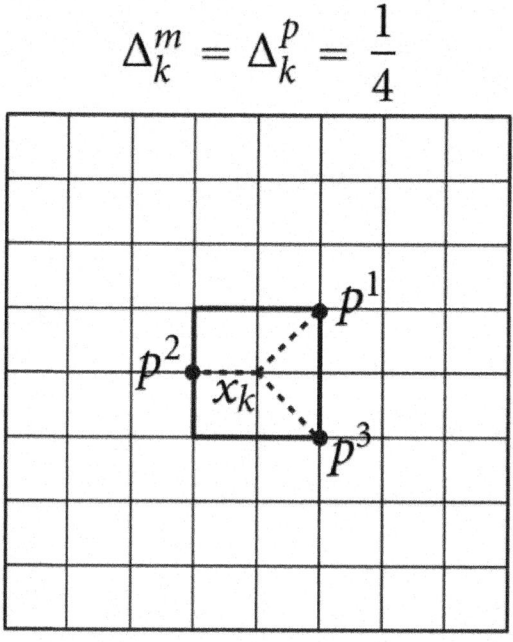

(a)

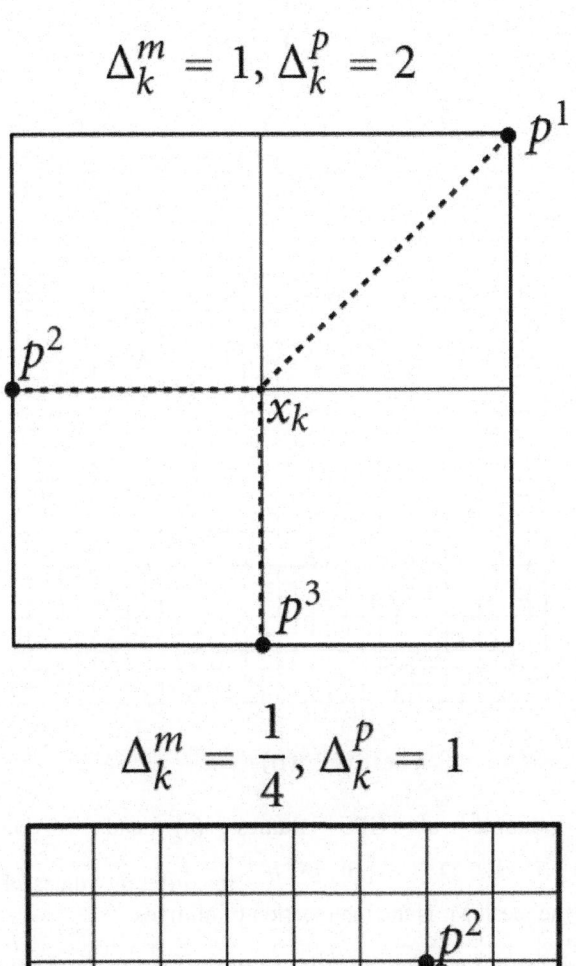

$$\Delta_k^m = \frac{1}{16}, \Delta_k^p = \frac{1}{2}$$

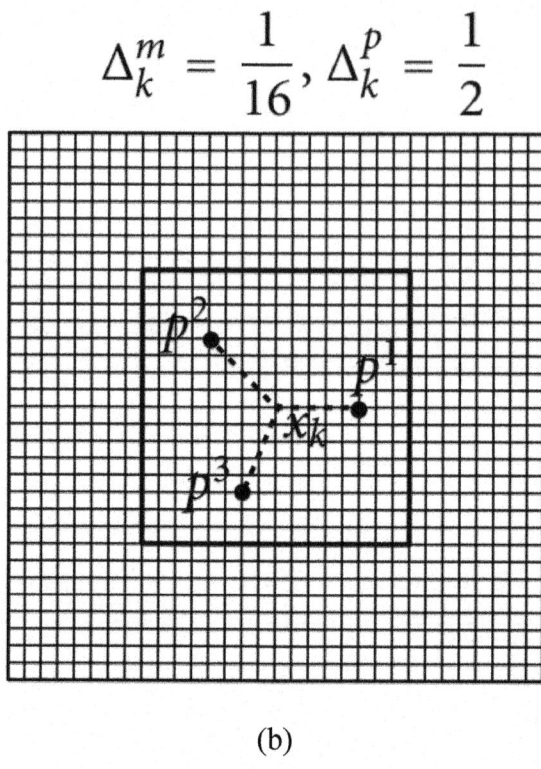

(b)

Figure 4: Example of GPS frames (a) and MADS frames (b) $P_k = \{x_k + \Delta_k^m d : d \in D_k\} = \{p^1, p^2, p^3\}$ for different values of $\Delta_k^m = \Delta_k^p$. In all six figures, the mesh M_k is the intersection of all lines.

In the economic-oriented multilayer optimization framework (see Figure 1) MADS is applied in the supervisory level to maximize the economic performance formalized as (1). The optimization problem is solved with respect to the the process constraints, (2), and the value of the probability of constraint violation, see (3)-(4), with varying setpoint signal (w). Since MADS needs a reduced number of runs of the augmented process simulator, the optimal value of the setpoint signals can be quickly obtained. The low number of iteration during optimization is necessary, since Monte Carlo simulation of the operative control level (augmented process simulator) is applied, which is highly computation demanding process.

In the following section the effective application of the proposed framework is going to be examined throughout the case studies of a benchmark, linear process, and an MPC controlled highly non-linear technology.

APPLICATION EXAMPLES

In this section, two application examples are presented to demonstrate the applicability of the proposed framework for enhancing the economic benefit of the operating technologies. The calculations for both examples are based on closed-loop data, generated using Matlab-Simulink. The uncertainties are presented in the examples as noise superimposed to inputs and outputs.

A SISO Process

Consider a SISO process, characterized by G_p shown in Figure 5 subject to disturbance dynamics G_d described by

$$y_k = G_p u_k + G_d \alpha_k = \frac{0.6299 z^{-1}}{1 - 0.8899 z^{-1}} u_{k-2}$$

$$+ \frac{1 - 0.8 z^{-1}}{1 - 0.8899 z^{-1}} \alpha_k \quad k = 1 \ldots q,$$

(60

where α_k is a normally distributed white noise sequence of mean 0 and variance 1. q signs the last time step of the considered simulation. The objective in the supervisory control level is to maximize the output ($\bar{y}$—mean of the output on the considered time horizon) with respect to the process constraints. The optimization problem can be formalized as

$$\max_{w} 2\bar{y}$$

(7)

subject to

$$-10 \leq y_k \leq 10$$

$$-5 \leq u_k \leq 5$$

$$k = 1 \ldots q.$$

(8)

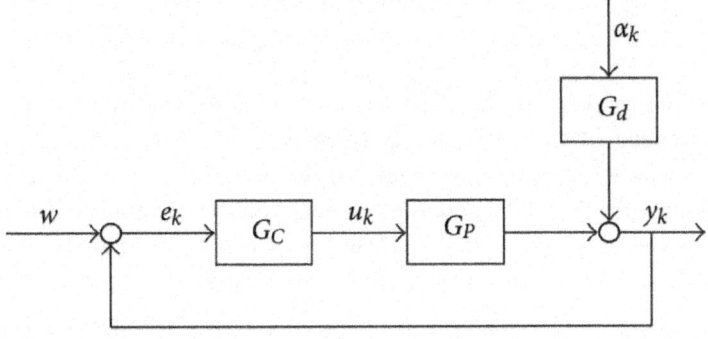

Figure 5: Block diagram of the SISO closed-loop system.

As base case a PI controller is designed. The controller parameters are K_c = 1.926T_I = 0.6. As previously shown, specifying the probability of not violating the constraint defined on the output variable defines a non-linear constraint for the optimization problem. During the presented studies this confidence level is assigned as 95% and 90%. In the literature [6] the same SISO process is utilized with the same probabilities. The means of the output are $\bar{y}$ = 1.49 and $\bar{y}$ = 2.72 confidence level of 95% and 90%, respectively. The output data in 95% confidence level is depicted in Figure 6.

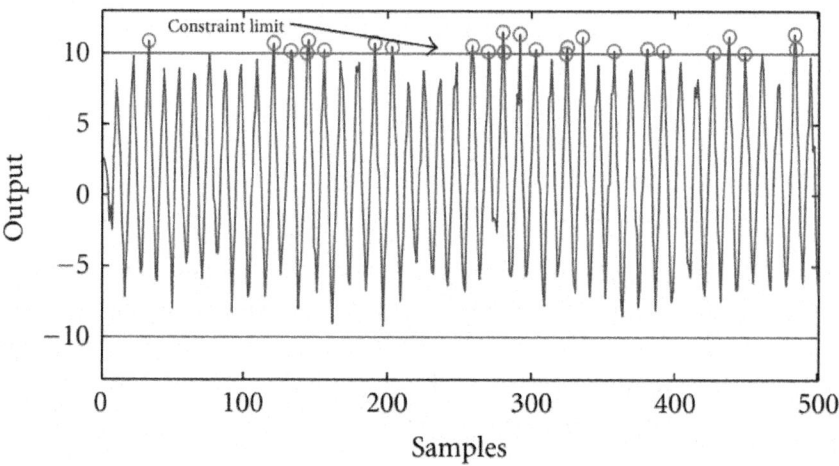

Figure 6: Base case operation with probability constraint level of 95%.

Since MPC is highly applicable for variance reduction purposes the PI controller has been replaced with a linear Dynamic Matrix Controller (DMC) [17]. DMC applies the linear convolution model of the process for predicting the effects of the considered manipulated variable sequence. With the application of DMC lower variance ($\bar{\sigma}_{PI}$ = 5.3 in contrast to $\bar{\sigma}_{DMC}$ = 1.05) and higher economic benefit might be expected. The control rule of the DMC has been proposed in (5). The tuning parameters of the applied DMC are H_m = 50, H_p = 20, and H_c = 10. The value of λ is chosen as 1000. By applying the previously proposed multilayer optimization framework significant improvement could be experienced in the economic performance (the number of Monte Carlo iterations was set to 100). The means of the output are $\bar{y}$ = 8.28 and $\bar{y}$ = 8.68 at confidence level of 95% and 90% respectively. The outputs in the improved operation are depicted in Figures 7 and 8. It can be clearly stated when the confidence level decreases the frequency of constraint violation increases.

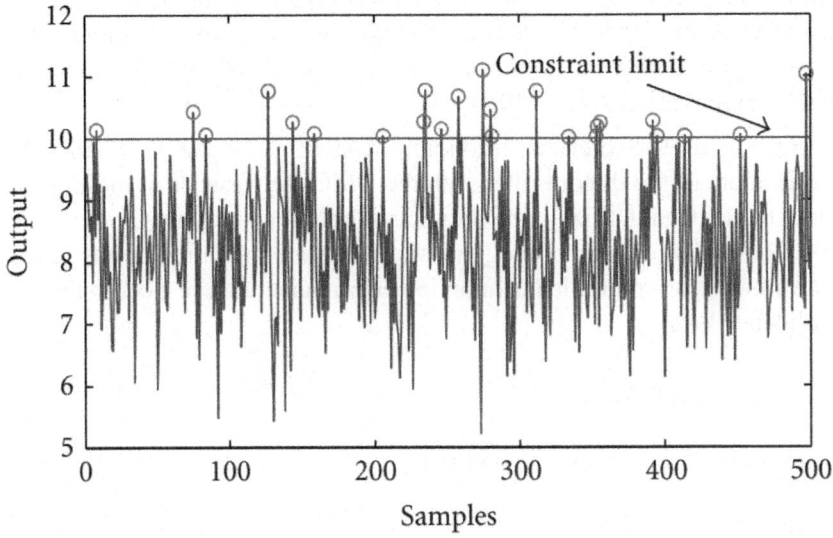

Figure 7: Improved case, optimal operation with probability constraint level of 95%.

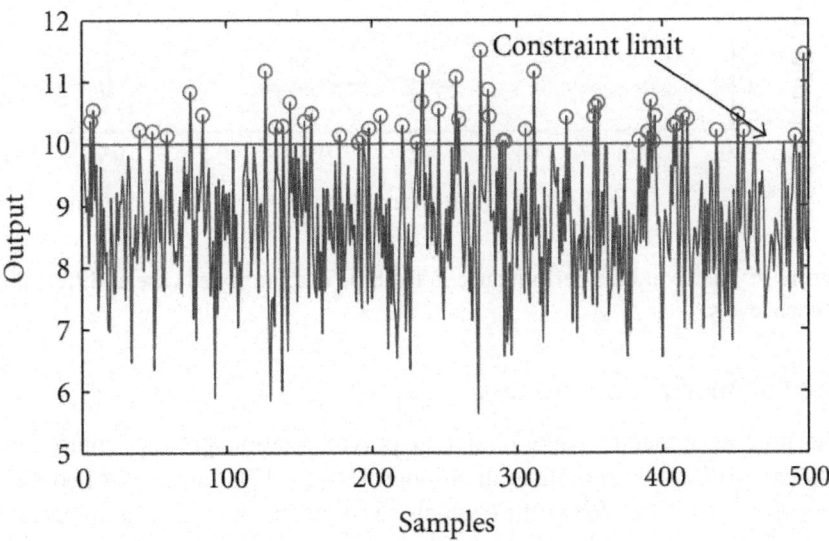

Figure 8: Improved case, optimal operation with probability constraint level of 90%.

The number of individual economic performance evaluations in the Monte Carlo-simulation has been set to 100. There has been an attempt to apply quadratic programming as optimization algorithm (utilizing Matlab, Optimization Toolbox), but the computational demand was extremely high, almost one hour even in this simple example. By applying MADS, the computation demand has been significantly decreased into 5 minutes. In both cases the initial setpoint for the optimizer was set equal to the upper constraint of the output variable, $w_0 = 10$.

In Figure 9 achievable economic benefit is depicted. Thanks to the replacement of the PI controller with the DMC the variance in the closed-loop could be reduced. Utilizing the previously introduced Monte Carlo simulation based optimization methodology new steady state operation points have been determined with multiplied economic performance with respect to the defined confidence level.

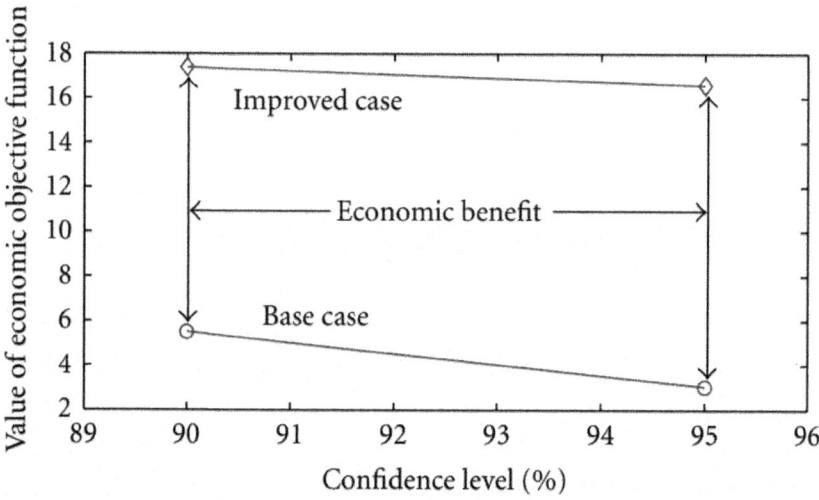

Figure 9: The economic performance in the base and improved case at different confidence levels.

The Polymerization Process

The process under consideration is a polymerization process controlled by a linear MPC, the previously mentioned DMC, [17]. The controlled system possesses all those difficulties which exist in an operating polymerization process.

Process Description

The reactor which has been studied is a CSTR where a free radical polymerization reaction of methyl-metacrylate is considered using azobisisobutyronitrile (AIBN) as initiator, and toluene as solvent. The aim of the process is to produce different kinds of product grades. The number average molecular weight is used for qualifying the product and process state. The polymerization process can be described by the following model equations [18]:

$$\frac{dC_m}{dt} = -\left(k_p + k_{fm}\right)C_m P_0 + \frac{F(C_{min} - C_m)}{V}$$

$$\frac{dC_I}{dt} = -k_I C_I + \frac{F_I C_{Iin} - F C_I}{V}$$

$$\frac{dD_0}{dt} = (0.5k_{tc} + k_{td})P_0^2 + k_{fm}C_m P_0 - \frac{FD_0}{V}$$

$$\frac{dD_1}{dt} = M_m\left(k_p + k_{fm}\right)C_m P_0 - \frac{FD_1}{V},$$

$$(9)$$

where

$$P_0 = \sqrt{\frac{2f^* C_I k_I}{k_{td} + k_{tc}}}$$

$$(10)$$

The notation for the equations can be seen in Table 1. The number average molecular weight (NAMW) is defined by the ratio of D_1/D_0. By assuming an isotherm operation model the process model consists of four states, represented by four differential equation (9) [19]. For the integration, the MATLAB's built-in ode45 function has been used, which is based on an explicit Runge-Kutta (4,5) formula. During simulations $T_s = 0.03$h is applied as sample time.

Table 1: Design parameters for MMA polymerization reactor

F	$1.0\,\text{m}^3/\text{h}$
C_{min}	$6.4678\,\text{kmol/m}^3$
C_{Iin}	$8\,\text{kmol/m}^3$
V	$0.1\,\text{m}^3$
M_m	$100.12\,\text{kg/kmol}$
f^*	0.58
$R\cdot$	$8.314\,\text{kJ/kmol}\cdot\text{K}$
k_p	$2.4952\cdot10^6\,\text{m}^3/\text{kmol}\cdot\text{h}$
k_I	$1.0224\cdot10^{-1}\,1/\text{h}$
k_{fm}	$2.4522\cdot10^3\,\text{m}^3/\text{kmol}\cdot\text{h}$
k_{tc}	$1.3281\cdot10^{10}\,\text{m}^3/\text{kmol}\cdot\text{h}$
k_{td}	$1.0930\cdot10^{11}\,\text{m}^3/\text{kmol}\cdot\text{h}$

MPC Controller Strategy and the Economic Performance Assessment

The qualification of the product and process operation is based on the number average molecular weight. Thanks to the non-linear model equations the development economic performance turns into a highly non-linear optimization problem.

The control objective on the supervisory control level is to maximize the economic performance of the process. The objective function is formalized as

$$\max_{w_{NAMW}} E = P_{onspec} \cdot Q_{onspec} - P_{offspec} \cdot Q_{offspec},$$

(11)

where P_{onspec} (with the value of 10000) and $P_{offspec}$ (with the value of 3500) are the prices of the polymer product which fulfill/not fulfill the product specifications; w_{NAMW} indicates the steady state setpoint. Q is the quantity of the polymer product, calculated with the following expression:

$$Q_{polymer} = F \cdot D_1.$$

(12)

During the economic-oriented optimization, the following process constraints have to be considered:

$$24000 \leq NAMW \leq 26000.$$

(13)

An important characteristic of the process is the increasing of product quantity when shifting the steady state operation closer to the lower limit, hence the optimal steady state operation point is expected near to the lower limit. The maximum probability of violating the process constraints is 1%, so the mentioned confidence level is 99%. The number of Monte Carlo simulations is 100, similarly to the previous case. The closed-loop variance of the process is caused by the noise added to the inlet monomer flowrate (F) with means of 0 and $\overline{\sigma} = 0.014$. Other source of the closed-loop variance is the noise added to the controlled variable with the mean of 0 and $\overline{\sigma} = 143$.

On the operative control level the previously proposed DMC is designed. The manipulated variable in the control strategy of the reactor is the initiator inlet flow rate. The tuning parameters of the applied DMC are $H_m = 30$, $H_p = 3$, and $H_c = 3$. The value of λ is chosen as 4.10^{12}.

As base case the safest steady state operation point has been chosen which is in the middle of the specified operation range ($w_{NAMW} = 25000$). As the result of the economic performance optimization the optimal steady state setpoint is $w_{NAMW} = 24300$. Thanks to this setpoint modification the quantity of the produced polymer has been increased with 5% and throughout this the economic performance also increased with 5%. The result of the closed-loop simulation with the optimal setpoint is depicted in Figure 10.

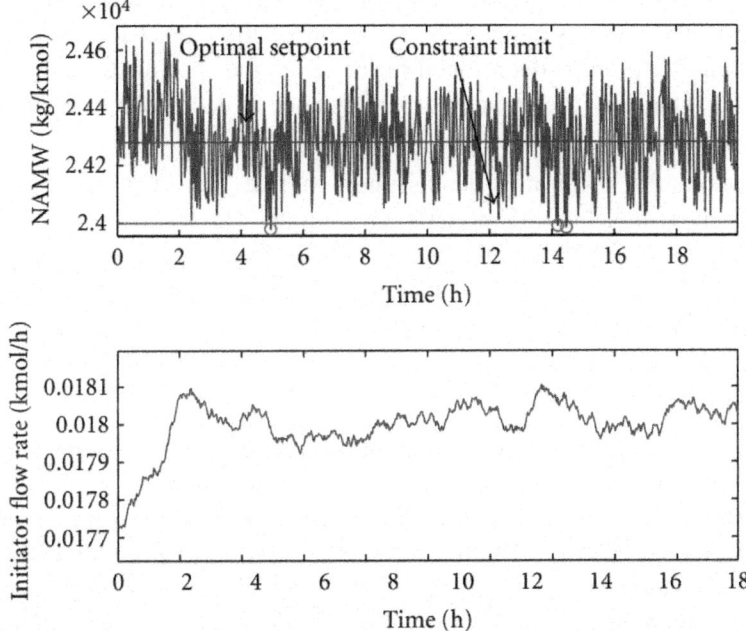

Figure 10: Improved case, optimal steady state operation point of PMMA reactor.

The number of individual economic performance evaluations in the Monte Carlo simulation has been set to 100. In this case study there has been an attempt to apply quadratic programming as optimization algorithm. The same result has been obtained with quadratic programming but the computational demand was extremely high, almost 10 hour. By applying MADS, the computation demand has been decreased into 1 hour. As it can be seen, the quadratic programming might be applicable but its computation demand is exaggerated. The initial setpoint for the optimizer was set equal to the lower constraint of the output variable, $w_0 = 24000$.

As Figure 10 shows that the frequency of constraint violation is conspicuously low, the process constraint has been violated only three times. It means 99.55% probability of not violating the limits, in contrast to the previously determined confidence level, which was 99%. It may happen since the off-specification product (products which do not fulfill the requirements) means extra outgoings in the economic objective function. Accordingly it is not worth to produce even just 1% off-specification product; however, this amount can be accepted technologically.

If the circumstances of the steady state operation change, there will be the need of redetermining the optimal operation point. This case is considered when the deviation of the noise on the monomer inlet flow rate has been increased to $\overline{\sigma} = 0.03$. Since the variation of the closed loop is increased, the optimal setpoint is determined further from the specification limit, w_{NAMW} = 24400. This way the economic performance is decreased compared to the previous case with 0.5%.

The results confirmed the assumption that the economically optimal operation is close to the process constraints. However, 99% was set as confidence level of limit violation; the way of formulating the economic cost function does not allow such a low quantity of off-specification product, since it causes extra outgoings during operation.

CONCLUSION

In this paper an economic-oriented optimization framework has been introduced to determine optimal operation regimes of complex chemical process systems. Situations where the economically optimal steady state operation point is close to the technological limits of the operation have been studied. Due to process variance caused by unmeasured disturbances and measurement noise the determination of the optimal values of the controller setpoints is rather difficult since shifting the operation point closer to the process limits results in risk of violation of constraints related to process safety and product quality requirements. By formulating an economic objective function the performance and the risk level of the operation can be quantitatively evaluated. Monte Carlo simulation is applied with multiple runs of process model (augmented with the model of the control system) with economic performance assessment to handle the stochastic phenomena of process variance. Integrating the Monte Carlo simulation-based economic performance assessment tool into the Mesh Adaptive Direct Search (MADS) methodology can take process constraints and desired risk/confidence levels into consideration. Since MADS is the one of the recent gradient-free optimization methodology with high efficiency its application in Monte-Carlo simulation is much more effective than the classical gradient-based quadratic programming (SQP) by using MADS the time demand of optimization can be shortened to one-tenth of SQP.

The efficiency of the proposed framework is demonstrated throughout benchmark examples. In the first benchmark problem significant economic benefit was realized by finding the optimal setpoint signal after variance reduction. The process operation has been optimized at different confidence levels. Thanks to the efficiency of MADS the optimization has taken only 5 minutes; however, 100 iterations have been set in the Monte Carlo simulation. In case of the non-linear process the increase of economic throughput is not as significant as in the previous case, but with application of the proposed framework 5% profit increase can be obtained. The conditions of the optimization were similar to the previous case, 100 individual runs hve been set in the Monte Carlo simulation. Since the process model is non-linear the time consumption of the optimization is longer, almost one hour.

The application of the proposed methodology requires an existing process model with the description of the control system and detailed analysis of the process uncertainties. These modeling and analysis tools are widely available in advanced technologies thanks to the increasing interest for APC and Operator Training Systems (OTS). Another inevitable condition of application is the availability of an economic objective function. Although economic performance measures are frequently missing in current technologies, nowadays online economic performance monitoring is more and more indispensable. The aim of the application of these economic oriented monitoring tools is to avoid operations that are not economically or energetically optimal. Utilizing these performance assessment tools and integrating them with optimization the proposed framework is resulted.

ACKNOWLEDGMENTS

This work was supported by the European Union and financed by the European Social Fund in the frame of the TAMOP-4.2.1/B-09/1/KONV-2010-0003 and GOP-1.1.1-11-2011-0045 projects.

REFERENCES

1. K. H. Lee, E. C. Tamayo, and B. Huang, "Industrial implementation of controller performance analysis technology," Control Engineering Practice, vol. 18, no. 2, pp. 147–158, 2010.

2. J. Oakland, Statistical Process Control, Butterworth-Heineman, 2007.

3. M. Bauer and I. K. Craig, "Economic assessment of advanced process control—a survey and framework," Journal of Process Control, vol. 18, no. 1, pp. 2–18, 2008.

4. X. Chen, M. Heidarinejad, J. Liu, and P. D. Christofides, "Distributed economic mpc: application to a nonlinear chemical process network," Journal of Process Control, vol. 22, pp. 689–699, 2012.

5. K. H. Lee, B. Huang, and E. C. Tamayo, "Sensitivity analysis for selective constraint and variability tuning in performance assessment of industrial MPC," Control Engineering Practice, vol. 16, no. 10, pp. 1195–1215, 2008.

6. C. Zhao, Y. Zhao, H. Su, and B. Huang, "Economic performance assessment of advanced process control with LQG benchmarking," Journal of Process Control, vol. 19, no. 4, pp. 557–569, 2009.

7. C. E. García, D. M. Prett, and M. Morari, "Model predictive control: theory and practice—a survey,"Automatica, vol. 25, no. 3, pp. 335–348, 1989.

8. R. Rubinstein and D. Kroese, Simulation and the Monte Carlo Method, Wiley-Interscience, 2008.

9. E. Borgonovo, M. Marseguerra, and E. Zio, "A Monte Carlo methodological approach to plant availability modeling with maintenance, aging and obsolescence," Reliability Engineering and System Safety, vol. 67, no. 1, pp. 61–73, 2000.

10. C. Audet and J. E. Dennis, "Mesh adaptive direct search algorithms for constrained optimization," SIAM Journal on Optimization, vol. 17, no. 1, pp. 188–217, 2006.

11. W. H. Ray, Advanced Process Control, McGraw-Hill, New York, NY, USA, 1981.

12. A. Prékopa, Stochastic Programming, Springer, 1995.

13. K. Marti, Y. Ermoliev, and G. Pflug, Dynamic Stochastic Optimization, Springer, 2004.

14. G. Kjellstrom, "Network optimization by random variation of component values," Ericsson Technics, vol. 25, no. 3, pp. 133–151, 1969.

15. M. A. Abramson, C. Audet, J. W. Chrissis, and J. G. Walston, "Mesh adaptive direct search algorithms for mixed variable optimization," Optimization Letters, vol. 3, no. 1, pp. 35–47, 2009.

16. V. Torczon, "On the convergence of pattern search algorithms," SIAM Journal on Optimization, vol. 7, no. 1, pp. 1–25, 1997.

17. N. L. Ricker, "The use of biased least-squares estimators for parameters in discrete-time pulse-response models," Industrial and Engineering Chemistry Research, vol. 27, no. 2, pp. 343–350, 1988.

18. A. Silva-Beard and A. Flores-Tlacuahuac, "Effect of process design/operation on the steady-state operability of a methyl methacrylate polymerization reactor," Industrial and Engineering Chemistry Research, vol. 38, no. 12, pp. 4790–4804, 1999.

19. B. R. Maner and F. J. Doyle, "Polymerization reactor control using autoregressive volterra-based mpc,"AIChE Journal, vol. 43, no. 7, pp. 1763–1784, 1997.

Chapter 5

STUDY ON THE COLOR LEVELNESS OF SILK FABRIC DYED WITH VEGETABLE DYES

Mohammad Gias Uddin

Department of Textile Engineering, Ahsanullah University of Science and Technology, Dhaka 1208, Bangladesh

ABSTRACT

At present, customers are more conscious about the quality of clothing items. As a result, it becomes more challenging for the manufacturers to maintain the consistent dyeing quality. Unlevelness is generally unaccepted in commercial dyeing, except some cases such as stone-washed, acid-washed or pigment-dyed textiles. The aim of the study was to evaluate the color levelness quality of fabric dyed with vegetable dyes. For this purpose, dyeing was carried out on silk fabric with the aqueous extracts of mango, guava and henna leaves. Then the levelness of colored fabric was evaluated by spectrophotometric measurements in terms of color difference, ΔE value. It was concluded that unmordanted dyed samples have better color levelness quality than the mordanted ones. The unmordanted samples dyed with mango and guava leaves extracts showed excellent levelness quality while the unmordanted ones dyed with henna leaves extracts showed good levelness quality. On the other hand, the mordanted dyed samples showed good leveling quality for the three vegetable dyes except ferrous sulphate mordanted sample dyed with henna leaves extract. Again, among the mardanted samples the levelness quality was comparatively better in case of alum mordanted dyed samples and the average ΔE value was increased gradually when moved from alum to tin and finally to ferrous sulphate.

BACKGROUND

The vegetable dyes are generally used for the coloration of the products in textile and fashion industries (For yarns, fabrics and garments), food industry (For sweets, confectionery and bakery products), and cosmetics industry

(For soap, cream, powder, lipstick etc.). There is a big potential for using the vegetable dyes in textile industry due to one and only reason i.e. eco-friendliness of the product [1, 2]. Whole plant (e.g. weld/marigold), barks (e.g. oak/maple), leaves (e.g. tea/henna), seeds (e.g. annatto), roots (e.g. madder/blood root), semen shell (e.g. Persian nut), flowers (e.g. dahlia/gul mohar), rhizome (e.g. water lily), fruits (e.g. elder berries), fruit rinds (e.g. pomegranate) or, skin (e.g. onion) etc. can be the sources of vegetable dyes [3]. The use of vegetable materials in textile coloration is a well-known way of utilizing renewable raw materials according to the technical and ecological requirements of the 21st Century [4]. For this purpose, traditional dye plants have also been cultivated in many countries. As an estimate 1 kg of plant material will be required to dye 1 kg of textile goods, this indicates that the use of vegetable dyes including secondary products will be cost effective [5]. Many dyes can be easily available from garden waste. In many of the world's developing countries, vegetable sources not only give final products with a great variety of shades but also the possibility of an income through sustainable harvest and sale of these plants [6]. Many of the vegetable dyes attract the most consumers with a special fragrance and many provide antibacterial and UV protective functions to fabrics [7, 8].

However, with the increased interest in vegetable dyes a greater importance was given on using mordant. Most of these dyes require some sort of mordant to set permanently in the fibers. Metal ions of mordants which are widely used act as electron acceptors for electron donors to form coordination bonds with the dye molecules, making them insoluble in water [9, 10]. This leads to improve dye uptake and retention, which result in a greater depth of shade and color fastness properties [11]. Common mordants used are alum, ferrous sulphate, stannous chloride etc. [12].

Day by day the customers are going to be more conscious about the quality of clothing items. As a result, it becomes more challenging for the manufacturer to maintain the consistent dyeing quality. Unlevelness is generally unaccepted in commercial dyeing, except for stone-washed, acid-washed or pigment-dyed textiles, where an effect is sought to comply with a fashion trend. Color levelness is actually a description of the uniformity of shade in different places of the fabric. Traditionally it is judged by visual assessment, or by comparing similar samples, e.g. from the same fabric batch, to determine color differences. The human eyes can detect small differences in color, but it is difficult to quantify color differences accurately. Textile manufacturing industries are mainly concerned with the appearance of color uniformity in terms of levelness parameters in dyed fabrics and/or union shade in dyed fabrics of more than one fiber type [13]. Since mordants are also used in the processing of vegetable dyes, unlevelness could be a major problem in textile dyeing and dyeing with the vegetable sources should have proper control over color levelness.

METHODS

Materials

The raw silk fabric (1/1 plain weave and 22 g/m^2) purchased from Sopura Silk Ltd., Dhaka was used in this study.

Methods

Pre-Treatment

The raw silk fabric was degummed in an aqueous soap solution (15 g/L soap) at pH 9, material to liquor ratio 1:50, and temperature 80°C for 60 min. The degummed fabric was washed with 2 g/L detergent at 65°C for 10 min. Then, the fabric was treated with 35% hydrogen peroxide (3 mL/L) solution, maintaining a material to liquor ratio of 1:50 at pH 9 and temperature 60°C for 60 min followed by wash with 2 g/L detergent at 65°C for 10 min. Mordanting was carried out using 5% (On fabric weight) of ferrous sulphate (FeSO$_4$), alum [KAl(SO$_4$)$_2$·12H$_2$O], tin (SnCl$_2$) mordants individually and using three different combinations of mordants such as ferrous sulphate–alum (2.5% + 2.5%), ferrous sulphate–alum–tin (2% + 2% + 1%) and alum–tin (2.5% + 2.5%) at 60°C for 60 min keeping material to liquor ratio 1:30.

Dyeing

Extraction of the coloring component from vegetable leaf sources—mango, guava and henna leaves was carried out in aqueous boiling process. It was done by extracting fixed quantity of powdered leaves with a material to liquor ratio of 1:10 (Weight of leaves powder in g; amount of water in mL) at boil for 60 min. Dyeing was carried out in exhaust method with the dye extracts as per standard dyeing conditions recommended for silk fabric [14] at 80°C for 60 min under pH 5, keeping material to liquor ratio 1:50 in a IR sample dyeing machine.

Color Difference, ΔE

Each dyed sample is measured in a dual beam reflectance spectrophotometer (Datacolor 650, USA) using D$_{65}$ illuminant and 10° observer settings considering the reading-1 as standard and other nines as sample batches. Data for each batch are analyzed with respect to color difference, ΔE value. ΔE is a single value that takes into account the differences between the L*, a* and b* values of the sample and standard in the CIE L*a*b* color system. The Eq. 1 was used to calculate the ΔE [15].

$$\Delta E = \sqrt{[(\Delta L^*)^2 + (\Delta a^*)^2 + (\Delta b^*)^2]} \tag{1}$$

ΔL^* = L* sample − L* standard, Δa^* = a* sample − a* standard, Δb^* = b* sample − b* standard; where standard refers to the reading-1 in dyed fabric, sample refers to other readings in the corresponding dyed fabric.

The ΔE value is an important parameter in the dyeing process, which can indicate degree of levelness of dyed fabric. For the study, the degree of levelness was described according to ΔE values as shown in Table 1.

Table 1: Suggested interpretation of ΔE values

ΔE value	Visual appearance of levelness	Extent of unlevelness
≤0.20	Excellent levelness	Unlevelness not detectable
0.21–0.50	Good levelness	Unlevelness noticeable under close examination
0.51–1.0	Poor levelness	Apparent unlevelness
>1.0	Bad levelness	Conspicuous unlevelness

RESULTS AND DISCUSSION

Levelness of the Dyed Samples

The ΔE values of silk fabric dyed with henna, guava and mango leaves extracts using different mordants were shown in Table 2.

Table 2: ΔE values of silk fabrics dyed with vegetable dyes

Dye sources	Mordant type	ΔE values of dyed samples										Average ΔE
		R-1	R-2	R-3	R-4	R-5	R-6	R-7	R-8	R-9	R-10	
		Batch readings										
Mango leaves	No mordant	Standard	0.19	0.17	0.07	0.22	0.10	0.21	0.09	0.12	0.10	0.14
	FeSO$_4$		0.45	0.24	0.37	0.42	0.56	0.22	0.32	0.41	0.61	0.40
	FeSO$_4$–Alum		0.28	0.42	0.32	0.26	0.61	0.37	0.35	0.27	0.22	0.34
	FeSO$_4$–Alum–Tin		0.29	0.24	0.13	0.34	0.32	0.22	0.55	0.14	0.56	0.31
	Alum		0.19	0.30	0.18	0.25	0.12	0.32	0.43	0.27	0.15	0.25
	Alum–Tin		0.39	0.23	0.25	0.19	0.45	0.34	0.27	0.13	0.17	0.27
	Tin		0.24	0.15	0.39	0.17	0.15	0.51	0.38	0.11	0.25	0.26
Guava leaves	No mordant		0.12	0.27	0.15	0.13	0.03	0.42	0.21	0.14	0.24	0.19
	FeSO$_4$		0.49	0.54	0.36	0.68	0.56	0.55	0.29	0.30	0.46	0.47
	FeSO$_4$–Alum		0.54	0.37	0.56	0.24	0.38	0.34	0.26	0.34	0.18	0.36
	FeSO$_4$–Alum–Tin		0.21	0.20	0.18	0.45	0.24	0.43	0.49	0.51	0.43	0.35
	Alum		0.08	0.24	0.14	0.21	0.13	0.33	0.05	0.38	0.51	0.23
	Alum–Tin		0.48	0.29	0.07	0.23	0.24	0.30	0.09	0.17	0.28	0.23
	Tin		0.31	0.64	0.34	0.17	0.18	0.23	0.54	0.43	0.22	0.34
Henna leaves	No mordant		0.29	0.22	0.11	0.20	0.16	0.17	0.33	0.18	0.24	0.21
	FeSO$_4$		0.75	0.23	0.45	0.57	0.48	0.62	0.79	0.33	0.55	0.53
	FeSO$_4$–Alum		0.34	0.37	0.24	0.23	0.42	0.37	0.30	0.13	0.54	0.33
	FeSO$_4$–Alum–Tin		0.35	0.26	0.58	0.40	0.41	0.43	0.13	0.43	0.42	0.38
	Alum		0.13	0.25	0.21	0.16	0.53	0.51	0.17	0.27	0.19	0.27
	Alum–Tin		0.46	0.35	0.12	0.34	0.18	0.23	0.24	0.25	0.63	0.31
	Tin		0.32	0.53	0.19	0.36	0.20	0.24	0.11	0.32	0.23	0.28

From Table 2, it was found that levelness was found highest when no mordant was used on the fabric as the average ΔE value was the lowest for all the three vegetable dyes in such cases.

In case of mango leaves dyed fabric, when no mordant was used, the minimum average ΔE value was found 0.14 among all the results and in this case, maximum levelness was found which can be expressed as excellent levelness i.e., unlevelness was not detectable (See Table 1). In addition, when mordants were used, all the average ΔE values were found within 0.21–0.50 range and can be described as good leveling i.e., unlevelness was noticeable under close examination. Among the mordanted samples dyed with mango leaves extracts, the order of levelness found was: alum (0.25) > tin (0.26) > alum–tin (0.27) > ferrous sulphate–alum–tin (0.31) > ferrous sulphate–alum (0.34) > ferrous sulphate (0.40).

Again, in case of guava leaves dyed fabric, the minimum average ΔE value was found 0.19 when no mordant was used and in this case, maximum levelness was found which can be expressed as excellent levelness. When mordants were used, all the average ΔE values of guava leaves dyed fabric were found within 0.21–0.50 range and can be described as good leveling. Among the mordanted samples dyed with guava leaves extracts, the order of levelness found was: alum and alum–tin (0.23) > tin (0.34) > ferrous sulphate–alum–tin (0.35) > ferrous sulphate–alum (0.36) > ferrous sulphate (0.47).

Moreover, in case of henna leaves dyed fabric, good levelness was found when no mordant was used, as the average ΔE value was found 0.21. But except the ferrous sulphate mordanted dyed sample, all the mordanted dyed samples showed the average ΔE value within 0.21–0.50 range i.e., good levelness. For ferrous sulphate mordanted dyed sample the average ΔE value was found highest (0.53). It can be interpreted as sample of poor levelness quality i.e., showing some kind of apparent unlevelness. Thus among the mordanted samples dyed with henna leaves extracts, the order of levelness found was: alum (0.27) > tin (0.28) > alum–tin (0.31) > ferrous sulphate–alum (0.33) > ferrous sulphate–alum–tin (0.38) > ferrous sulphate (0.53).

Thus, among the fabrics dyed with the three vegetable dyes, the noticeable trend found was unmordanted dyed sample showed better levelness than mordanted dyed samples and among the mardanted samples the levelness quality was comparatively better in case of alum mordanted dyed samples and the average ΔE value was found increased gradually when moved from alum to tin and finally to ferrous sulphate.

The phenomenon of levelness or unlevelness can be interpreted mainly by the homogeneity of the dye distribution on the fabrics dyed with the vegetable dyes. The reason behind the better levelness of the unmordanted samples was due to homogeneity of the dye distribution on the fabric, while non-homogenous distribution of the dyes can result in the greater ΔE values. In addition, ferrous sulphate as a transition metal mordant forms a large number of complexes with the dye molecules, mostly octahedral ones with coordination number 6. As a result, ferrous sulphate salts can form a ternary complex on one site with the fiber and in the other site with the dye, which resulted in slightly higher color difference.

CONCLUSIONS

This study was carried out to evaluate the color levelness of silk fabric dyed with mango, guava and henna leaves extracts and the differences among ΔE values obtained from different areas of fabric were used for such evaluation. It was concluded that unmordanted dyed samples have better color levelness quality than the mordanted ones. The unmordanted samples dyed with mango and guava leaves extracts showed excellent levelness quality while the unmordanted ones dyed with henna leaves extracts showed good levelness quality. On the other hand, in case of mordanted samples dyed with mango leaves extracts, the order of color levelness found was: alum > tin > alum–tin > ferrous sulphate–alum–tin > ferrous sulphate–alum > ferrous sulphate. Among the mordanted samples dyed with guava leaves extracts, the order found was: alum and alum–tin > tin > ferrous sulphate–alum–tin > ferrous sulphate–alum > ferrous sulphate. Thus among the mordanted samples dyed with henna leaves extracts, the order found was: alum > tin > alum–tin > ferrous sulphate–alum > ferrous sulphate–alum–tin > ferrous sulphate. It was found that all the mordanted dyed samples showed good leveling quality for the three vegetable dyes except ferrous sulphate mordanted sample dyed with henna leaves extract. Except this one, it was concluded that good to excellent leveling quality was achieved on silk fabric dyed with three vegetable leaf dyes.

REFERENCES

1. Adeel S, Fazal-ur-Rehman, Hanif R, Zuber M, Ehsan-ul-haq, Munir M (2014) Ecofriendly dyeing of UV-irradiated cotton using extracts of acacia nilotica bark (kikar) as source of quercetin. Asian J Chem 26(3):830–834

2. Naz S, Bhatti IA, Adeel S (2011) Dyeing properties of cotton fabric using un-irradiated and gamma irradiated extracts of Eucalyptus camaldulensis bark powder. Indian J Fibre Text Res 36(2):132–136

3. Ahmed WYW, Rahim R, Ahmad MR, Kadir MIA, Misnon MI (2011) The application of Gluta Aptera wood (Rengas) as natural dye on silk and cotton fabrics. Univers J Environ Res Technol 1(4):545–551

4. Tayade PB, Adivarekar RV (2013) Dyeing of silk fabric with Cuminum L as a source of natural dye. Int J ChemTech Res 5(2):699–700

5. Bechtold T, Mussak R (eds) (2009) Handbook of natural colorants. Wiley, West Sussex

6. Jothi D (2008) Extraction of natural dyes from African marigold flowers (*Tagetes Ereectal*) for textile coloration. AUTEX Res J 8(2):49–53

7. Khan AA, Iqbal N, Adeel S, Azeem M, Batool F, Bhatti IA (2014) Extraction of natural dye from red calico leaves: gamma ray assisted improvements in color strength and fastness. Dyes Pigm 103:50–54

8. Ajmal M, Adeel S, Azeem M, Zuber M, Akhtar N, Iqbal N (2014) Modulation of pomegranate peel colorant characteristics for textile dyeing using energy radiations. Ind Crops Prod 58:188–193

9. Uddin MG (2014) Effects of different mordants on silk fabric dyed with onion outer skin extracts. J Textiles 2014:1–8

10. Zarkogianni M, Mikropoulou E, Varella E, Tsatsaroni E (2010) "Color and fastness of natural dyes: revival of traditional dyeing techniques", Society of Dyers and Colorists. Color Technol 127:18

11. Win ZM, Swe MM (2008) Purification of the natural dyestuff extracted from Mango bark for the application on protein fibers. World Acad Sci Eng Technol 22:536–540

12. Saravanan P, Chadramohan G (2011) Dyeing of silk with ecofriendly natural dye obtained from Barks of *Ficus Religiosa. L.* Univ J Environ Res Technol 1(3):268–273

13. Yang Y, Li S (1993) Instrumental measurement of the levelness of textile coloration. Text Chem Color 25(9):75–78

14. Clariant manual. Recommendations for dyeing silk. Sandoz Chemicals, Basel. (ID-05543.00.94)

15. Millward S (2009) Color difference equations and their assessment. Test Target J 9:19–26

Chapter 6

RECENT DEVELOPMENTS IN MALEIC ACID SYNTHESIS FROM BIO-BASED CHEMICALS

Robert Wojcieszak[1], Francesco Santarelli[1,2,3], Sébastien Paul[1,2], Franck Dumeignil[1,4], Fabrizio Cavani[3] and Renato V Gonçalves[5]

[1]CNRS UMR 8181, Unité de Catalyse et Chimie du Solide (UCCS), Université Lille 1 Sciences et Technologies, 59655 Villeneuve d'Ascq Cedex, France

[2] Ecole Centrale de Lille, ECLille, Cité Scientifique, 59650 Villeneuve d'Ascq, France

[3]Dipartimento di Chimica Industriale "Toso Montanari", Università di Bologna, Viale Risorgimento 4, 40136 Bologna, Italy

[4] Institut Universitaire de France, IUF, Maison des Universités, 103 Boulevard Saint-Michel, Paris 75005, France

[5] Institute of Chemistry, USP, Av. Professor Lineu Prestes, 748, São Paulo 05508-000, SP, Brazil.

ABSTRACT

This review paper presents the current state of the art on maleic acid synthesis from biomass-derived chemicals over homogeneous or heterogeneous catalysts. It is based on the most recent publications on the topic, which are discussed in details with respect to the observed catalytic performances. The recent developments and the technical drawbacks in the gas and the liquid phases are also reported. In addition, recent results on the mechanistic aspect are discussed giving insights into the probable reaction mechanisms depending on the starting molecule (furan, furfural and 5-hydroxymethylfurfural).

GRAPHICAL ABSTRACT

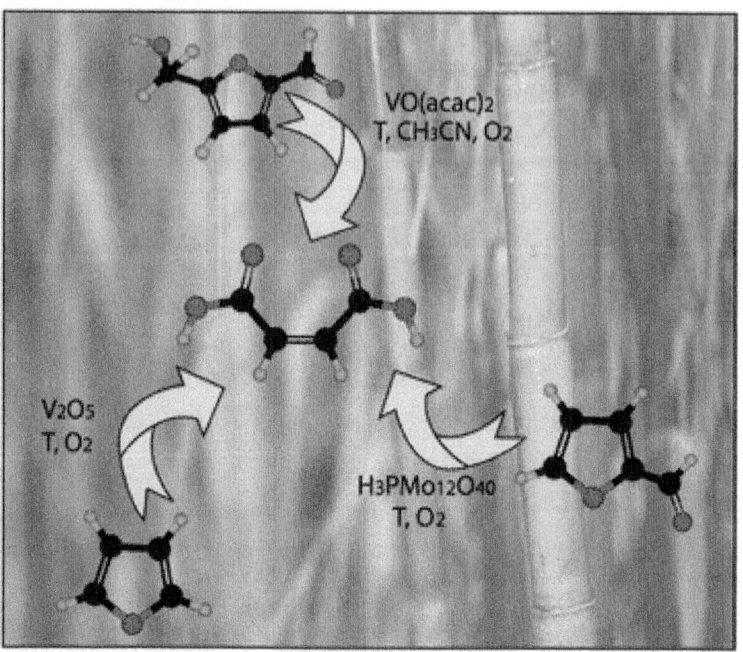

BACKGROUND

Maleic acid (MA) and fumaric acid (FA) are very important chemical intermediates that find applications in nearly every field of industrial chemistry. Maleic acid is an important raw material used in the manufacture of lubricant additives, unsaturated polyester resins, surface coatings, plasticizers, copolymers and agricultural chemicals [1–5]. Fumaric acid is naturally present in many plants and its name originates from Fumaria officinalis, a climbing annual plant, from which it was first isolated [6]. It is used as a food acidulent and as a raw material in the manufacture of unsaturated polyester resins, quick setting inks, furniture lacquers, paper sizing chemicals, and aspartic acid [7, 8].

Maleic acid and fumaric acid are dicarboxylic acid isomers that have the same carbon skeleton. They both yield succinic acid (SA) by hydrogenation. Each can be converted to the identical anhydride by heat treatment, but maleic acid reacts much more rapidly. This, coupled with the fact that mild hydrolysis of maleic anhydride (MAnh) leads to maleic acid, is linked to the cis structure of maleic acid and to the transstructure of fumaric acid (Scheme 1). They contain two acid carbonyl groups and a double bond in the α,β position. These functional groups are very reactive, which makes the control of the selectivity of their synthesis reactions a key parameter.

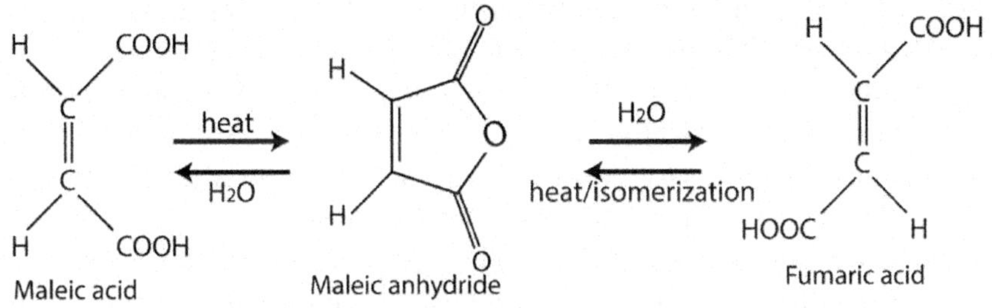

Scheme 1: Chemical structures of maleic acid, fumaric acid and maleic anhydride.

Historically, these two acids were first prepared in the 1830's, [9] but their commercial manufacture did not begin until almost one century later. Maleic acid was commercially available in 1928 and fumaric acid production began in 1932 using an acid-catalyzed isomerization of maleic acid process.

Maleic anhydride can be commercially produced by the vapor-phase oxidation of benzene or butene/butane using O_2 as an oxidant [10–15]. This later reaction is very exothermic and CO and CO_2 thus constitute the main by-products. The catalyst used in the production of maleic anhydride from butane is vanadium–phosphorus-oxide (VPO). There are several routes to prepare this catalyst, but the industrial way involves the reaction of vanadium (V) oxide and phosphoric acid to form vanadyl hydrogenophosphate, $VOHPO_4 \oplus 0.5H_2O$, which is then treated thermally to produce $(VO)_2P_2O_7$. The catalyst used for the conversion of benzene to maleic anhydride consists of supported vanadium oxide [16]. The vanadium oxide on the surface of the support is often modified with molybdenum oxides. The support is an inert oxide such as alumina or silica of a relatively low surface area. The conversion of benzene to maleic anhydride is a less complex oxidation than the conversion of butane, which enables obtaining very good conversions together with high selectivities [9].

Fumaric acid is generally produced by fermentation [17–20]. Many aspects such as the applied microbial strain and its morphology, the use of a neutralizing agent, and the applied feedstock play a crucial role in the fermentation process. This latter is very interesting because it involves CO_2 fixation. Indeed, it is known that the mechanism involving CO_2 fixation and catalyzed by pyruvate carboxylase enables obtaining higher yields in FA production. In case of the maximum theoretical yield, two moles of CO_2 could be fixed per mole of glucose consumed [19].

The quest for sustainable and environmentally benign sources of energy and, more recently, of chemicals has attracted much attention in the recent years [6]. The production and the use of chemicals and fuels from biomass [9, 21–23] seem to be an ideal solution to tackle environmental issues and fossil resources progressive depletion, if correct measures are taken. In that context, biomass-derived platform molecules, such as maleic acid, fumaric acid or maleic anhydride have been identified as top value-added chemicals. Therefore, a highly effective method to produce these dicarboxylic acids from biomass is necessary. Moreover, from the industrial application point of view, this method should be simple and environmentally friendly. In this context, heterogeneous catalysis could bring new economic and environmental solutions. Indeed, nowadays, new synthesis techniques permit to control the morphology and physical and chemical properties of the catalysts. This yields in higher conversion rates and selectivities. In some cases they are as good as for the enzymatic or homogeneous catalysts. As a matter of fact, a better understanding of catalytic nanomaterials is essential for the synthesis of fine chemicals.

In this review, we present the current state of the art on maleic acid synthesis by upgrading biomass-derived molecules (furfural and 5-hydroxymethylfurfural) using heterogeneous and homogeneous catalytic processes. This paper is based on the most recent publications, and we put emphasis on the factors that have to be considered to understand the catalytic activity of the nanomaterials in maleic acid synthesis.

Platform Molecules used for MA, FA and MAnh Synthesis

Production of high value added chemicals from biomass sources remains one of the greatest contemporary challenges for heterogeneous catalysis. A very important point to be analyzed is related to the choice of the substrate and to its availability in the future. The major sources of this kind of raw material are agricultural residues and wastes, such as rice straw, wheat straw, wood (hardwood), byproducts left over from the corn milling process (corn strover), annual and perennial crops, waste paper and sweet sorghum. These raw materials comprise three types of main biopolymers: cellulose, hemicellulose, and lignin [24].

Due to its numerous advantages for growth and production, biomass raw materials has been identified as a suitable source of chemical energy for biofuels [25]. However, in order to synthesize fine chemicals of the desired size and properties, catalytic C–C bond formation is required. To this respect, 5-hydroxymethylfurfural (HMF), furan and furfural obtained by transformation of carbohydrates, have been widely identified as useful platform molecules [26, 27]. The respective chemical structures of these molecules are represented in Scheme 2.

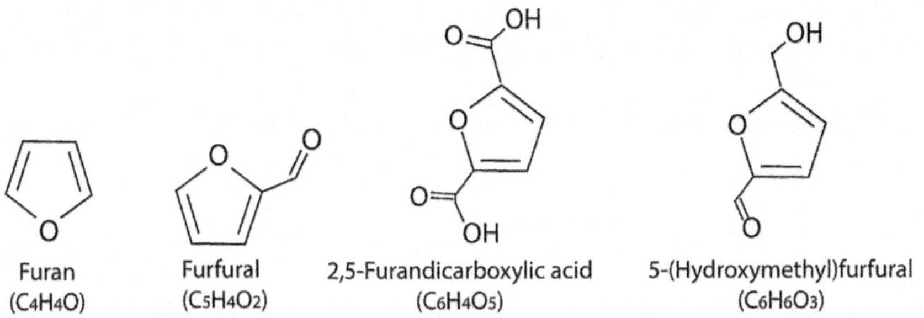

Scheme 2: Chemical structures of furan, furfural, 2,5-furandicarboxylic acid and 5-hydroxymethylfurfural (HMF).

The list of the most important building block chemicals (platform molecules) that can be produced from sugars via biological or chemical conversions is now well established [28]. These building blocks can be subsequently converted to a number of high-value bio-based chemicals or materials. These molecules have multiple functional groups that possess the potential to be transformed into new families of useful molecules.

Other very important platform molecules are HMF and furfural. HMF is a versatile platform chemical. HMF could be easily transformed into maleic anhydride because it contains a furyl ring in its basic structure. This transformation could be achieved via oxidative C–C bond cleavage of HMF. Furfural is an important renewable, non-petroleum based, chemical feedstock. It could be easily transformed into furfuryl alcohol (FAlc), via hydrogenation, which is a very useful chemical intermediate [precursor of tetrahydrofurfuryl alcohol (THFA)]. It could be also transformed into maleic anhydride or maleic acid, via oxidation, as we will discuss later in the present paper. The by-product of furfural oxidation in gas phase is furan. This molecule is also one of the important intermediates in chemical industry. Moreover, as indicated below, furan is also the first intermediate in the mechanism of the furfural oxidation. Even if it cannot be produced directly from biomass (it can be produced from furfural) due to its presence in the reaction mechanism we have decide to include it into the present review.

Biomass feedstocks are highly reactive by nature, and, consequently, high temperature is normally not required to achieve their transformation. However, reactions carried out in the liquid phase increases the possibility of leaching issues. For this reason, rather than thermal stability, the new catalysts for biomass conversion should be designed so as to be resistant to leaching [29]. This is one of the most important challenges in liquid phase heterogeneous catalysis. However, taking into account the huge number of paper on leaching issues, we think that this subject needs a separate review. That is the reason why we do not discuss on leaching phenomenon in details in this work.

Liquid Phase Oxidation: Homogeneous and Heterogeneous Catalysis

Guo and Yin [30] studied aerobic oxidation of furfural into maleic acid using phosphomolybdic acid catalysts. They performed the reaction in a biphasic aqueous/organic medium. The oxidation takes place in the aqueous phase and the organic phase plays the role of a reservoir, which gradually releases the substrate, which is unstable in the aqueous phase, through phase equilibrium. They studied the influence of the co-solvent addition on the distribution of furfural between both phases, which influences its overall conversion. Without the organic co-solvent, the yield of maleic acid was 38.1% with 44.2% of selectivity, and the conversion of furfural was as high as 86.2%. Addition of an organic co-solvent generated a biphasic system, and improved the selectivity to maleic acid (up to 61% in the case of tetrachloroethane) with a concomitant reduction of the furfural conversion in most cases.

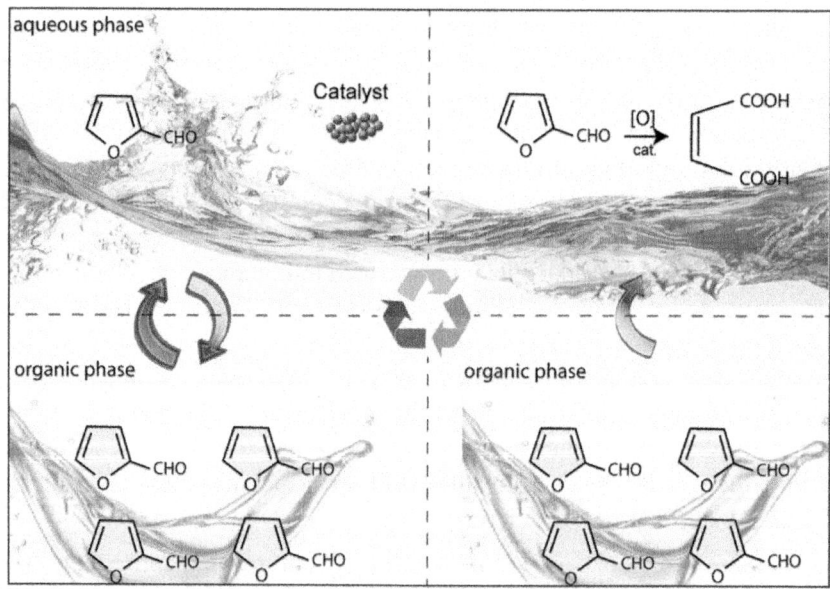

Figure 1: Simplified flow line of catalytic oxidation of furfural in biphasic system adapted from [31].

The authors also studied the influence of the reaction temperature on catalytic activity and they found that when the reaction temperature was increased, the furfural conversion, the maleic acid yield, and the selectivity to maleic acid first increased, and at a temperature higher than 383 K, the furfural conversion and the maleic acid yield increased, but the selectivity to maleic acid then decreased. At 383 K, the conversion of furfural, the yield and selectivity to maleic acid were 50.3, 34.5, and 68.6%, respectively, whereas they were 87.6, 47, and 53.3% at 403 K. Moreover, at this latter temperature 3% of fumaric acid was observed, which reflects competitive polymerization (Figure 1).

Du et al. [31] studied the selective oxidation of HMF to maleic anhydride using $VO(acac)_2$ catalysts. They also observed a crucial role of the solvent in this reaction. 52% yield of MA was obtained in acetonitrile at 363 K, but a 14% yield of 2,5-diformylfuran (DFF) was also observed. Besides acetonitrile, acetic acid was preferred as a solvent for the formation of MA (50% MA yield). In contrast, MA yields of about 7% were observed in the case of N,N-dimethyl-formamide (DMF) and α,α,α-trifluorotoluene (TFT). The authors also observed that the reaction temperature strongly affected the oxidation process in terms of products distribution. When the reaction was performed at 343 K, the HMF conversion was no more than 50%. However, when the temperature was raised from 363 to 403 K, the yield of MA gradually decreased from 52 to 20% while the yield of DFF increased from 14 to 52%. Transition metals complexes such as $FeSO_4$, $CuSO_4$, $Mn(acac)_2$, $MoO_2(acac)_2$, $Co(acac)_2$ and $Co(OAc)_2$ studied by the authors in the same conditions were found much less effective for catalytic oxidation of HMF into MA, and the yield of MA was then <2%.

Quite similar results were observed in the oxidation of furfural in the presence of redox metal salt catalysts [32]. It was shown that copper acetate and iron sulfate enabled respectively 18.6 and 12.1% yields in the expected maleic acid. On the other hand, the catalysts based on other metal sources such as $Mn(OAc)_2$, $RuCl_3$ and $NiCl_2$ are substantially much less efficient for maleic acid formation. Interestingly, a $Pd(OAc)_2$ catalyst showed a 15.9% yield to furoic acid, while only traces of this compound were observed over other metal catalysts. However, the major competitive process for the selective oxidation is the polymerization of furfural to generate resins under the oxidative conditions. The authors observed that in some experiments in the absence of a catalyst, the conversion of furfural was 56.3% with only a 7.2% yield of maleic acid. These results were explained by the polymerization of furfural to resins.

Table 1 shows the recent catalytic results obtained with different catalysts in liquid phase oxidation. The best results were obtained with heteropolyacids such as $H_3PMo_{12}O_{40}$, $H_3PMo_{12}O_{40}$ modified with $Cu(NO_3)_2$ and $VO(acac)_2$ (up to 52% of MA yield and 56% of selectivity to MA). This showed the influence of acidity of the catalysts on the overall catalytic activity in this reaction.

Table 1: Summary of catalytic results obtained in liquid phase

Catalyst	Conditions	Furfural or HMF conversion	MA yield	MA selectivity	References
$H_3PMo_{12}O_{40}$	383 K, O_2 20 bar, H_2O + nitrobenzene	67 (furfural)	38	56	[30]
$H_3PMo_{12}O_{40}$	383 K, O_2 20 bar, H_2O + toluene	73 (furfural)	37	50	[30]
$H_3PMo_{12}O_{40}$	383 K, O_2 20 bar, H_2O + p-xylene	66 (furfural)	35	53	[30]
$H_3PMo_{12}O_{40}$	383 K, O_2 20 bar, H_2O + cyclohexane	85 (furfural)	38	45	[30]
$H_3PMo_{12}O_{40}$	383 K, O_2 20 bar, H_2O + tetradecane	82 (furfural)	38	46	[30]
$H_3PMo_{12}O_{40}$	383 K, O_2 20 bar, H_2O	86 (Furfural)	38	44	[30]
$VO(acac)^2$	363 K, O_2 10 bar, CH_3CN	99 (HMF)	52	52	[31]
Amberlyst 15	353 K, H_2O_2, H_2O, 24 h	99 (furfural)	11	11	[33]
Nafion NR50	353 K, H_2O_2, H_2O, 24 h	99 (Furfural)	11	11	[33]
Nb_2O_5	353 K, H_2O_2, H_2O, 24 h	99 (furfural)	4	5	[33]
ZrO_2	353 K, H_2O_2, H_2O, 24 h	99 (furfural)	5	5	[33]
$H_6PV_3MO_9O_{40}$	383 K, O_2 20 bar, CH_3CN	99 (furfural)	12	12	[44]
$VO(acac)^2$	363 K, O_2 10 bar, DMF	96 (HMF)	7	7	[31]
$H_5PV_2Mo_{10}O_{40}$ + $Pd(OAc)_2$ (1/1)	383 K, O_2 20 bar, CH_3CN	94 (furfural)	14	15	[44]
$VO(acac)^2$	363 K, O_2 10 bar, TFT	96 (HMF)	7	7	[31]
$VO(acac)^2$	363 K, O_2 10 bar, CH_2Cl_2	99 (HMF)	16	16	[31]
$VOSO_4$	363 K, O_2 10 bar, CH_3CN	NC (furfural)	34	NC	[31]
$Co(OAc)_2$	363 K, O_2 10 bar, CH_3CN	NC (furfural)	2	NC	[31]
$Co(NO_3)_2$	371 K, O_2 20 bar, H_2O	69 (furfural)	4	6	[32]
$FeSO_4$	371 K, O_2 20 bar, H_2O	90 (furfural)	12	13	[32]
V_2O_5	371 K, O_2 20 bar, H_2O	72 (Furfural)	6	8	[32]
$CuSO_4$	371 K, O_2 20 bar, H_2O	67 (furfural)	19	29	[32]
$Cu(NO_3)_2$	371 K, O_2 20 bar, H_2O	86 (furfural)	24	28	[32]
$Cu(OAc)_2$	371 K, O_2 20 bar, H_2O	71 (furfural)	19	26	[32]
$H_3PMo_{12}O_{40}$ + $Cu(NO_3)_2$ (2/1)	371 K, O_2 20 bar, H_2O	95 (furfural)	49	52	[32]

NC Not communicated.

Some authors studied also the influence of the nature of counter-ions of the copper (II) cation, which could modulate the redox properties of the Cu^{2+} [26]. It was found that the acetate, sulfate and nitrate counterparts could significantly improve the yield of maleic acid. In contrast, using the chloride anion did not yield any improvement, and using carbonate even led to a detrimental effect on catalytic activity [32]. However, the extra addition of nitrate to the reaction mixture did not improve the catalytic activity at all, which suggests that the nitrate anion does not independently promote the conversion of furfural to maleic acid. However, it may play the role of modulator of the redox potential of Cu^{2+} to increase the catalytic oxidation efficiency [32].

Very interesting results were observed in the case of the combination of copper nitrate with phosphomolybdic acid. The mixture of these two reagents yielded a drastic enhancement of the reaction of furfural oxidation to maleic acid. For a mixture of 0.8 mmol of phosphomolybdic acid and 0.4 mmol of copper nitrate, the yield of maleic acid could be improved up to 49.2% with a selectivity of 51.7%, and the conversion of furfural was 95.2% (as compared to the test without $Cu(NO_3)_2$: 38.4, 43.3 and 88.7%, respectively) [32].

Muzyczenko et al. have studied the oxidation of furfural with hydrogen peroxide in water and in absolute ether containing small amounts of water [34]. They found that the presence of water inhibits the formation of furfuryl hydroxyhydroperoxide but promotes the formation of acids. Moreover, the acids formed during the reaction catalyzed the other reaction steps. In the case of an excess of water, water molecules blocked the carbonyl group of furfural. The formation of hydrogen bonds with the carbonyl group and water prevails forming the furfural–H_2O complex, which hinders the access of H_2O_2 molecules to the furfural. In the case of the absence of water molecules, the formation of a polar furfural–H_2O_2 complex prevails [34].

Recently, Fagundez et al. studied the selective liquid-phase oxidation of furfural to maleic acid using hydrogen peroxide as an oxidant and titanium silicalite (TS-1) as a catalyst [35]. The highest yield of 78 mol.%, was obtained using an H_2O_2/furfural molar ratio of 7.5 at 323 K after 24 h of reaction (furfural/catalyst ratio of 1). However, Ti leaching was observed, especially during the first run, and became much less important in the subsequent successive cycles. In addition, the leaching affected both anatase and Ti species within a silicalite framework. Moreover, the authors observed that, when using pure furfural, the catalyst could be reused for five runs without noticeable deactivation, whereas when using furfural directly derived from biomass, visible deactivation occurred. It was explained by the presence of some organic impurities in biomass-derived furfural.

Gas Phase Oxidation: Heterogeneous Catalysis

Gas phase oxidation reactions are one of the most important processes in the chemical industry. The first catalytic tests to transform furfural into maleic acid were performed in the first half of the 20th century. In 1949, Nielsen reported on gas phase furfural oxidation to maleic acid based on iron molybdates materials and carried out in nickel-tube reactors as shown in Figure 2 [36].

In this study, three different materials were used as a reactor tube: nickel, iron and aluminum. It was shown that nickel enabled obtaining higher conversion rates as compared to Fe or Al tubes. The authors did not explain this phenomenon but assumed that nickel should be a good catalyst for one of the steps in furfural oxidation. The authors compared also two types of iron catalysts: iron molybdate impregnated on alumina and catalysts prepared by mixing iron nitrate and ammonium molybdate. The best catalytic results were obtained for the former, indicating the influence of method of preparation on catalytic activity (acidity of the catalyst). The authors claimed 95% or higher furfural conversion and over 80% of MA yield. These results were much better that those reported earlier in the literature [36–39].

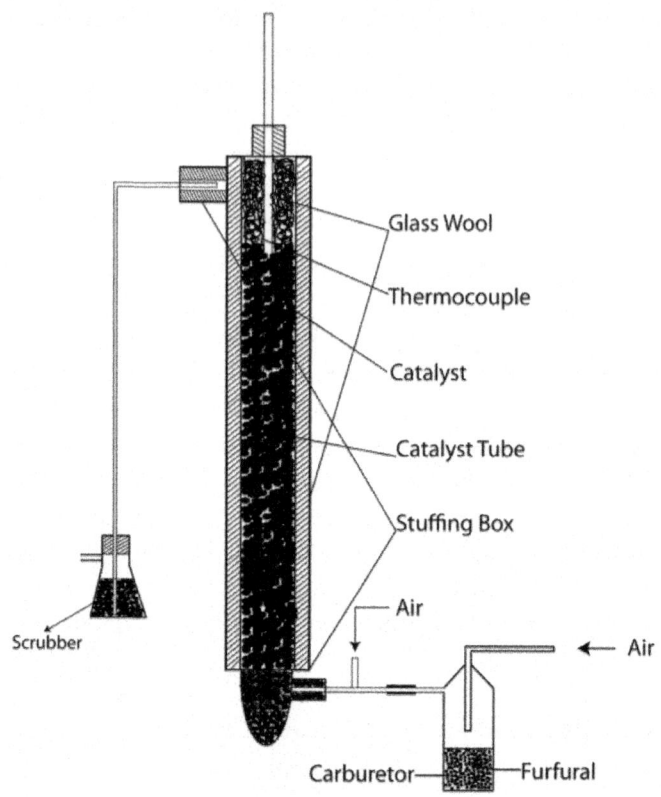

Figure 2: Nickel-tube reactor apparatus for gas phase furfural oxidation (reproduced with permission from American Chemical Society, Ref. [36]).

Recently, Ojeda et al. reported that furfural could be converted into maleic anhydride (73% yield) through selective gas phase oxidation at 593 K with O_2 over VO_x/Al_2O_3 catalysts [40]. They were the first to undertake detailed investigations on the effects of VO_x structures on maleic anhydride formation by furfural oxidation. They found that at low surface VO_x densities (<2.5 at $V \cdot nm^{-2}$), highly dispersed vanadia species (monovanadates, VO_4) with one short $V = O$ terminal bond and three anchoring $V–O–Al$ bonds are formed. With increasing vanadium surface density (>2.5 at $V \cdot nm^{-2}$), polyvanadates ($V–O–V$ bonds) are expected to form gradually on the catalyst surface. The presence of various active species should then be related to different catalytic activities. Indeed, polyvanadates exhibited the highest normalized rate as compared to monovanadates and V_2O_5, which showed similar activities. Moreover, they found that increasing O_2 pressure exhibited a positive effect on furfural conversion. However, higher concentrations of O_2 led to less selective furfural oxidation and formation of furan (up to 9%).

Wang et al. [41] have studied V_2O_5–MoO_3–P_2O_5/γ–Al_2O_3 catalysts prepared by impregnation in the oxidation of furfural to maleic anhydride. The catalytic behavior of the catalysts was evaluated in a continuous fixed-bed reactor, using a feeding stream comprising 2% furfural in air at 578 K. The authors found that the activity depends on the MoO_3/V_2O_5 ratio. When this ratio was equal 0.4, the conversion of furfural reached 82% and the yield of maleic anhydride was close to 50%. The authors indicated that the introduction of MoO_3 adjusts the interaction between the support and the active phase. This enables the rapid exchange of crystal lattice oxygen in the catalysts and then accelerates the overall activity and improves the selectivity to maleic anhydride [41].

Shimanskaya et al. studied vapor-phase oxidation of furan to maleic anhydride in the presence of metal-containing catalysts [42]. High yields of maleic anhydride could be obtained. However, oxidation by-products such as carbon monoxide and dioxide with small amounts of acetic and oxalic acids were also observed. It was established that the vapor-phase oxidation of furan, in contrast to its derivatives, is more selective. The yield of maleic anhydride is higher, and complete transformation of the furan is achieved in a shorter period.

Mechanism of Maleic Acid Formation

The maleic acid formation mechanism involves the removal of one or two carbon atoms from the considered starting reactant (furfural or HMF, respectively). Meanwhile, there is still no consensus on the role of the nature of the active phase and on the mechanism of the reaction. It has been claimed that MA can form via different pathways depending on the starting molecule (furan, furfural or HMF), which seems a reasonable assumption.

The oxidation of furan in the presence of vanadium (V) oxide takes place through a route involving the formation of an endoperoxide, [43] which is then oxidized to maleic anhydride, as described below (Scheme 3).

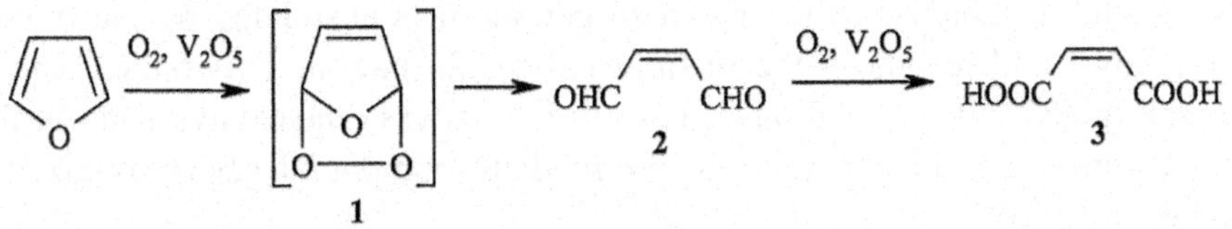

Scheme 3: Oxidation of furan and its homologs with molecular oxygen (reproduced with permission from Springer Science + Business Media, Ref. [43]).

In the case of the furan molecule, the formation of maleic acid goes through a first step of ring opening before a subsequent step of oxidation of the aldehyde groups to acid groups.

In the oxidation of furfural to maleic acid, one atom of carbon must be removed. There are two potential pathways to lose one carbon atom through decarboxylation. The first consists on decarboxylation and then furan ring-opening while the second one starts with ring-opening and then proceeds through decarboxylation. Guo and Yin studied furfural oxidation mechanism on phosphomolybdic acid catalysts. They observed that neither furoic acid nor furan were the intermediates in the MA formation reaction. These results confirmed a two-step mechanism with furan ring opening as a first step and decarboxylation to MA as the final step. Moreover, they proposed a plausible mechanism based on a radical pathway as shown in Scheme 4.

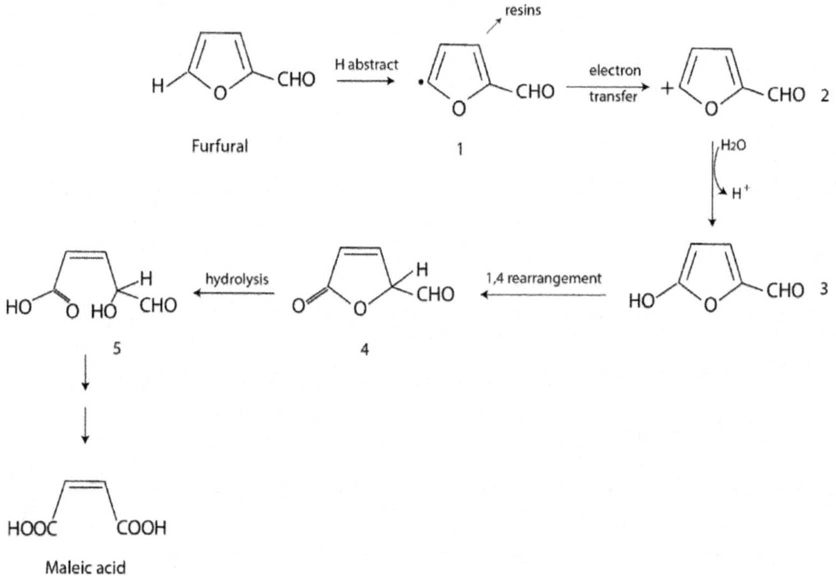

Scheme 4: Catalytic oxidation of furfural on phosphomolybdic acid catalyst. Adapted from [30].

In this mechanism, the first hydrogen atom is abstracted by the oxygen or catalyst and results in the furfural radical formation. This radical can be transformed to furfural cation (via electron transfer to the catalyst) and further react with water to generate an intermediate compound easily oxidized to maleic acid [30].

A different mechanism of maleic acid formation from furfural in the gas phase was proposed by the Ojeda's group [40] (Scheme 5).

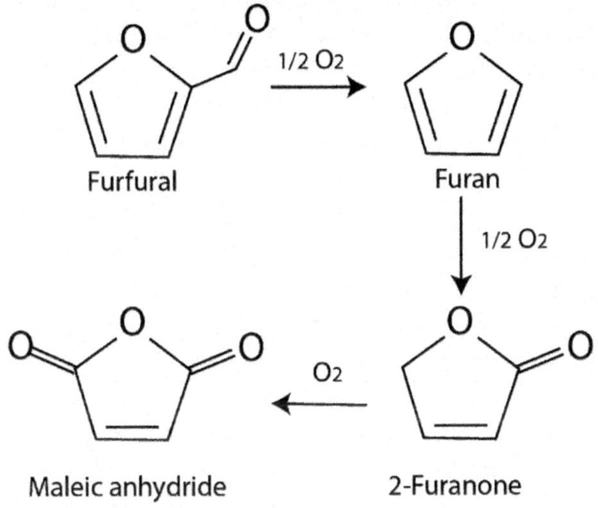

Scheme 5: Reaction pathways proposed for the selective oxidation of furfural to furan and maleic anhydride over VOx/Al_2O_3 catalysts [40].

As described by these authors, first, furfural decarbonylation yields furan, which is the main reaction intermediate (Scheme 5). Subsequently, furan is oxidized in two steps, first to 2-furanone and then to maleic anhydride. CO_2 and H_2O could be formed through total oxidation of furfural and/or intermediate compounds. The authors carried out also kinetic studies of furfural oxidation. They found that neither furfural nor O_2 saturates the catalytic surface under the studied reaction conditions. Moreover, taking into account that furfural oxidation to maleic anhydride differs from zero-order kinetics towards O_2, it implies a different reaction pathway compared to the Mars–van-Krevelen mechanism. The authors then proposed a Langmuir–Hinshelwood model for the furfural oxidation over VO_x/Al_2O_3.

Lan et al. [44] proposed that the formation of maleic anhydride could be initiated from the furfural radical intermediate formed after the first hydrogen abstraction by oxygen. This radical can then attack the $C = O$ bond of another furfural molecule to start polymerization or it can also initiate the formation of maleic anhydride as shown in Scheme 6.

In the most plausible pathway, the furfural radical generates the furfural cation intermediate, which is then attacked by a H_2O molecule to form the intermediate 4-hydroxyfurfural. This intermediate follows then a 1,4-rearrangement and decarbonylation to finally form maleic anhydride (Scheme 6). Moreover, as expected, the mechanism of maleic anhydride formation in the liquid phase is different from that observed in gas phase oxidation of furfural. The same authors claimed that furan is not the intermediate in the maleic anhydride formation over the $H_5PV_2Mo_{10}O_{40}$ catalyst in the liquid phase [44].

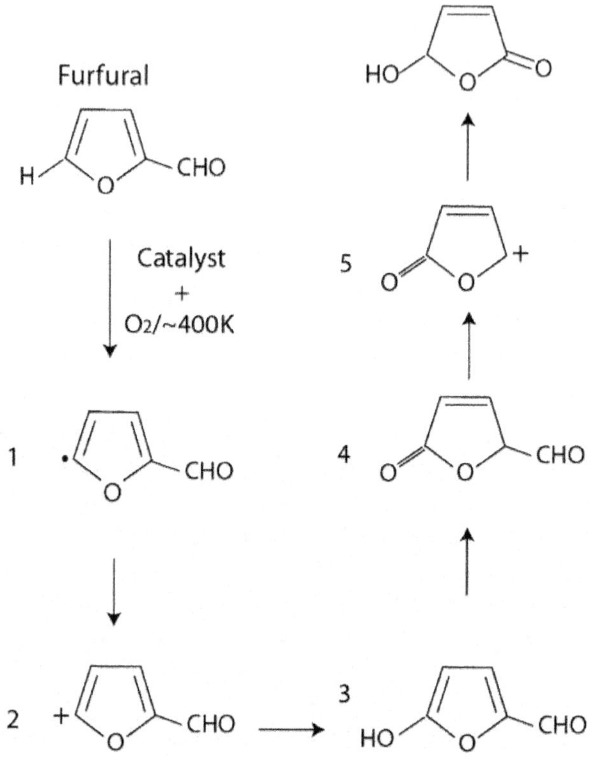

Scheme 6: Plausible pathway of furfural oxidation to maleic anhydride [44].

The formation of CO_2 in vapor phase oxidation of furfural to maleic anhydride could originate from both direct oxidation of furfural or successive oxidation of maleic anhydride. Murthy compared three possible reactions schemes over vanadium pentoxide catalysts: parallel consecutive mechanism (parallel formation of CO_2 from furfural and maleic anhydride), parallel reaction (CO_2 originating from maleic anhydride is negligible) and consecutive mechanism (CO_2 formation from furfural is negligible). He found that the oxidation of furfural at 493-553 K followed the parallel reaction scheme. The experimental data were satisfactorily explained by the well-known two-stage redox mechanism of Mars and van Krevelen. The re-oxidation of the catalyst was found to be the rate-determining step in the investigated temperature range. Moreover, the order of the reaction was found to be one with respect to furfural and oxygen partial pressures [45].

The formation of maleic acid from HMF involves a C–C bond cleavage and the removal of two carbon atoms. The mechanism of this reaction was studied by Du et al. [31] (Scheme 7). They showed that the aldehyde group of HMF was relatively stable under oxygen atmosphere, and that MA was probably not formed via a decarboxylation reaction. Indeed, intermediate compounds such as DFF or FDCA were not involved in the reaction mechanism. Moreover, they claimed that the formation of MA from HMF was related to the hydroxymethyl group of HMF rather than the aldehyde group, which was not necessary for C–C bond cleavage.

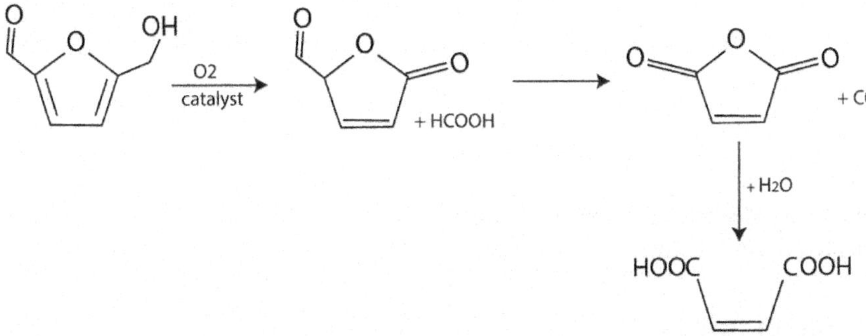

Scheme 7: Possible mechanism of 5-hydroxymethylfurfural (HMF) oxidation with O_2 in liquid phase. Adapted from [31].

Choudhary et al. [33, 46] have studied oxidation of furfural with H_2O_2 to succinic acid (SA) on acid catalysts. High conversions of furfural and good selectivity to succinic acid were reported (74%). However, in some cases, relatively good selectivity to MA was also observed (up to 16% in the case of Amberlyst-15 catalyst). It was shown that the acid catalyst enhanced the efficiency of H_2O_2. FA was formed in higher yields than SA in the initial stages of the reaction. Thereafter, a decrease in the FA yield concomitantly with an increase in the MA yield was observed. This trend was explained by the low solubility of FA in water, which then converts to a highly soluble MA via isomerization. Interestingly, it is worth noting that such isomerization was enhanced by the formed formic acid in the presence of Amberlyst-15. The authors proposed two reaction pathways involving the Baeyer–Villiger oxidation. In the first step, the furan ring was opened up to undergo oxidation by H_2O_2. The pathway was well supported by experiments and revealed that MA is obtained from furan-2(5H)-one intermediate (Scheme 8).

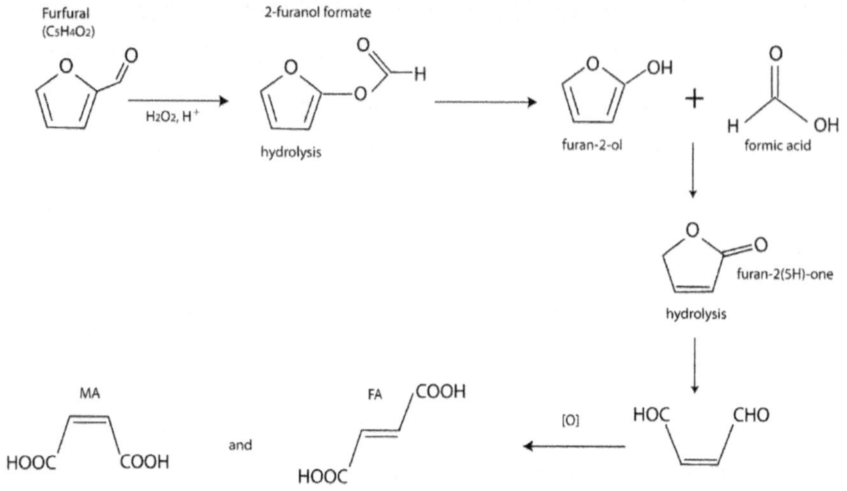

Scheme 8: Mechanism of furfural oxidation with H_2O_2 via Baeyer–Villiger oxidation. Adapted from [33, 46].

CONCLUSION

In this paper, we provide a close look at the existing literature on the preparation of maleic acid from biomass-based feedstocks via oxidation of HMF, furfural and furan. The oxidation can be carried out in gas or liquid phase. We report the best ways to obtain MA, MAnh or FA using O_2 or H_2O_2 as oxidants over widely available catalysts. The best results in term of MA selectivity were obtained in liquid phase using homogeneous $H_3PMo_{12}O_{40}$ and VO(acac)$_2$ catalysts. In the case of heterogeneous catalysis, the best results reported so far are obtained by vapor phase oxidation using vanadium oxide or vanadium-molybdenum mixed oxides supported on Al_2O_3. The liquid phase oxidation using heterogeneous catalysts still need improvement. Especially, the leaching of active phase should be avoided.

In this review, we put emphasis on the different mechanisms reported in the literature because fundamental studies are now essential in the design of the catalysts and optimization of conditions of the synthesis of these valuable chemicals. However, there is still no consensus on the exact mechanism of the reaction and the role of the nature of the active phase. It has been claimed in the literature that MA can form from furfural via different pathways: (1) abstraction of hydrogen to form furfural radical intermediate that then gives 4-hydroxyfurfural; (2) decarboxylation of furfural to furan, which then gives a 2-furanone intermediate; or by (3) first Baeyer–Villiger oxidation of furfural to furan-2-ol and then to MA through furan-2(5H)-one intermediate. The next challenge in the synthesis of maleic acid will be the elaboration and the use of efficient heterogeneous catalytic systems working under «green conditions», i.e., using low temperature and environmentally friendly solvents.

AUTHORS' CONTRIBUTIONS

RW, FS, SP, FD, FC, RVG have been involved in drafting of this manuscript and in revising it critically for important intellectual content. All authors read and approved the final manuscript.

REFERENCES

1. Hoffman RL (1980) Monsanto Company. US Patent, US 4238572 A. http://www.google.com/patents/US4238572

2. Sabu T, Kuruvilla J, Malhotra SK, Goda K, Sreekala MK (eds) (2012) Polymer Composites, Macro- and Microcomposites, vol 1. ISBN 978-3-527-32624-2

3. Otto G (2008) Piringer ALB: plastic packaging materials for food: barrier function, mass transport, quality assurance, and legislation. Wiley-VCH, New York

4. Yang CQ, Chen D, Guan J, He Q (2010) Cross-linking cotton cellulose by the combination of maleic acid and sodium hypophosphite. 1. fabric wrinkle resistance. Indus Eng Chem Res 49:8325–8332

5. Klaus Weissermel H-JA (2008) Industrial organic chemistry, 3rd, Completely Revised Edition. Wiley, New York

6. Mabee W, Gregg D, Saddler J (2005) Assessing the emerging biorefinery sector in Canada. Appl Biochem Biotechnol 123:765–778

7. Lohbeck K, Haferkorn H, Fuhrmann W, Fedtke N (2000) Maleic and fumaric acids. In: Ullmann's encyclopedia of industrial chemistry. Wiley, GmbH & Co. KGaA

8. Doscher CK, Kane JH, Cragwall GO, Staebner WH (1941) Industrial applications of fumaric acid. Indus Eng Chem 33:315–319

9. Felthouse TR, Burnett JC, Horrell B, Mummey MJ, Kuo Y-J (2001) Maleic anhydride, maleic acid, and fumaric acid. In: Kirk-Othmer Encyclopedia of Chemical Technology, Wiley

10. Skinner WA, Tieszen D (1961) Production of maleic acid by oxidizing butenes. Ind Eng Chem 53:557–558

11. Beach LK (1951) Cis-butene oxidation. US Patent, US 2537568 A. http://www.google.fr/patents/US2537568

12. Slotterbeck OC, Tribit SW (1941) Preparation of maleic acid. Google Patents

13. Harry O, Mitchell B, Russell JL (1964) Preparation of fumaric acid by catalytically isomerizing maleic acid produced by the oxidation of butene-2. Google Patents

14. Weiss Downs (1925) Catalytic oxidation of benzene to maleic acid. J Chem Educ 2:1178

15. Bhattacharyya SK, Venkataraman N (1958) Catalytic vapour-phase oxidation of benzene to maleic acid. J Appl Chem 8:728–737

16. Cornils B (2004) Handbook of commercial catalysts. In: Rase HF (eds) Heterogeneous catalysts. Angew Chem Int Ed 43:2324–2325

17. Xu Q, Li S, Huang H, Wen J (2012) Key technologies for the industrial production of fumaric acid by fermentation. Biotechnol Adv 30:1685–1696

18. Das RK, Brar SK, Verma M (2014) Enhanced fumaric acid production from brewery wastewater by immobilization technique. J Chem Technol Biotechnol

19. Roa Engel C, Straathof AJ, Zijlmans T, van Gulik W, van der Wielen LM (2008) Fumaric acid production by fermentation. Appl Microbiol Biotechnol 78:379–389

20. Podgorska E, Kasprzak M, Szwajgier D (2004) Fumaric acid production by rhizpous nigricans and rhizopus oryzae using apple juice. Pol J Food Nutr Sci 13/54:47–50

21. Lichtenthaler FW (2000) Carbohydrates as organic raw materials. In: Ullmann's encyclopedia of industrial chemistry. Wiley, GmbH & Co. KGaA

22. Milbrandt A, Kinchin C, McCormick R (2013) The feasibility of producing and using biomass-based diesel and jet fuel in the United States. National Renewable Energy Laboratory, US, Department of Energy, Technical Report NREL/TP-6A20-58015. http://www.nrel.gov/docs/fy14osti/58015.pdf

23. Shahbazali E (2013) Biorefinery: from biomass to chemicals and fuels. In: Volker H, De Gruyter (eds) Green Processing and Synthesis, vol 2. No 1 pp 87–88

24. Visioli L, Enzweiler H, Kuhn R, Schwaab M, Mazutti M (2014) Recent advances on biobutanol production. Sustain Chem Process 2:15

25. Somerville C, Youngs H, Taylor C, Davis SC, Long SP (2010) Feedstocks for lignocellulosic biofuels. Science 329:790–792

26. Alonso DM, Bond JQ, Dumesic JA (2010) Catalytic conversion of biomass to biofuels. Green Chem 12:1493–1513

27. Corma A, delaTorre O, Renz M, Villandier N (2011) Production of high-quality diesel from biomass waste products. Angew Chem Int Edn 50:2375–2378

28. Aresta M, Dibenedetto A, Dumeignil F (eds) from biomass to chemicals and fuels. De Gruyter, ISBN: 978-3-11-026028-1

29. Triantafyllidis K, Lappas A, Stöcker M (2013) The role of catalysis for the sustainable production of bio-fuels and bio-chemicals. Elsevier

30. Guo H, Yin G (2011) Catalytic aerobic oxidation of renewable furfural with phosphomolybdic acid catalyst: an alternative route to maleic acid. J Phys Chem C 115:17516–17522

31. Du Z, Ma J, Wang F, Liu J, Xu J (2011) Oxidation of 5-hydroxymethylfurfural to maleic anhydride with molecular oxygen. Green Chem 13:554–557

32. Shi S, Guo H, Yin G (2011) Synthesis of maleic acid from renewable resources: catalytic oxidation of furfural in liquid media with dioxygen. Catal Commun 12:731–733

33. Choudhary H, Nishimura S, Ebitani K (2012) Highly efficient aqueous oxidation of furfural to succinic acid using reusable heterogeneous acid catalyst with hydrogen peroxide. Chem Lett 41:409–411

34. Muzychenko GF, Badovskaya LA, Kul'nevich VG (1972) Role of water in the oxidation of furfural with hydrogen peroxide. Chem Heterocycl Compd 8:1311–1313

35. Fagundez NA, Agirrezabal-Telleria I, Arias P, Fierro JLG, Mariscal Lopez R, Lopez Granados M (2014) Aqueous-phase catalytic oxidation of furfural with H_2O_2: high yield of maleic acid by using titanium silicalite-1. RSC Adv 4:54960–54972

36. Nielsen ER (1949) Vapor phase oxidation of furfural. Indus Eng Chem 41:365–368

37. Sessions WV (1928) Catalytic oxidation of furfural in the vapor phase. J Am Chem Soc 50:1696–1698

38. Zumstein F (1929) A process for the preparation of maleic acid. Patent DE478726 C. http://www.google.je/patents/DE478726C?cl=en

39. Zurnetein F (1934) Process for preparing maleic anhydride and maleic acid. American Patent 1956482

40. Alonso-Fagúndez N, Granados ML, Mariscal R, Ojeda M (2012) Selective conversion of furfural to maleic anhydride and furan with VO_x/Al_2O_3 catalysts. Chem Sustain Chem 5:1984–1990

41. Chun WSLYLFHCY (2009) Catalytic oxidation of furfural in vapor-gas phase for producing maleic anhydride. Chem Ind Eng Prog 28:10–19

42. Kreile DR, Slavinskaya VA, Shimanskaya MV, Lukevits EY (1969) The reactivity of furan compounds in vapor-phase catalytic oxidation. Chem Heterocycl Compd 5:429–430

43. Badovskaya LA, Povarova LV (2009) Oxidation of furans (Review). Chem Heterocycl Comp 45:1023–1034

44. Lan J, Chen Z, Lin J, Yin G (2014) Catalytic aerobic oxidation of renewable furfural to maleic anhydride and furanone derivatives with their mechanistic studies. Green Chem 16:4351–4358

45. Murthy MS, Rajamani K (1974) Kinetics of vapour phase oxidation of furfural on vanadium catalyst. Chem Eng Sci 29:601–609

46. Choudhary H, Nishimura S, Ebitani K (2013) Metal-free oxidative synthesis of succinic acid from biomass-derived furan compounds using a solid acid catalyst with hydrogen peroxide. Appl Catal A General 458:55–62

Chapter 7

MULTI-STEP ENZYME-ORGANOCATALYST C–C BOND FORMING REACTIONS IN DEEP-EUTECTIC-SOLVENTS: TOWARDS IMPROVED PERFORMANCES BY ORGANOCATALYST DESIGN

Christoph R Müller[1], Andreas Rosen[1] and Pablo Domínguez de María[1,2]

[1]Institut für Technische und Makromolekulare Chemie (ITMC), RWTH Aachen University, Worringerweg 1, 52074 Aachen, Germany

[2]Sustainable Momentum, SL, Ap. Correos 3517, 35004 Las Palmas De Gran Canaria, Canary Islands, Spain.

ABSTRACT

Background

Deep eutectic solvents (DES) have recently emerged as promising non-hazardous environmentally-friendly solvents. In this respect, the use of DES as media for multi-step enzyme-organocatalysis (C–C bond formation via aldol-type reactions) represents a promising sustainable option. Being soluble in DES, organocatalysts may be retained in the DES phase during biphasic extractive work-up (e.g. with biogenic 2-methyl-tetrahydrofuran), enabling product recovery and organocatalyst recycling within the DES phase simultaneously.

Main Results

Herein, the proof-of-concept of designing organocatalysts—sspecifically tailored for DES—that may be properly retained in the DES phase (*immobilized*) among extractive cycles is demonstrated for the first time. To this end, the incorporation of novel hydrogen-bond donor groups (e.g. −OH) in the organocatalyst structure appears as a promising option to achieve improved results, leading to 1.5-fold higher conversions and yields, together with excellent chemoselectivities (>90%) for the new organocatalyst. Reactions are conducted using different bio-based DES, showing the broad applicability and possibilities that these processes may have.

Conclusions and Implications

In this work it is demonstrated that organocatalysts can be tuned to be used in different DES. This first proof-of-concept may trigger new research and applications of DES as sustainable solvents for enantioselective C–C bond forming reactions, whereby the organocatalyst design can play an important role for optimized integrated process set-up.

BACKGROUND

The quest for novel bio-based and environmentally-friendly solvents for synthetic processes is presently an important trend in chemistry, as solvents typically account for a significant part of the pollution produced through chemical reactions [1]. In this area, very recently Deep Eutectic Solvents (DES) have emerged as promising neoteric solvents for different chemical segments, e.g. using them as solvents for organic synthesis, as additives, or as extractive phases, among some relevant uses [2–5]. Previously, eutectic mixtures comprising substrates (e.g. composed of amino acids for peptide synthesis) had been reported to perform solvent-free biocatalytic processes [6]. In the novel applications, DES are typically formed by combination of a halide salt (e.g. choline chloride) with hydrogen-bond-donor molecules, such as (bio-based) alcohols, carboxylic acids, amines, etc. DES may be cost-effective solvents, environmentally-friendly, tunable and biodegradable, and thus their use in different synthetic reactions is gaining particular attention, stimulated by the above-described potential advantages [7]. With regard to organic compounds solubilities, DES enable often the dissolution of hydrogen-bonding molecules, such as alcohols, carboxylic acids, amines, etc., whereas non-hydrogen bonding compounds tend to form a second phase. That feature has further allowed several promising strategies, ranging from biphasic forming systems with many commonly used extractive solvents (e.g. ethyl acetate, 2-MeTHF) [7] to unique extraction properties towards alcohol separation from esters [8, 9]. Due to the high tuneability of DES, many options and creative innovations are possible.

In recent years, the set-up of combined multi-step enzyme-organocatalytic reactions has also emerged as a promising branch of catalysis, synergistically using biocatalysis together with small molecules as catalysts for many (enantioselective) processes, e.g. efficient C–C bond formations under rather mild reaction conditions [10, 11]. In this area, we have successfully reported the first chemo-enzymatic cascade reaction in DES recently. Using immobilized lipase B from *Candida antarctica* (*i*CALB), in combination with vinyl acetate and 2-propanol, an transesterification occurs, acetaldehyde is produced in situ [12] and subsequently undergoes an organocatalytic cycle—based on enamine-iminium intermediates—to afford the final aldol product in an enantioselective fashion. For this purpose diaryl prolinols catalysts showed the best results, giving the final product in high yields and excellent enantioselectivities (Scheme 1) [13].

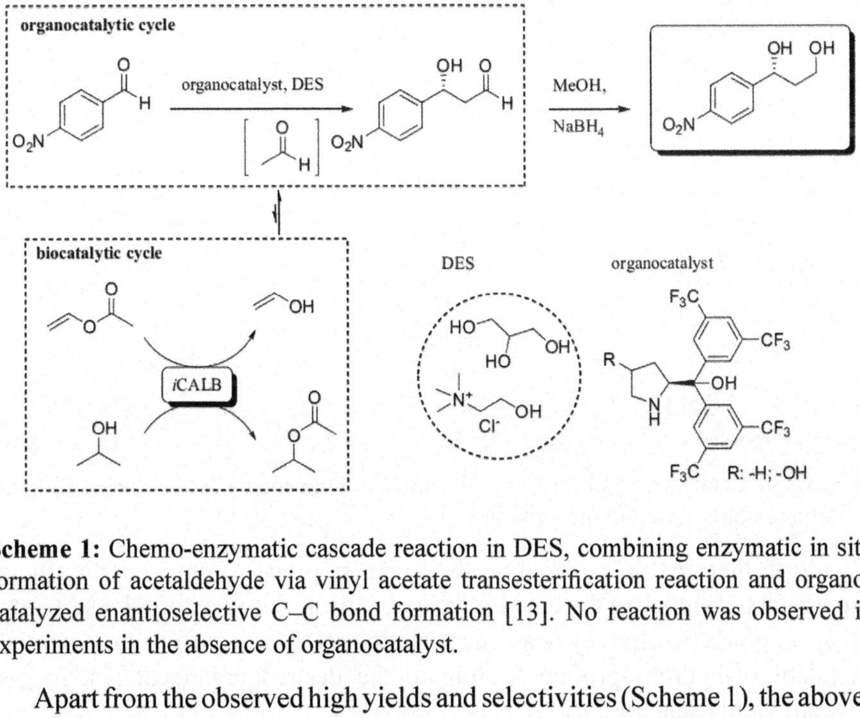

Scheme 1: Chemo-enzymatic cascade reaction in DES, combining enzymatic in situ formation of acetaldehyde via vinyl acetate transesterification reaction and organo-catalyzed enantioselective C–C bond formation [13]. No reaction was observed in experiments in the absence of organocatalyst.

Apart from the observed high yields and selectivities (Scheme 1), the above-reported process was also promising in terms of organocatalyst recycling, as the used diaryl prolinol organocatalyst bears different hydrogen-bond donor groups, which lead to strong interactions with the DES. When ethyl acetate was used for extractive purposes as second phase, the organocatalyst remained partly *immobilized* in the DES phase [13]. Thus, the DES phase could be used for two cycles without the need of adding fresh organocatalyst. Given the typically used high organocatalyst loadings, the recycling process may improve the overall economics.

Based on these promising observed results in terms of yield and enantioselectivity, and on the prognosis on ecological footprints and recycling [13], it was hypothesized that the incorporation of further hydrogen-bond donor groups along the organocatalyst structure might lead to higher DES-organocatalyst interactions and hence a better catalyst *immobilization*, while keeping yields and enantioselectivities in the same level. In this article, a proof-of-concept in that direction is reported for the first time.

RESULTS AND DISCUSSION

Starting from the diaryl prolinol as successful catalyst (Fig. 1) [13], the strategy consisted on incorporating an extra –OH moiety, leading to an organocatalyst with a further hydrogen-bond-donor group for the envisaged selective interaction with DES (Fig. 1).

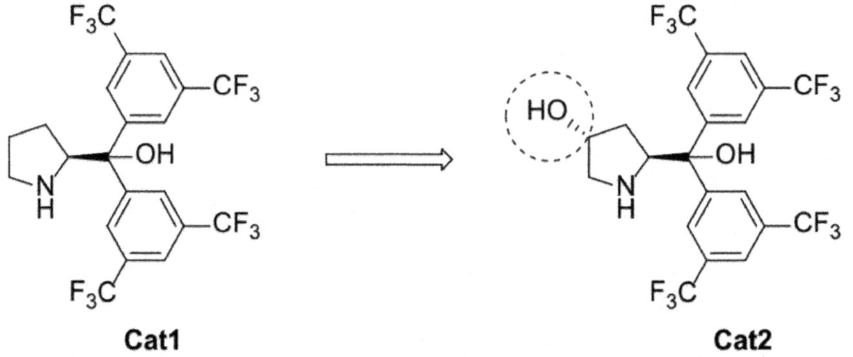

Figure. 1: Catalyst design for immobilization in DES media, adding an extra hydrogen-bond donor group in the structure.

With that goal in mind, the subsequent step was to envision an efficient approach to synthetize the desired organocatalyst. Based on previous literature [14], a synthetic strategy was successfully set, starting from commercially available 4-hydroxy-proline leading to the desired organocatalyst in good overall yield (Scheme 2).

Once the organocatalysts, Cat1 and Cat2 were synthetized, the multistep enzyme-organocatalyst enantioselective aldol reaction (Scheme 1) was assessed to compare the performance of the two organocatalysts, and thus evaluate the influence of the extra –OH moiety in the performance and in the recyclability. As DES, the same 1:2 (mol:mol) ChCl:Gly was used. Yields, conversions and the respective chemoselectivity (yield over conversion) were studied along different cycles. Among each cycle, the DES-reactive phase was extracted with bio-based 2-MeTHF [15]—to recover the aldol product—, and the DES phase was reused without addition of fresh organocatalyst or enzyme. The overall intended process concept is depicted in Fig. 2.

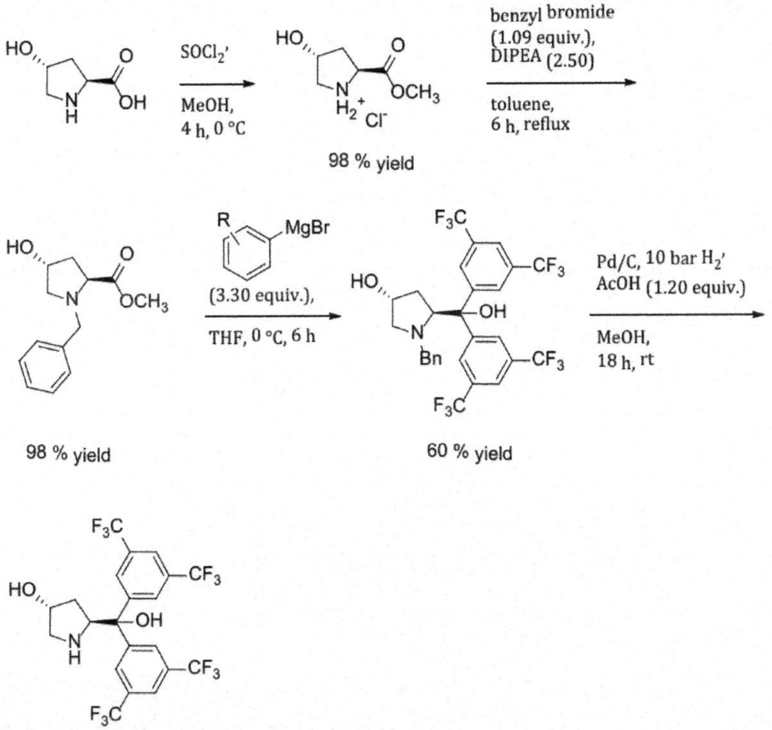

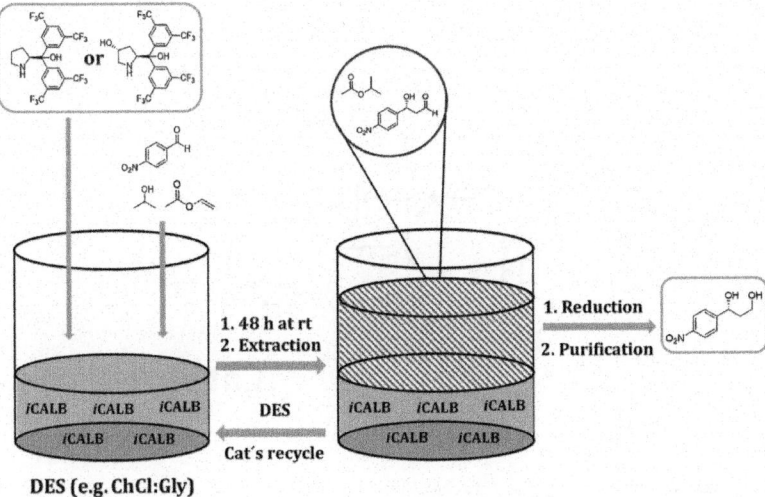

Scheme 2: Synthetic approach for the design of more DES-interactive organocatalysts. See "Experimental" part for conditions.

Figure. 2: Overall concept of the tandem enzyme-organocatalyst process combined with extraction among cycles and further chemical reduction. Immobilized CAL-B (iCALB) is suspended in the DES. Less than 1 wt% water is assumed in the reactive media.

Upon addition of substrates (vinyl acetate, *iso*propanol and *p*-nitrobenzaldehyde) and organocatalysts to the DES phase (containing iCALB), the reaction was conducted for 48 h, after which the addition of 2-MeTHF followed. While organocatalyst and enzymes remained in the DES phase, the organic phase extracted the aldol product, together with other side-products such as isopropyl acetate. Overall results of the reaction are shown in Fig. 3.

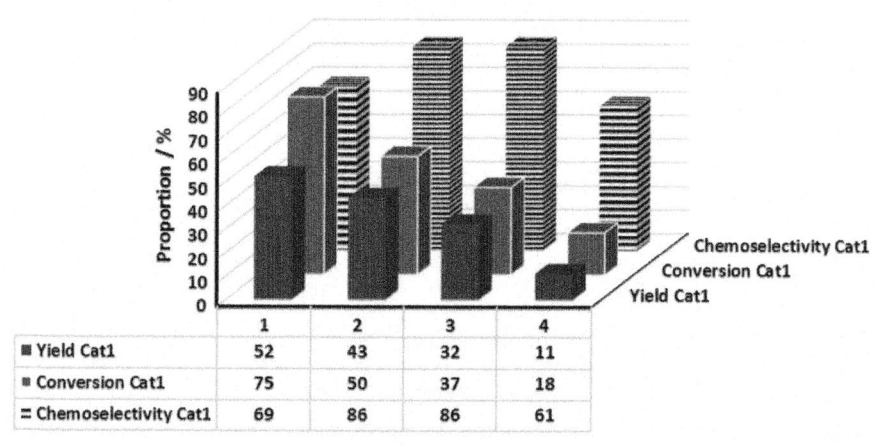

	1	2	3	4
■ Yield Cat1	52	43	32	11
▪ Conversion Cat1	75	50	37	18
≡ Chemoselectivity Cat1	69	86	86	61

Cycles

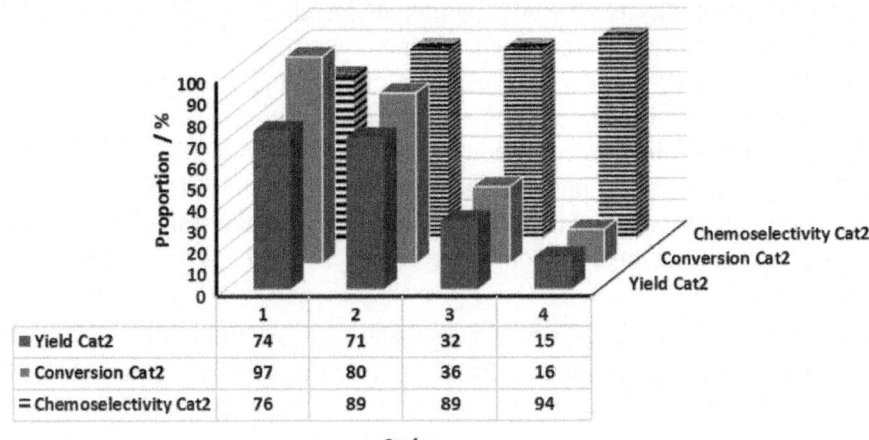

	1	2	3	4
■ Yield Cat2	74	71	32	15
▬ Conversion Cat2	97	80	36	16
═ Chemoselectivity Cat2	76	89	89	94

Cycles

Figure. 3: DES and catalyst recycling. Reaction conditions: 1.00 mmol p-nitrobenzaldehyde, 3.00 mmol vinyl acetate, 3.00 mmol i-PrOH, 3 mgiCALB, 0.20 mmol catalyst, 1 mL 1:2 ChCl/glycerol DES, 48 h, RT; Extraction with 3 × 2 mL 2-MTHF; Reduction of the organic phase phase with 2.0 mL MeOH, 6.00 mmol $NaBH_4$, 1 h, 0°C to yield the reduced alcohol. Less than 1 wt% water is assumed in the reactive media.

As it can be observed (Fig. 3), the envisioned Cat2—bearing an extra hydrogen-bond donor group in the structure—led not only to better recycling profiles than Cat1, but also to higher yields, conversions and chemoselectivities. This finding clearly shows that the subtle incorporation of an extra −OH group exerts a beneficial effect in the organocatalytic process performed in DES media. The effect is particularly relevant within the first two reaction cycles, where Cat2 virtually led to 1.5-fold improved performance compared to that of Cat1 (yields of 43–52% for Cat1 and 71–74% for Cat2). After the third cycle, however, both catalysts showed a significant decrease in yield and conversion, albeit chemoselectivity remained outstanding, especially in the case of Cat2 (>90% along cycles), suggesting a cleaner and more efficient overall synthetic process. Despite results are still non-optimized, to our knowledge this represents the first proof-of-concept of organocatalyst specifically tailored to DES media. As it can be envisaged, the incorporation of further hydrogen-bond donor groups—or other beneficial moieties—, in the structure might lead to improved and tailored outcomes for proper *immobilized* organocatalysts in DES.

Apart from envisaging organocatalyst-design as an option to improve multi-step organocatalytic reactions in DES, another line of research is to broaden the type of DES used as non-hazardous, environmentally-friendly media for multi-step reactions. In this case, many halide salts, as well as hydrogen-bond donor (HBD) compounds can be envisaged. Along this line, different choline-chloride-based DES combined with a range of different HBDs—covering polyols, urea, sugar-based alcohols—, were subsequently used for the intended multi-step reaction (Scheme 1). Results are depicted in Fig. 4.

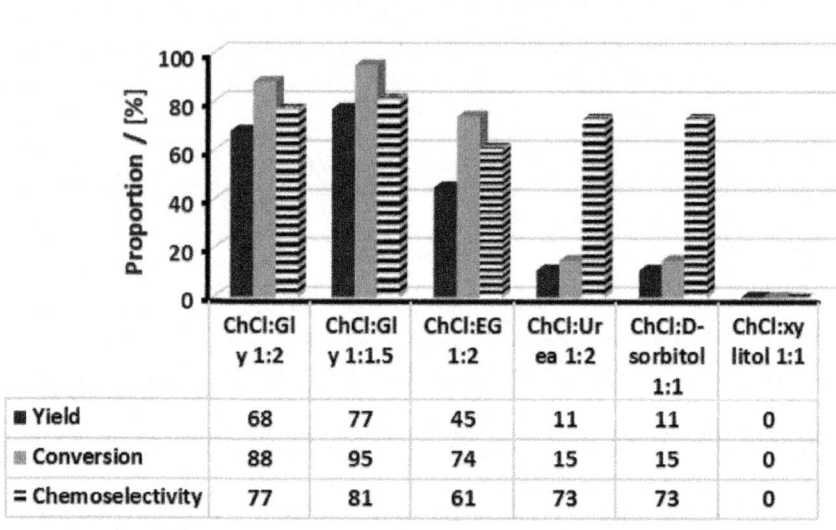

	ChCl:Gly 1:2	ChCl:Gly 1:1.5	ChCl:EG 1:2	ChCl:Urea 1:2	ChCl:D-sorbitol 1:1	ChCl:xylitol 1:1
■ Yield	68	77	45	11	11	0
▨ Conversion	88	95	74	15	15	0
≡ Chemoselectivity	77	81	61	73	73	0

Figure. 4: DES screening. Reaction conditions: 1.00 mmol *p*-nitrobenzaldehyde, 3.00 mmol vinyl acetate, 3.00 mmol *i*-PrOH, 3 mg *i*CALB, 0.20 mmol Cat1, 1 mL DES, 48 h, room temperature; Reduction with 2.0 mL MeOH, 6.00 mmol $NaBH_4$, 1 h, 0°C. *Gly* glycerol, *EG* ethylene glycol. Less than 1 wt% water is assumed in the reactive media.

As observed (Fig. 4), the reaction media played an important role in the outcomes of the enzyme-organocatalyst reaction. Glycerol-based DES (comprising 1:2 and 1:1.5) led to excellent results, and even the subtle change of proportion (from 2 to 1.5 mol of glycerol) led to improved results in yields and chemoselectivities. The substitution of glycerol (as hydrogen-bond donor) by ethylene–glycol gave, however, lower yields and chemoselectivities, emphasizing the interactions that DES have with the catalytic process. Likewise, detrimental effects were more dramatic when using ChCl:urea and sugar-based DES, presumably due to enzyme denaturation. In the case of xylitol as hydrogen-bond donor, the reaction mixture froze within the first minutes of reaction, thus rapidly invalidating this option for a combined enzyme-organocatalyst process. Overall, the choice of DES media appears as a relevant matter as well for the setup of these enzyme-organocatalyst multi-step combined reactions.

CONCLUSIONS

In this paper the use of Deep Eutectic Solvents (DES) for multi-step enzyme-organocatalyst processes—leading to selective C–C bond forming aldol-type reactions—has been assessed. The design of organocatalysts specifically designed for DES media has been envisioned for the first time, leading to the proof-of-concept of an organocatalyst with improved performances (yield, conversion and chemoselectivities) in DES, by the incorporation of novel hydrogen-bond donor groups. Hence, triggering the catalyst immobilization via hydrogen bond interactions with the DES phase. Likewise, different DES have been tested for the reaction successfully. Overall, results suggest that the high tuneability of DES, combined with tailored organocatalyst-design, may lead to powerful synergies to perform selective organic reactions in non-hazardous environmentally-friendly media under rather mild reaction conditions. More research and design is needed to first understand the molecular interactions between DES and organocatalysts, and to ultimately exploit the tremendous potential that this multi-disciplinary field may have.

EXPERIMENTAL SECTION

Chemicals

All chemicals were purchased from Sigma-Aldrich and used without further purification. Immobilized form of *Candida antarctica* Lipase B (iCAL-B) was purchased from c-LEcta GmbH (trade name CALB-immo).

DES Preparation

The components were mixed in the desired molar ratio and stirred at 60°C until a clear solution was obtained. After cooling down to room temperature, the DES was directly used. DESs were stored for maximum 1 month in a closed vessel.

Analytics

All NMR spectra were measured on a 400 MHz (^{1}H-NMR: 400 MHz, ^{13}C-NMR: 101 MHz), and 300 MHz (^{1}H-NMR: 300 MHz, ^{13}C-NMR: 75 MHz) Bruker device from BioSpin GmbH at 20°C. Chemical shifts are relative to the used solvents (CDCl$_3$: ^{1}H: δ = 7.26 ppm, ^{13}C: δ = 77.16 ppm), indicated in ppm. Following abbreviations were used for the signal patterns: s = singlet, bs = broad signal, d = doublet, t = triplet, q = quartet, m = multiplet, dd = doublet of doublets etc. for the ^{1}H-spectra.

Cat2 Synthesis

(2S,4R)-4-hydroxy-2-(methoxycarbonyl)pyrrolidin-1-ium chloride

Freshly distilled methanol (125 mL) was added to a round bottom flask and *trans*-4-hydroxy-l-proline (5.00 g, 38.13 mmol) was suspended. The slurry was cooled down to 0°C. After drop wise addition of thionyl chloride (2.78 mL, 38.2 mmol, 1.00 equiv.) the reaction mixture was stirred for 4 h at RT. Removal of the solvent under reduced pressure gave the product (6.80 g, 37.4 mmol, 98% yield) as a white solid. ^{1}H-NMR (CDCl$_3$, 300 MHz): δ = 2.08–2.18 (2H, m), 3.06 (1H, d, J = 12.0 Hz), 3.36 (1H, dd, J = 12.0, 4.4 Hz), 3.75 (3H, s), 4.41–4.48 (2H, m), 5.62 (1H, br-s), 9.91 (2H, br-s) *ppm*. ^{13}C-NMR (CDCl$_3$, 75 MHz): δ = 37.0, 53.0, 57.3, 68.4, 68.5, 169.0 *ppm*. Data are fully consistent with previous literature [16].

Methyl (2S,4R)-1-benzyl-4-hydroxypyrrolidine-2-carboxylate

Toluene (30 mL) was added to a round bottom flask and (2S,4R)-4-hydroxy-2-(methoxycarbonyl)pyrrolidin-1-ium chloride (5.30 g, 29.2 mmol) was suspended. The mixture was cooled down to 0°C. Firstly N,N-Diisopropylethylamine (DIPEA, 12.60 mL, 72.34 mmol, 2.48 equiv.) was added drop wise and then followed by benzyl bromide (3.80 mL, 31.92 mmol, 1.09 equiv.). The reaction mixture was refluxed for 6 h and quenched with saturated aqueous ammonium chloride solution (30 mL). The aqueous solution was extracted with ethyl acetate (3 × 150 mL). The combined organic layers were washed with brine (150 mL) and dried over sodium sulfate. Filtering of the drying agent and solvent removal under reduced pressure gave the product (6.74 g, 28.7 mmol, 98% yield) as yellowish oil. ^{1}H-NMR (CDCl$_3$, 300 MHz): δ = 1.75 (1H, br-s), 2.02–2,13 (1H, m), 2.18–2.33 (1H, m), 2.48 (1H, dd, J = 10.2, 3.7 Hz), 3.33 (1H, dd, J = 10.2, 5.6 Hz), 3.58–3.70 (2H, m), 3.66 (3H, s), 3.90 (1H, d, 12.9 Hz), 4.45 (1H, br-s), 7.20–7.38 (5H, m) ppm. ^{13}C-NMR (CDCl$_3$, 75 MHz): δ = 39.6, 51.9, 58.2, 61.3, 63.7, 70.4, 127.36, 128.4, 129.2, 138.2, 174.1 ppm. Analytic data agree with literature [17].

(3R,5S)-1-benzyl-5-(bis(3,5-bis(trifluoromethyl)phenyl)(hydroxy)methyl)pyrrolidin-3-ol

Dry THF (20 mL) was added to a round bottom flask under argon atmosphere and magnesium turnings (1.70 g, 70.00 mmol, 5.00 equiv.) were suspended. A solution of 1-bromo-3,5-bis(trifluoromethyl) benzene (7.97 mL, 46.20 mmol, 3.30 equiv.) in dry THF (180 mL) was added drop wise at 0°C. The reaction mixture was refluxed for 2 h, transferred into a dropping funnel and added drop wise to a solution of methyl (2S,4R)-1-benzyl-4-hydroxypyrrolidine-2-carboxylate (3.29 g, 14.00 mmol) in dry THF (60 mL) at 0°C over 30 min. The mixture was stirred 4 h at RT, before it was quenched by the addition of saturated aqueous ammonium chloride solution (100 mL) and subsequently neutralized with aqueous 10% HCl solution. The aqueous layer was extracted with ethyl acetate (3 × 100 mL). The combined organic layers were washed with brine (100 mL) and dried over sodium sulfate. Filtering of the drying agent and solvent removal under reduced pressure gave the crude product, which was purified via flash column chromatography (eluent: EtOAc/PE 1:6; R_f = 0.22). The final product (5.29 g, 8.37 mmol, 60% yield) was obtained as yellowish oil. ^{1}H-NMR (CDCl$_3$, 400 MHz): δ = 1.67–1.75 (1H, m), 1.76–1.85 (1H, m), 2.68–2.76 (1H, m), 3.12–3.26 (2H, m), 3.52 (1H, d, J = 13.0 Hz), 3.49–3.55 (1H, m), 4.53 (1H, dd, J = 8.7, 7.5 Hz), 5.55 (1H, br-s), 6.97–7.00 (2H, m), 7.19–7.29 (3H, m), 7.75 (2H, d, J = 11.6 Hz) 8.06 (2H, s), 8.24 (2H, s) *ppm*. ^{13}C-NMR (CDCl$_3$, 101 MHz): δ = 38.9, 61.8, 62.7, 70.8, 71.1, 76.2, 121.5, 121.7, 121.9, 124.6, 125.7, 126.1, 127.7, 128.3, 128.7, 132.2, 138.5, 147.2, 149.4 *ppm*. Analytic data agree with literature results [18].

(3R,5S)-5-(bis(3,5-bis(trifluoromethyl)phenyl)(hydroxy)methyl) pyrrolidin-3-ol

To a suspension of acetic acid (106 μL, 1.86 mmol, 1.20 equiv.) and palladium on carbon (10 wt% Pd/C, 100 mg), and dry methanol (3.0 mL) (3R,5S)-1-benzyl-5-(bis(3,5-bis(trifluoromethyl)phenyl)(hydroxy)   methyl) pyrrolidin-3-ol (1.00 g, 1.55 mmol) was added. The reaction was stirred in a hydrogen atmosphere (18 bar) for 18 h at RT. Filtering over Celite and removal of the solvent under reduced pressure gave an oil which was solubilized in ethyl acetate (15 mL). Saturated aqueous NaHCO$_3$ solution (15 mL) was added and neutralized to pH 7. The aqueous layer was extracted with ethyl acetate (3 × 15 mL). The combined organic layers were washed with brine (15 mL) and dried over sodium sulfate. Filtering of the drying agent and solvent removal under reduced pressure gave the product (748 mg, 1.38 mmol, 89% yield) as yellowish solid. ^{1}H-NMR (CDCl$_3$, 300 MHz): δ = 1.40–1.52 (1H, m), 1.67–1.77 (1H, m), 3.05-3.22 (2H, m), 4.41–4.47 (1H, m), 4.72 (1H, dd,J = 10.1, 6.3 Hz), 7.76 (1H, s), 7.79 (1H, s), 7.94 (1H, s), 8.06 (1H, s) *ppm*. ^{13}C-NMR (CDCl$_3$, 75 MHz): δ = 36.5, 55.6, 63.3, 72.6, 76.5, 125.7, 126.3, 132.2, 146.1, 149.2 *ppm*. Analytic data agree with previous literature [19].

DES Screening

4-Nitrobenzaldehyde (1.00 mmol) was added with (*S*)-α,α-Bis[3,5-bis(trifluoromethyl)phenyl]-2-pyrrolidinemethanol (cat 1, 0.20 mmol), vinyl acetate (276 µL, 3.00 mmol, 3.00 equiv.), *iso*-propanol (230 µL, 3.00 mmol, 3.00 equiv.) and iCALB (3 mg) to 1 mL of 1:2 ChCl/glycerol DES in a G15 vial, equipped with a 15 mm × 4 mm magnetic stirring bar. The vessel was closed with a cap and gasket. After stirring (300 rpm) for 24 h at RT, methanol (2.00 mL) was added and the reaction mixture was transferred into a round bottom flask and cooled down to 0°C. After slow addition of sodium borohydride (226 mg, 6.00 mmol over 30 min), the reaction was allowed to stir another hour at 0°C. Quenching was conducted by the addition of aqueous saturated ammonium chloride solution (15 mL) followed by the addition of ethyl acetate (15 mL). The phases were separated and the aqueous phase was extracted with ethyl acetate (2 × 15 mL). After washing of the combined organic layers with brine (15 mL), drying was proceed over sodium sulfate. Filtration of drying agent and solvent removal under reduced pressure led to the crude product which was purified via flash chromatography (eluent: 4:1 EtOAc/PE; R$_f$ = 0.33) to give a colorless oil. ^{1}H-NMR (CDCl$_3$, 300 MHz): δ = 1.97 (q, J = 5.8 Hz, 2H), 2.18 (br-s, 1H), 3.50 (d, J = 2.9 Hz, 1H), 3.90–3.95 (m, 2H), 5.10 (t, J = 6.00 Hz, 1H), 7.55 (d, J = 8.8 Hz, 2H), 8.21 (d, J = 8.8 Hz, 2H) ppm. ^{13}C-NMR (CDCl$_3$, 75 MHz): δ = 40.6, 61.7, 73.7, 124.1, 126.8, 147.6, 152.1 ppm.

Recycling System

4-Nitrobenzaldehyde (1.00 mmol) was added with organocatalyst (0.20 mmol), vinyl acetate (276 µL, 3.00 mmol, 3.00 equiv.), *iso*-propanol (230 µL) and iCALB (3.00 mg) to 1 mL of 1:2 ChCl/glycerol DES in a G15 vial, equipped with a 15 mm × 4 mm stirring bar. The vessel was closed with a cap and gasket. After stirring for the 48 h at 300 rpm, the DES phase was extracted with the 2-MTHF (3 × 2 mL). The combined organic layers were separated and methanol was added, while the mixture was cooled down to 0°C. After slow addition of sodium borohydride (6.00 mmol over 30 min), the reaction was allowed to stir another hour at 0°C. Quenching was conducted by the addition of aqueous saturated ammonium chloride solution (15 mL) followed by the addition of ethyl acetate (15 mL). The phases were separated and the aqueous phase was extracted with ethyl acetate (2 × 15 mL). After washing of the combined organic layers with brine (15 mL), drying was proceed over sodium sulfate. Removal of drying agent and solvent under reduced pressure led to the crude product which was purified via flash chromatography (eluent: 4:1 EtOAc/PE; R_f = 0.33) to give a colorless oil. The extracted DES was directly used in the next cycle.

AUTHORS' CONTRIBUTIONS

CRM and AR performed the experimental work. CRM and PDdM designed the experiments, interpreted the results, and wrote the manuscript. All authors read and approved the final manuscript.

ACKNOWLEDGEMENTS

Financial support from DFG training group 1166 "BioNoCo" ("Biocatalysis in Non-conventional Media") is gratefully acknowledged.

REFERENCES

1. Gu Y, Jerome F (2013) Bio-based solvents: an emerging generation of fluids for the design of eco-efficient processes in catalysis and organic chemistry. Chem Soc Rev 42:9550–9570

2. Zhang Q, De Oliveira Vigier K, Boyer S, Jerome F (2012) Deep eutectic solvents: syntheses, properties and applications. Chem Soc Rev 41:7108–7146

3. Carriazo D, Serrano MC, Gutiérrez MC, Ferrer ML, del Monte F (2012) Deep eutectic solvents playing multiple roles in the synthesis of polymers and related materials. Chem Soc Rev 41:4996–5014

4. Maugeri Z, Domínguez de María P (2012) Novel choline chloride based deep eutectic solvents with renewable hydrogen bond donors: levulinic acid and sugar-based polyols. RSC Adv 2:421–425

5.   Domínguez de María P, Maugeri Z (2011) Ionic Liquids in biotransformations: from proof-of-concept to deep eutectic solvents. Curr Opin Chem Biol 15:220–225

6.   Gill I, Vulfson E (1994) Enzymic catalysis in heterogeneous eutectic mixtures of substrates. Trends Biotech. 12:118–122

7.   Smith EL, Abbott AP, Ryder KS (2014) Deep eutectic solvents and their applications. Chem Rev 114:11060–11082

8.   Maugeri Z, Leitner W, Domínguez de María P (2012) Practical separation of alcohol-ester mixtures using deep-eutectic-solvents. Tetrahedron Lett 53:6968–6971

9.   Krystof M, Pérez-Sánchez M, Domínguez de María P (2013) Lipase-catalyzed (trans)esterification of 5-hydroxymethylfurfural and separation of HMF-esters using deep eutectic-solvents. Chem Sus Chem 6:630–634

10. Groeger H, Hummel W (2014) Combining the "two worlds" of chemocatalysis and biocatalysis towards multi-step one-pot processes in aqueous media. Curr Opin Chems Biol 19:171–179

11. Heidlindemann M, Rully G, Berkessel A, Hummel W, Groeger H (2014) Combination of asymmetric organo- and biocatalytic reactions in organic media using immobilized catalysts in diferente compartments. ACS Catal 4:1099–1103

12. Majumder AB, Ramesh NG, Gupta MN (2009) A lipase catalysed condensation reaction with a tricyclic diketone: yet another example of biocatalytic promiscuity. Tetrahedron Lett 50:5190–5193

13. Mueller CR, Meiners I, Domínguez de María P (2014) Highly enantioselective tandem enzyme-organocatalyst crossed aldol reactions with acetaldehyde in deep eutectic solvents. RSC Adv 4:46097–46101

14. Lin Q, Meloni D, Pan Y, Mia M, Rodgers J, Shepard S et al (2009) Enantioselective synthesis of Janus Kinase inhibitor INCB018424 via an organocatalytic Aza-Michael reaction. Org Lett 11:1999–2002

15. Pace V, Hoyos P, Castoldi L, Domínguez de María P, Alcántara AR (2012) 2-methyltetrahydrofuran (2-MeTHF): A biomass-derived solvent with broad application in organic chemistry. Chem Sus Chem 5:1369–1379

16. Pickering L, Malhi BS, Coe PL, Walker RT (1995) 4'-Methyloxycarbamyl-3'-deoxy-5-methyluridine; synthesis of a novel nucleoside analogue. Tetrahedron 51:2719–2728

17. Alza E, Sayalero S, Kasaplar P, Almasi D, Pericas MA (2011) Polystyrene supported diarylprolinol ethers as highly efficient organocatalysts for Michael-type reactions. Chem Eur J 17:11585–11595

18. Maltsev OV, Kucherenko AS, Chimishkyan AL, Zlotin SG (2010) α, α-Diarylprolinol-derived chiral ionic liquids: recoverable organocatalysts for the domino reaction between α, β-enals and N-protected hydroxylamines. Tetrahedron Asymmetry 21:2659–2670

19. Itoh T, Ishikawa H, Hayashi Y (2009) Asymmetric Aldol reaction of acetaldehyde and isatin derivatives for the total syntheses of ent-convolutamydine E and CPC-1 and a half fragment of madindoline A and B. Org Lett 11:3854–3857

Chapter 8

EXPERIMENTAL AND MODELLING STUDIES ON THE UNCATALYSED THERMAL CONVERSION OF INULIN TO 5-HYDROXYMETHYLFURFURAL AND LEVULINIC ACID

A Fachri[1,3], R M Abdilla[1] , C B Rasrendra[2] and H J Heeres[1]

[1]Chemical Engineering Department, University of Groningen, Nijenborgh 4, 9747 AG Groningen, The Netherlands

[2] Chemical Engineering Department, Faculty of Industrial Technology, Institut Teknologi Bandung, Ganesha 10, Bandung, Indonesia

[3] Faculty of Engineering, University of Jember, Kalimantan 37, Jember, Indonesia.

ABSTRACT

Background

5-Hydroxymethylfurfural (HMF), an important biobased platform chemical, is accessible by the acid catalysed conversion of biopolymers containing hexoses (cellulose, starch, inulin) and monomeric sugars derived thereof. We here report an experimental study on the uncatalysed, thermal conversion of inulin to HMF in aqueous solutions in a batch set-up.

Results

The reactions were conducted in a temperature range of 153–187°C, an inulin loading between 0.03 and 0.12 g/mL and batch times between 18 and 74 min using a central composite experimental design. The highest experimental HMF yield in the process window was 35 wt% (45 mol%), which is 45% of the theoretical maximum (78 wt%). The HMF yields were modeled using a statistical approach and good agreement between experiment data and model was obtained. The possible autocatalytic role of formic acid (FA) and levulinic acid, two main byproducts, was probed by performing reactions in the presence of these acids and it was shown that particularly FA acts as a catalyst.

Conclusions

Inulin is an interesting feed for the synthesis of HMF in water. A catalyst is not required, though autocatalytic effects of FA play a major role and also affect reaction rates and product yields.

GRAPHICAL ABSTRACT

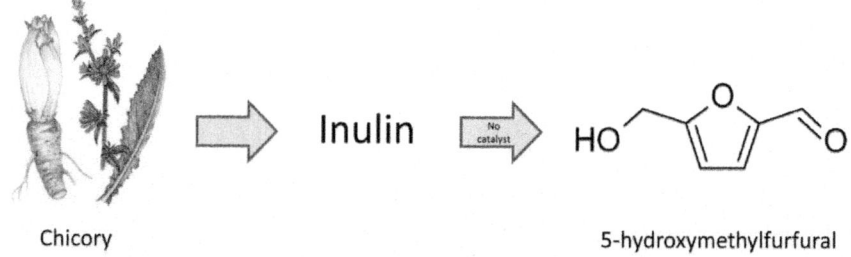

Chicory   5-hydroxymethylfurfural

BACKGROUND

Biomass has been identified as an attractive alternative for crude oil, natural gas and coal to produce fuels and chemicals. When considering biobased chemicals, particularly the carbohydrate fraction (cellulose and hemicellulose) of lignocellulosic biomass shows high potential. An example of a biobased chemical is 5-hydroxymethylfurfural (HMF), which has been classified as one of the top 12 biobased chemical from biomass by the US department of Energy (DOE) [1, 2]. HMF is a very versatile building block for biofuels or biofuel additives [3, 4], solvents, surface-active agents, fungicides [5] and for interesting monomers for the plastics industry [1, 6, 7].

5-Hydroxymethylfurfural may be synthesised from of a wide range of hexoses by elimination of three water molecules (Scheme 1, example ford-fructose).

$$CH_2OH \; CH_2OH \quad \xrightarrow[H^+]{-3H_2O} \quad HOH_2C \cdots CHO \quad \xrightarrow[H^+]{+2H_2O} \quad H_3C \cdots OH \quad + \quad H \cdots OH$$

fructose   HMF   levulinic acid   formic acid

Scheme 1: A simplified reaction scheme for HMF formation.

Conventionally, the reaction is carried out in water using a Brönsted acid with d-fructose as the preferred carbohydrate source [8–20]. Hydrochloric acid and sulfuric acid are the most commonly used Brönsted acids. Kuster [21], reported that d-fructose is more reactive and selective towards to HMF than d-glucose and a HMF yield from d-fructose of 68% was reported using HCl as the catalyst (C_{fruct} = 9 wt%, 1.5 h). Yields in water are less than quantitative due to the formation of insoluble byproducts (humins) and a subsequent reaction of HMF to levulinic acid (LA) and formic acid (FA, Scheme 1). Two main reaction mechanisms have been proposed for the reaction in water, one involving acyclic intermediates and another with cyclic intermediates [22].

Considerable progress has been made to enhance HMF yields, among others by using solvents other than water and the use of advanced catalysts [8]. Typical homogeneous Brönsted catalysts (mineral acids) have the advantage that they are relatively cheap, though recycling is often cumbersome. As such, heterogeneous catalysts could be beneficial and have been tested in detail. However, catalyst lifetime needs to be established and a major concern is the deposition of humin substances on the heterogeneous catalysts leading to irreversible catalyst deactivation. Examples of solvents other than water tested for the reaction include acetone [17], methanol [17], toluene [23] and dimethyl sulfoxide (DMSO) [24]. Among them, reactions in DMSO give essentially quantitative HMF yields [8], though down-stream processing is considerably more complicated than with water and is a critical issue.

Despite these advances, efficient, economically viable routes to produce HMF from d-fructose have not yet been commercialized. Bicker et al. [17] estimated an HMF cost price of 2 €/kg, using the assumption that d-fructose is available at a price of 0.5 €/kg. Torres et al., [25] reported that the manufacturing costs of HMF are between 1.97 and 2.43 $/kg, depending on the solvent used, based on a d-fructose price of 0.55 $/kg and 7000 ton/year HMF production unit. An HMF price of around 1.00 $/kg is considered a good starting point to allow its use in bulk-scale chemical applications.

The major variable cost item in the economic evaluations for HMF manufacture is the cost of the d-fructose feed. As such, the identification of low priced d-fructose alternatives is of high importance for the development of techno-economically viable routes to HMF. Alternative feeds are biopolymers enriched in d-fructose units, which are likely cheaper than purified d-fructose. An interesting biopolymer is inulin, an oligosaccharide consisting of mainly fructose units, in some cases capped with a glucose unit [6, 15, 26]. Inulin can be extracted from plants, examples are jerusalem artichoke [27–31], chicory [28–31] and dahlia tubers [28–31]. Of particular interest are the Jerusalem artichoke and chicory, which are reported to have a high inulin content of up to 20% on fresh weight [32].

A number of studies have been performed on the catalysed inulin conversion to HMF in water, the solvent of choice in the current investigation (Table 1). Temperatures are typically between 80 and 200°C, reaction times vary between minutes and 3 h, and intakes of inulin are between 5 and 10 wt% on solvent. A range of catalysts has been applied, varying form homogeneous (H_2CO_3, formed in situ by CO_2 addition) to heterogeneous catalysts. Reactions not only have been performed in water but also in water-organic solvent mixtures and ionic liquids. Highest HMF yields (88 mol% on inulin) were reported by Hu et al. [33] using DMSO as the solvent.

Table 1: Overview of HMF yield data for the catalysed conversion of Inulin to HMF in water

C_{inulin} (wt%)	T (°C)	t	Catalyst	Catalyst loading	HMF yield (%)[a]	References
Water/homogeneous catalyst						
5	160	4 min	CO_2	6 MPa	45[b]	[34]
5	160	4 min	CO_2	9 MPa	42[b]	[34]
5	180	2 min	CO_2	4 MPa	45[b]	[34]
5	180	2 min	CO_2	6 MPa	50[b]	[34]
5	180	2 min	CO_2	11 MPa	52[b]	[34]
5	200	45 min	CO_2	6 MPa	53[b]	[34]
5	200	45 min	CO_2	9 MPa	49[b]	[34]
Water/solid catalyst						
6	100	0.5 h	Cubic ZrP_2O_7	0.6 g	26 (wt%)	[35]
6	100	1 h	Cubic ZrP_2O_7	0.6 g	35 (wt%)	[35]
6	100	2 h	Cubic-ZrP_2O_7	0.6 g	36 (wt%)	[35]
6	100	1 h	$Ti(PO_4)(H_2PO_4) \cdot 2H_2O$	0.6 g	41 (wt%)	[35]
6	100	2 h	$Ti(PO_4)(H_2PO_4) \cdot 2H_2O$	0.6 g	65 (wt%)	[35]
6	80	2 h	FeVOP	5 wt%	35 (mol%)	[36]
10	155	18 min	HNb_3O_8	SCR = 50	43 (mol%)	[37]
6	100	3 h	Niobium phosphate	SCR = 1.6	31 (mol%)	[38]

SCR substrate to catalyst ratio.
[a] wt or mol in brackets after entries.
[b] wt% or mol% not provided.

We here report a study on the conversion of inulin to HMF using water as the solvent in the absence of a catalyst. In this way, catalyst recycle procedures are eliminated and drawbacks of heterogeneous catalysts (among others deactivation by humin deposition) are avoided. Water was selected as the solvent of choice, as it is environmentally benign, is a good solvent for many carbohydrates and it is cheap, nontoxic, and nonflammable [39–42].

A number of studies have been reported on the uncatalysed conversion of d-fructose to HMF (Table 2). For inulin, very limited information is available in the literature. The only paper is by Wu et al. [34] who reported an HMF yield of 48% (200°C, inulin loading of 5 wt% and 1 h reaction time).

Table 2: Overview of studies for the uncatalysed reaction of d-fructose and inulin to HMF in water

Substrate	Substrate intake (wt%)	T (°C)	t	Yield (%)[b]	References
ᴅ-Fructose	4.5	175	45 min	56[c]	[14]
	5	140	1 h	4 (mol%)	[16]
	9	200	5 min	41 (mol%)	[18]
	30	160[a]	5 min	1 (mol%)	[43]
	30	190[a]	5 min	36 (mol%)	[43]
	2	200[a]	5 min	13 (mol%)	[44]
	30	170	3 h	43 (mol%)	[45]
	0.05[d]	240	3 min	20[c]	[54]
	5	125	5 min	0.8[c]	[56]
	11	200	30 min	51[c]	[57]
Inulin	5	160	4 h	38[c]	[34]
	5	180	2 h	41[c]	[34]
	5	200	1 h	48[c]	[34]

[a] Heating by microwave irradiation.
[b] wt or mol in brackets after entries.
[c] wt% or mol% not provided.
[d] In molar.

Thus, it can be concluded that a detailed study on the effect of process conditions on HMF yields for the thermal decomposition of inulin in water has not been reported to date. We here report a systematic study on the effect of process conditions like reaction time, inulin intake and reaction temperature on the HMF yield. A total of 24 batch experiments were performed using a composite design. The data were analysed statistically and a model was developed to describe the HMF yield versus process conditions. Finally, possible autocatalytic effects of organic acids (FA and LA) on the rates of reactions and HMF yields were explored.

EXPERIMENTAL SECTION

Chemicals

Inulin from Dahlia tubers was purchased from Acros Organic (Geel, Belgium). d-Fructose (99%) and LA (≥97%) were obtained from Acros Organic (Geel, Belgium). Formic acid (≥95%) and d-glucose (≥99.5%) were purchased from Merck KGaA (Darmstadt, Germany). HMF (≥99%) was obtained from Sigma Aldrich (Steinheim, Germany). 2,5-Dihydroxybenzoic acid (DHB)

(≥99%) was purchased from Fluka (Deisenhofen, Germany). All chemicals were used without purification. De-ionized water was used to prepare the solutions.

Experimental Procedures

The experimental procedures are based on previous research by our group (Girisuta et al. [19]). In a typical experiment, the pre-determined amount of inulin and de-ionized water (4 mL) were loaded to glass ampoules with an internal diameter of 5 mm, a length of 15 cm and thickness of 1.5 mm. The ampoules were sealed with a torch.

For the exploratory experiments, a series of ampoules was placed in a rack in a heating oven (Heraeus Instruments Type UT6060) at constant temperature. At different reaction times, an ampoule was taken from the oven and quickly quenched in cold water to stop the reaction. The experiments carried out in the framework of the experimental design were individually performed in an oven (Heraeus Instruments Type UT6060) at the pre-set temperature. For the autocatalytic experiments with LA and FA, a series of ampoules were filled with inulin (0.1 g/mL) and the appropriate amount of each acid (0 or 0.1 M) and placed in an oven [Binder, APT Line™ FD (E2)] at 180°C for a predetermined reaction time. The experiments were repeated at different reaction times allowing construction of an HMF yield versus time plot for each individual organic acid and the blank experiment. All experiments were conducted in duplicate and the average value is taken. The outcome of the autocatalytic experiments cannot be compared directly with that of the screening and experimental design experiments as the heating up profile (temperature versus time) for both ovens differs.

After reaction, the ampoules were opened and the reaction mixture was taken out, and centrifuged for about 10–30 min to remove the solids. The liquid product was diluted with demi water before analysis.

The composition of the inulin sample and particularly the type and amount of C_6-sugars was determined by an acid-catalysed hydrolysis reaction. For this purpose, inulin (2.5 g) was dissolved at 70°C in 150 mL of water under stirring. The pH was adjusted to 1.4–1.6 by adding HCl (12 M). Then, the solution was placed in a water-bath for 30 min at 90°C. A liquid sample was taken and analysed by HPLC.

Analysis

Matrix assisted laser desorption ionization-time of flight mass spectrometry (MALDI-TOF MS) on a Voyager-DE PRO was used to determine the molecular weight of the inulin sample. DHB was used as the matrix.

HPLC was used to identify and quantify the liquid product from the reactions. The HPLC system consisted of a Hewlett Packard 1050 pump, a Bio-Rad organic acid column Aminex HPX-87H and a Water 410 differential refractive index detector. A sulfuric acid solution (5 mM) was used as the eluent with a constant flow rate of about 0.55 cm^3/min. The column was operated at 60°C. The HPLC was calibrated with solutions of the pure compounds at a range of concentrations. Using the chromatogram peak area and the external calibration curves, the concentrations of components in the liquid phase were determined.

Definitions

The HMF yield (y_{HMF}) is defined according to Eq. (1) and reported on a weight basis.

$$y_{HMF} = \frac{C_{HMF} \times M_{HMF} \times V}{W_{In}} \times 100\% \ (wt\%)$$

(1)

here, C_{HMF} represents HMF concentration (mol/L) at a certain time t, M_{HMF} is the molecular weight of HMF (g/mol), V represents the liquid volume (L) and w_{in} represents the intake of inulin (g).

The yield of HMF was converted from wt% to mol% by assuming that inulin consists of linked glucose/fructose units ($C_6H_{10}O_5$) which react to HMF according to the following stoichiometry:

$$C_6H_{10}O_5 \rightarrow C_6H_6O_3 + 2 \ H_2O \ C_6H_{10}O_5 \rightarrow C_6H_6O_3 + 2 \ H_2O$$

(2)

As such, the maximum yield of HMF is 78 wt%. Thus, the yield of HMF in mol% may be calculated from the yield in wt% by dividing the latter by 0.78.

Statistical Modelling

The optimisation experiments were modeled using Design-Expert 7 software (Stat-Ease). The yield of HMF (Y_{HMF}) was modeled using a standard expression as given in Eq. (3):

$$y_{HMF} = b_0 + \sum_{i=1}^{3} b_i x_i + \sum_{i=1}^{3} \sum_{j=1}^{3} b_{ij} x_i x_j$$

(3)

The independent variables (inulin intake, temperature and reaction time) are represented by the indices 1–3. The regression coefficients were obtained by statistical analyses of the data. Significance of factors was determined by their p value in the ANOVA analyses. A factor was considered significant if the p value was lower than 0.05, meaning that the probability of noise causing the correlation between a factor and the response is lower than 5%. Insignificant factors were eliminated using backward elimination, and the significant factors were used to model the data.

RESULT AND DISCUSSION

Inulin Characterization

The molecular weight distribution of the inulin used in this study (Dahlia tubers) was determined by MALDI-TOF/MS, a technique particularly suited for molecular weight determinations of oligosaccharides and polysaccharides [46, 47]. The M_n was found to be 2560, the M_w 3680, indicating an average degree of polymerization (DP) of about 16. Roberfroid [31] reported that the DP of inulin varies according to plant species, weather conditions, and the physiological age of the plant. In chicory, DP values range from 2 to 65, with 15 reported as an average. For inulin from onions, the DP is in the range of 2–12, for Jerusalem artichoke the DP is reported to be <40. Thus, the experimentally determined value of the DP of the inulin sample used in this study is within the ranges reported in the literature.

The d-fructose and d-glucose content of the inulin used in this study were determined by a mild hydrolysis of the samples followed by HPLC analyses of the liquid phase. The fructose content was 94 mol%, the remainder being glucose, giving a fructose to glucose ratio of 15–1. The fructose content is in close agreement with the literature for Dahlia tubers (94.1–96.7 mol%) [48]. Thus, the inulin sample contains mainly of oligomers with d-fructoside units, with each oligomer chain on average capped with a d-glucose molecule, in line with literature data [31].

HMF Synthesis from Inulin in the Absence of a Catalyst

Exploratory experiments on the thermal conversion of inulin were performed at 170°C, using an inulin intake of 0.1 g/mL. A typical concentration profile is given in Figure 1. Six water soluble components were identified in the reaction mixtures (HPLC). Three show a clear maximum in the course of the reaction, viz. d-fructose, d-glucose and HMF, and are as such intermediates in the reaction sequence. The final products are LA, FA and acetic acid (AA). In addition, some brown–black insolubles were observed, known as humins, which are always formed during the acid catalyzed conversions of carbohydrates in water, either in monomeric or polymeric form [8].

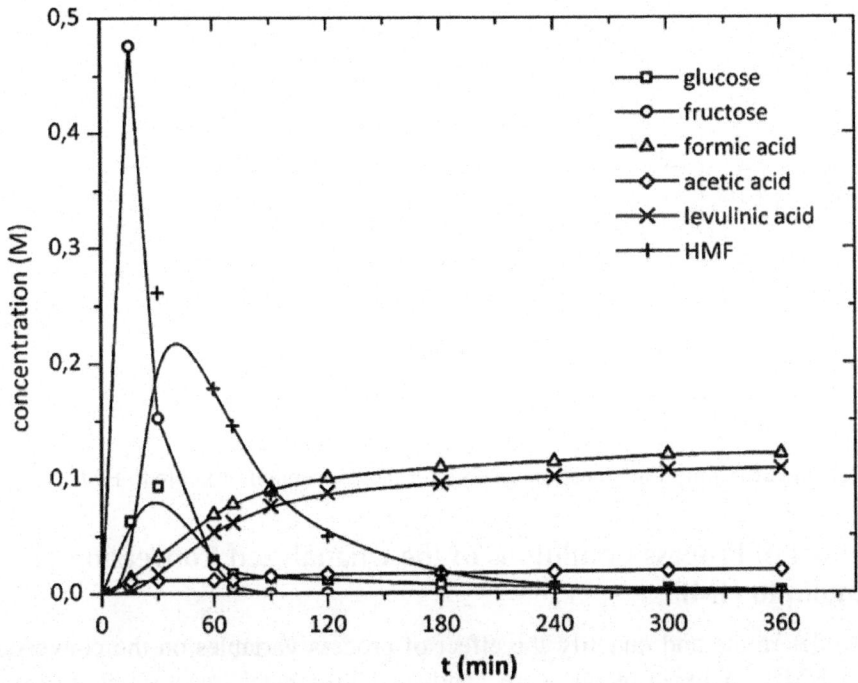

Figure 1: Concentration-time profiles for various compounds during the thermal conversion of inulin in water (170°C, C_{In} = 0.1 g/mL).

The product composition versus time is in agreement with the reaction network provided in Scheme 2. It involves the saccharification of inulin to the monomeric building blocks (d-fructose and some d-glucose), followed by the reaction of these C6 sugars to HMF. The latter is not inert under reaction conditions and reacts further to LA and FA [18, 19, 49–52].

Besides these products, acetic acid was detected in small amounts, though not quantified. AA is likely formed by hydrolysis of acetyl side groups in the inulin [53] or alternatively, from further degradation reactions of the intermediates. For instance, Asghari et al. [54] reported the formation of acetic acid (5 mol%) from fructose in subcritical water (200–320°C, residence time of 120 s, C_{fruct} of 0.05 M, no catalyst).

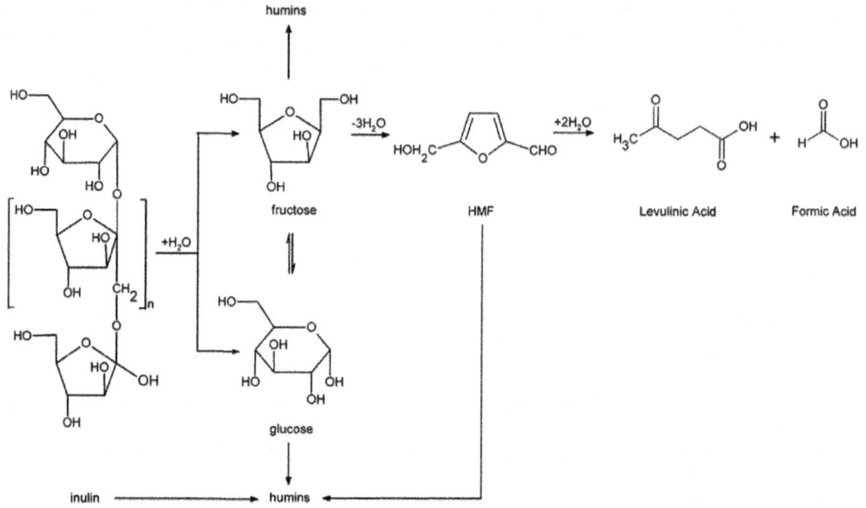

Scheme 2: A simplified reaction scheme for the conversion of inulin to HMF.

Effect of Process Conditions of the Unanalyzed Conversion of Inulin to HMF

To determine and quantify the effect of process variables on the conversion of HMF, 24 experiments were conducted in a batch reactor set up (glass ampoules) using a central composite design. Three independent variables, the temperature (153–187°C), inulin intake (0.03–0.12 g/mL) and reaction time (18–74 min), were explored and the yields of HMF and LA were taken as the dependent variables. The results are provided in Table 3.

Table 3: Overview of experiments for the thermal, unanalyzed conversion of inulin to HMF and LA

Run	C_{in} (g/mL)	$T_{reaction}$ (°C)	t (min)	y_{HMF} (wt%)	y_{LA} (wt%)
1	0.08	170	39	22.8	4.0
2	0.05	160	18	2.7	–
3	0.08	170	39	23.1	3.1
4	0.08	170	39	23.9	3.2
5	0.05	180	18	32.1	5.6
6	0.05	180	18	35.0	6.1
7	0.12	170	39	17.6	6.1

8	0.08	170	39	23.1	3.3
9	0.08	187	39	14.6	13.4
10	0.08	170	39	22.5	1.1
11	0.08	170	39	21.8	1.5
12	0.08	170	39	19.4	1.8
13	0.08	170	39	16.6	2.2
14	0.08	170	74	8.9	1.9
15	0.1	180	18	20.9	5.5
16	0.03	170	39	24.8	3.6
17	0.05	160	60	25.7	3.1
18	0.1	160	60	23.3	3.6
19	0.1	160	18	3.2	–
20	0.05	180	60	10.1	9.4
21	0.05	180	60	10.4	9.0
22	0.1	180	60	3.1	9.5
23	0.1	180	60	2.2	9.8
24	0.08	153	39	12.4	0.3

The center point of the central composite design was measured six times and the HMF yield was found to be on average 21.4 wt% with a standard deviation of 2.9 wt%. Thus, the reproducibility of the experimental procedure appears to be good.

The highest experimental HMF yield was about 35 wt% (45 mol%, entry 6 in Table 3) and was achieved at 180°C, an inulin intake of 0.05 g/mL and a reaction time of 18 min. Wu et al. [34] reported an HMF yield of 41% at 180°C, an inulin intake of 0.05 g/mL and a reaction time of 2 h. Comparison is cumbersome as conditions are not exactly similar and different starting materials were used (among others, inulin source and related properties like DP).

It is also of interest to compare the results with those obtained for inulin in water with sulphuric acid as the catalyst [55]. In the latter case, the highest experimental yield of HMF was 39.5 wt% (50.6 mol%), obtained at 170°C, 0.17 g/mL inulin intake and an acid concentration of 0.006 M and 20 min reaction time. Thus, the maximum HMF yield for the uncatalysed, thermal reaction within the window of process conditions is only slightly lower than when using a Brönsted acid. The reaction time is in general much longer for the thermal reaction than for the sulphuric acid catalysed reaction. As such, the use of a Brönsted acid catalyst has advantages in terms of reaction rates and as such a smaller reactor will be needed to achieve a certain production rate (kg/h). However, sulphuric acid recycle, complicating the work-up section, is not required for the uncatalysed reaction. Detailed process design studies will be required for both options to evaluate the best approach and these are beyond the scope of the current study.

Statistical Modeling

The HMF yield (y_{HMF}) as a function of the independent variables was statistically modelled using the Design-Expert 7 software package. All independent variables (temperature T, intake inulin C_{in} and reaction time t) were shown to be statistically significant and to have an effect on the HMF yield. The best model for HMF yield is given in Eq. (4) and includes both quadratic and interaction terms. An extended version with more significant numbers for the coefficients to be used for among others reactor engineering studies is given in the Additional file 1. The R-squared of the model is 0.9664, an indication that the model fits the experimental data well.

$$y_{HMF} = (1466.7)C_{in} + (11.8)T + (9.3)t - (9.2)C_{in}T$$
$$- (0.05)tT - (0.03)T^2 - (0.009)t^2 - 1223.9 \quad (4)$$

Analysis of the model variance is given in Table 4. A good agreement between the model and the experimental data was observed, as is shown in the parity plot provided in Figure 2.

Table 4: Analysis of variance of the HMF model

Source	Sum of squares	df	Mean square	F value	p value Prob > F
Model	1,878.14	7	268.31	65.66	<0.0001
C_{in} (A)	85.20	1	85.20	20.85	0.0003
T (B)	50.63	1	50.63	12.39	0.0028
t (C)	2.27	1	2.27	0.56	0.4665
AB	55.60	1	55.60	13.61	0.0020
BC	1,156.24	1	1156.24	282.96	<0.0001
B^2	107.51	1	107.51	26.31	0.0001
C^2	159.51	1	159.51	39.04	<0.0001
Residual	65.38	16			

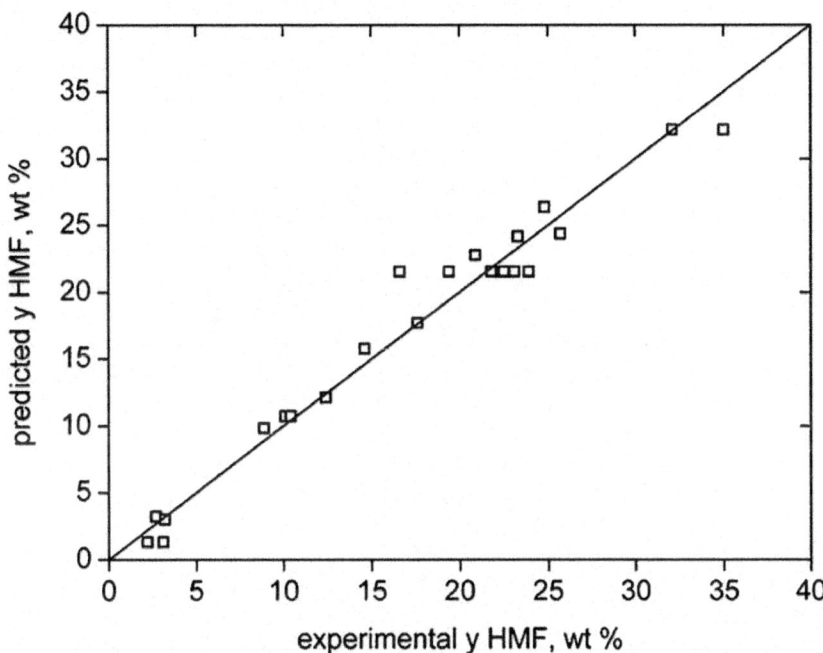

Figure 2: Parity plot between the experimental and modelled HMF yields.

With the statistical model available, it is possible to determine the effects of the process conditions on the HMF yield within the design window. To illustrate this, the model predictions for HMF yield for batch times of 30 min (Figure 3, left) and 40 min (Figure 3, right) are given versus the temperature and inulin intake. The yield of HMF is a complex function of the independent variables and it is difficult to draw general conclusions. Though, as expected, the HMF yields after 40 min are lower than for 30 min. This is due to the subsequent reaction of HMF to LA and FA (Scheme 2), leading to a lowering in the HMF yield. The effect of the inulin intake on the HMF yield is also complex in nature and temperature depending. At low temperatures, the HMF yield is slightly higher at higher inulin intakes, whereas the opposite trend is observed at higher temperatures, where a lower inulin intake is favored.

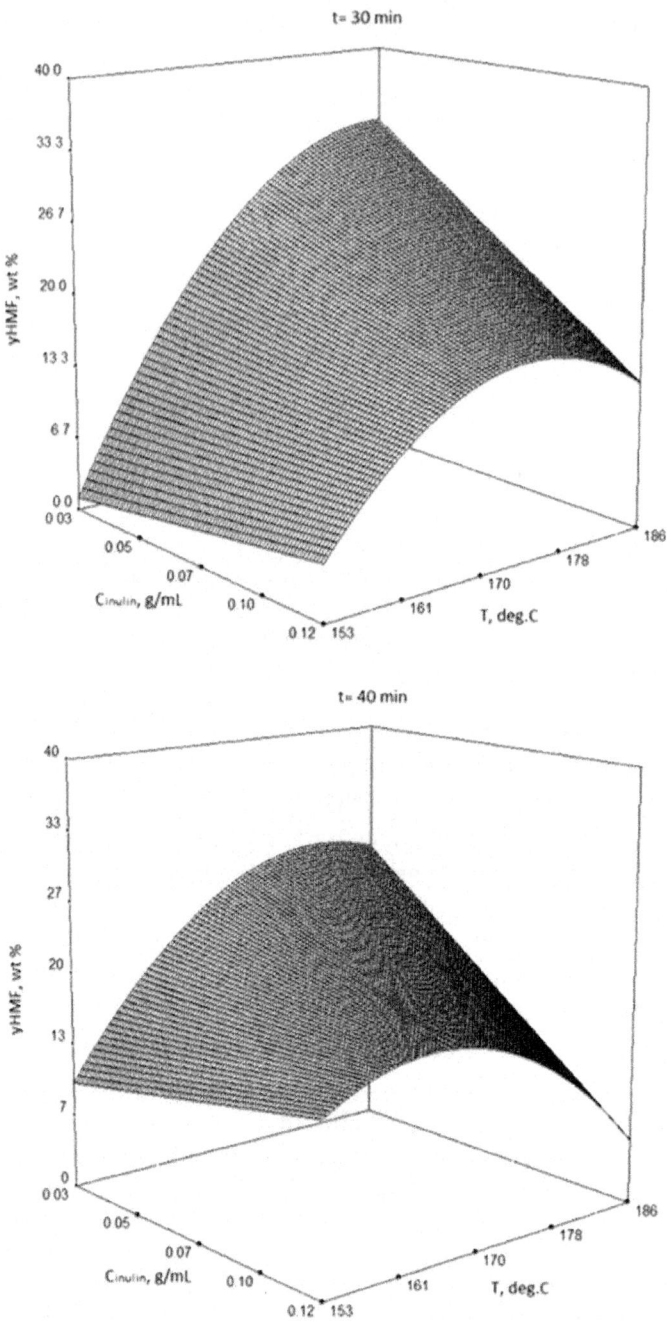

Figure 3: Modeled HMF yield versus temperature and inulin intake for two batch times (*left* 30 min; *right* 40 min).

The occurrence and importance of the consecutive reaction of HMF to LA was also confirmed by considering the LA yield (y_{LA}) versus the process conditions (Table 3). The LA yield was modeled and the results are given in Figure 4 and Table 5. The model equation with an R-square of 0.9464 is given in Eq. (5). An extended version with more significant numbers for the coefficients to be used for among others reactor engineering studies is given in the Additional file 1. The parity plot (Figure 5) reveals good agreement between the model and the experimental data.

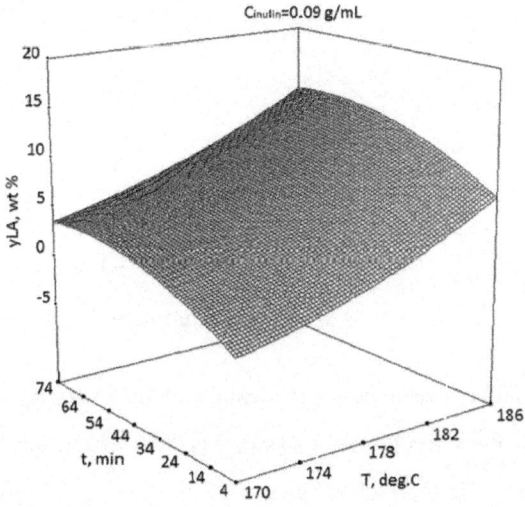

Figure 4: Modeled LA yield versus temperature and reaction time at a fixed inulin intake of 0.09 g/mol.

Table 5: Analysis of variance of the LA model

Source	Sum of squares	df	Mean square	F value	p value Prob > F
Model	273.91	9	30.43	27.44	<0.0001
C_{in} (A)	1.29	1	1.29	1.16	0.2989
T (B)	157.95	1	157.95	142.42	<0.0001
t (C)	25.28	1	25.28	22.79	0.0003
AB	8.95×10^{-3}	1	8.95×10^{-3}	8.07×10^{-3}	0.9297
AC	0.08	1	0.09	0.079	0.7829
BC	0.15	1	0.15	0.14	0.7151
A^2	14.29	1	14.29	12.88	0.0030
B^2	43.84	1	43.84	39.53	<0.0001
C^2	4.97	1	4.97	4.48	0.0527
Residual	15.53	14	1.11		

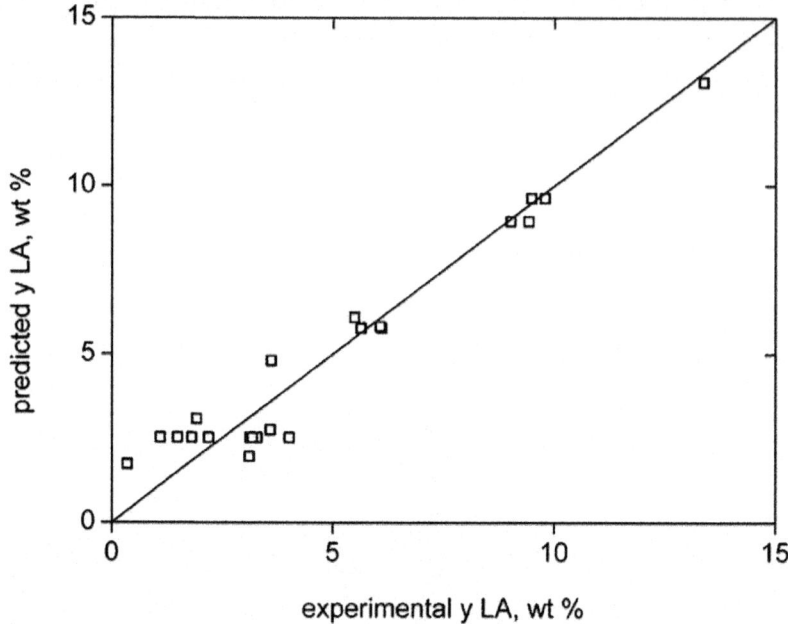

Figure 5: Parity plot between the experimental and modeled LA yield.

$$y_{LA} = 437.51 + (0.09)t + (0.17)C_{in}t + \left(5.88.10^{-4}\right)Tt$$
$$+ (1422.29)C_{in}^2 + (0.02)T^2 - (188.89)C_{in}$$
$$- (5.41)T - (0.12)C_{in}T - \left(1.69.10^{-3}\right)t^2 \qquad (5)$$

As expected on the basis of the reaction network proposed in Scheme 2, the LA yield is higher when using a longer batch time and higher temperatures, see Figure 4 for details.

Possible Role of Acid Formation on HMF Yields: Autocatalytic Effects

During the course of the reactions, organic acids such LA and FA are formed. These acids may act as catalysts for all reactions in the proposed reaction network (Scheme 2). To gain insight in the effect of these organic acids on the reaction rates and product distributions, a number of additional batch experiment was performed using inulin as the starting material and with one of the individual acids (LA or FA) and a combination of LA and FA present at the start of the reaction. The experiments were carried out temperature of 180°C, an acid concentration of 0.1 M, and an inulin loading of 0.1 g/mL. The results are given in Figure 6.

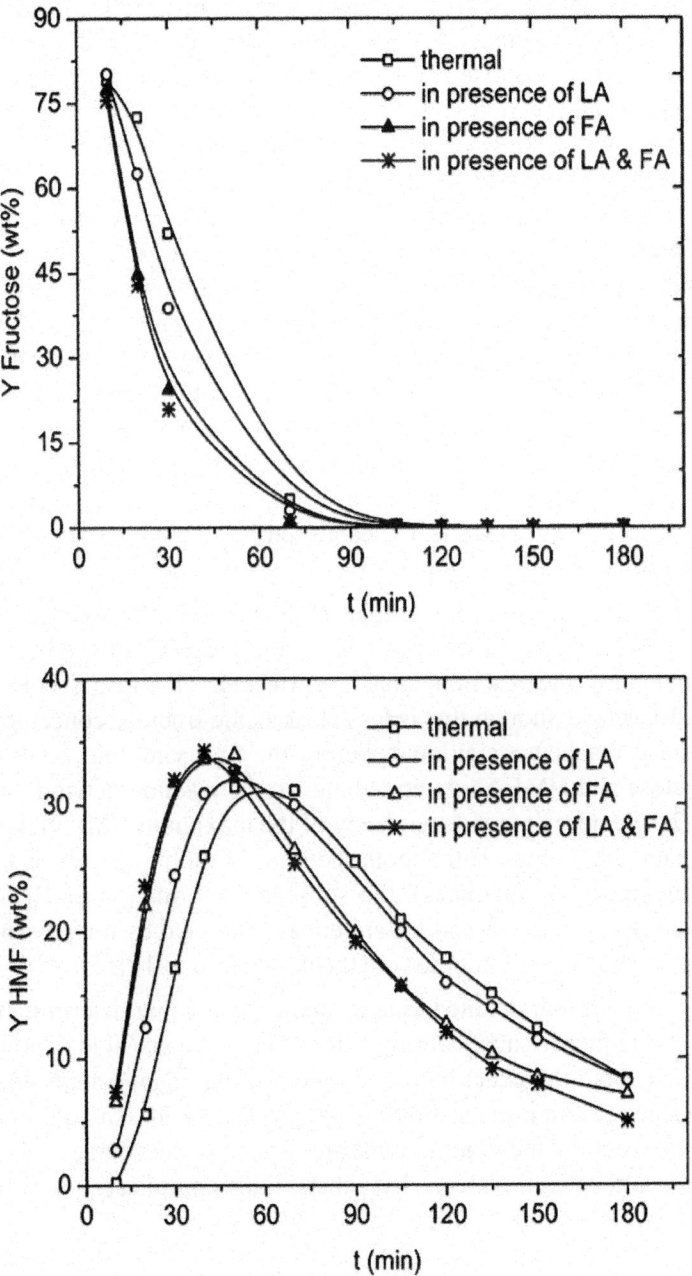

Figure 6: Autocatalytic effect of LA and formic acid on the fructose and HMF yield when using inulin as the feed (180°C, an acid concentration of 0.1 M, and an inulin loading of 0.1 g/mL).

When considering the fructose yield versus time profiles, it is evident that the presence of particularly FA has a positive effect on the rate of reaction of fructose and all fructose reacts away considerably faster when compared to the unanalyzed reaction and the reaction in the presence of LA. A similar trend was found for the HMF yield versus time profiles. Here the presence of FA also leads to a reduction of the time required to reach the maximum HMF yield. The autocatalytic effect of LA is by far less pronounced than for FA and the profiles are essentially similar to the unanalyzed reaction. This, we may conclude that particularly FA has a positive effect on the reaction rates. These findings may be explained by considering the pKA values of both acids, 3.75 for FA and 4.6 for LA, indicating that a stronger Brönsted acid leads to higher reaction rates due to a higher Brönsted acidity of the solution. Thus, the formation of d-fructose and HMF is mainly due to autocatalytic effects of FA. These findings are in line with studies reported in the literature using fructose as the substrate (Table 2). For instance Ranoux et al. [45], showed that the reaction of fructose to HMF is autocatalysed by FA and that the role of LA is by far less pronounced.

An interesting observation is that the maximum HMF yield in the yield-time profile is only slightly dependent on the presence of the organic acids, though evidently the reaction rates are affected. This also seems to be the case for d-fructose, though this is less clear as the fructose concentration may already have reached a maximum before the first sampling point (15 min). Both fructose and HMF are intermediates in the reaction network as given in Scheme 2. The near time independency of the maximum HMF yields as given in Figure 6 (right) suggests that both the rates of inulin hydrolysis to fructose (with oligomeric intermediates), the subsequent formation of HMF and the subsequent reaction to LA and FA are equally affected by the presence of FA, implying a similar reaction order in H^+ (likely close to 1) [55].

However, without detailed kinetic studies (including determination of the amount of oligomeric sugars during reaction) it is not possible to draw definite conclusions about the exact mode of action of the organic acids and the way they influence reaction rates and thus the product yields. In addition, also for the thermal reaction, the organic acids are formed in the course of the reaction, resulting in autocatalytic effects that substantially complicate a detailed kinetic analysis.

CONCLUSIONS

The unanalyzed, thermal conversion of inulin to HMF in water was studied for a wide range of reaction conditions, including variations in temperature (153–187°C), inulin intake (0.03–0.12 g/mL) and reaction time (18–74 min).

Inulin was selected as it contains mainly d-fructose units, which are known to be among the best sugars for HMF synthesis. In addition, inulin has found a wide range of applications, among others, in the food and beverage industry and as such is abundantly available. The global inulin market was 246.5 kilo tons in 2013 and is expected to grow to exceed over 400 kilo tons by 2020 [58].

The highest HMF yield was 35 wt%, corresponding to a 45% yield on a molar basis (180°C, inulin intake of 0.05 g/mL and a reaction time of 18 min). The experimental data were modelled using a statistical approach. The model shows a good fit with the experimental data and allows estimation of the HMF yield as a function of temperature, inulin intake and reaction time. This model may be used for reactor engineering purposes, for instance to optimize the HMF yield in continuous reactors with different degrees of mixing.

Autocatalysis of reaction products (FA and LA) and particularly by FA, the strongest acid, occurs to a significant extent as was shown by separate experiments with inulin in the presence of organic acids. These findings indicate that it may be advantageous regarding HMF yield to perform the reaction in a buffer solution at neutral pH values. These experiments are in progress and will be reported in due course.

Outlook: Besides the use of purified inulin, it also of interest to study the unanalyzed reaction with a real feed containing a high inulin content in order to further reduce the manufacturing costs of HMF. In this respect, the use of Chicory roots as a starting material for HMF production in water is of particular interest as this root has a reported inulin content of up to 20% on dry matter [32]. In addition, it is considered the best crop for inulin production as it has a high inulin yield per hectare (reported between 2.6 and 13.1 ton/ton dry matter chicory roots per hectare [59]) and a regular root facilitating mechanic harvesting. The catalyzed version was reported in a patent from the Süddeutsche Zucker-Aktiengesellschaft, who showed an HMF yield of 13% when using chicory roots for a reaction in water with sulphuric acid as the catalyst (pH of 1.8 at 140°C for 2 h [60]. Further studies will be required to show the proof of concept for the unanalyzed reaction.

AUTHORS' CONTRIBUTIONS

BAF and RMA carried out the experiments and drafted the manuscript. CBR and HJH supervised the entire study. All authors read and approved the final manuscript.

ACKNOWLEDGEMENT

B.A. Fachri and R.M. Abdilla would like to thank the Directorate General of Higher Education, Ministry of Education and Culture, Indonesia for funding of their PhD program.

ADDITIONAL FILES

Additional file 1: Extended model equations of both the HMF and LA yield versus process parameters.

yHMF   =

-1223.87339

1466.69731	* inuline intake
11.75492	* T
9.35425	* time
-9.19413	* inuline intake * T
-0.05071	* T * time
-0.02628	* T^2
-0.00945	* time^2

yLA   =

+437.51303

-188.89308	* inuline intake
-5.41277	* T
+0.092913	* time
-0.11725	* inuline intake * T
+0.17390	* inuline intake * time
+5.88156E-004	* T * time
+1422.29493	* inuline intake^2
+0.016862	* T^2
-1.69004E-003	* time^2

REFERENCES

1.   Werpy T, Petersen G (2004) Top value added chemicals from biomass. US Department of Energy, Springfield

2. Bozell JJ, Petersen GR (2010) Technology development for the production of biobased products from biorefinery carbohydrates—the US department of energy's "top 10" revisited. Green Chem 12:539–554

3. Li C, Zhang Z, Zongbao ZK (2009) Direct conversion of glucose and cellulose to 5-hydroxymethylfurfural in ionic liquid under microwave irradiation. Tetrahedron Lett 50:5403–5405

4. Roman-Leskhov Y, Barret J, Liu ZY, Dumesic JA (2007) Production of dimethylfuran for liquid fuels from biomass-derived carbohydrates. Nature 447:982–986

5. Lewkowski J (2001) Synthesis, chemistry and applications of 5-hydroxymethylfurfural and its derivative. ARKIVOC (i):17–54

6. Corma A, Velty A, Sara I (2007) Chemical routes for the transformation of biomass into chemicals. Chem Rev 107:2411–2502

7. Chheda JN, Huber GW, Dumesic JA (2007) Liquid-phase catalytic processing of biomass-derived oxygenated hydrocarbons to fuels and chemicals. Angew Chem 46:7164–7183

8. Boisen A, Christensen T, Fu W, Gorbanev YY, Hansen TS, Jensen JS et al (2009) Process integration for the conversion of glucose to 2,5-furandicarboxylic acid. Chem Eng Res Des 87:1318–1327

9. van Putten RJ, van der Waal JC, de Jong E, Rasrendra CB, Heeres HJ, de Vries JG (2013) Hydroxymethylfurfural, a versatile platform chemical made from renewable resources. Chem Rev 113:1499–1597

10. Mednick ML (1962) The Acid-base-catalyzed conversion of aldohexose into 5-(hydroxymethyl)-2-furfural. National Meeting of the American Chemical Society, Chicago

11. McKibbins S, Harris J, Saeman J (1962) Kinetics of the acid catalyzed conversion of glucose to 5-hydroxymethyl-2-furadehyde and levulinic acid. For Prod J Part V 12:17–23

12. Kuster BFM (1977) The influence of water concentration on the dehydration of d-fructose. Carbohydr Res 54:177–183

13. Kuster BFM, Baan HSVD (1977) The influence of the initial and catalyst concentrations on the dehydration of d-fructose. Carbohydr Res 54:165–176

14. Kuster BFM, Temmink HMG (1977) The Influence of pH and weak-acid anions on the dehydration of d-fructose. Carbohydr Res 54:185–191

15. Rapph K (1988) Process for preparing pure 5-hydroxymethylfuraldehyde. Patent No. US 4740605, Germany

16. Seri K, Inoue Y (2001) Catalytic activity of lanthanide(iii) ions for the dehydration of hexose to 5-hydroxymethyl-2-furaldehyde in water. Bull Chem Soc Jpn 74:1145–1150

17. Bicker M, Hirth J, Vogel H (2003) Dehydration of fructose to 5-hydroxymethylfurfural in sub and supercritical acetone. Green Chem 5:280–284

18. Watanabe M, Aizawa Y, Iida T, Aida TM, Levy C, Sue K et al (2005) Glucose reactions with acid and base catalysts in hot compressed water at 473 K. Carbohydr Res 340:1925–1930

19. Girisuta B, Janssen LM, Heeres HJ (2006) A Kinetic study on the conversion of glucose to levulinic acid. Chem Eng Res Des 84(A5):339–349

20. Tarabanko V, Chernyak MY, Nepomnyashchiy I, Smirnova MA (2006) High temperature 5-hydroxymethylfurfural synthesis in a flow reactor. Chem Sustain Dev 14:49–53

21. Kuster BFM (1990) 5-Hydroxymethylfurfural. A review focussing on its manufacturing. Starch/Starke 42:314–321

22. Newth FH (1951) The formation of furan compounds from hexoses. Adv Carbohydr Chem 6:83–106

23. Musau RM, Munavu RM (1987) The preparation of 5-hydroxymethyi-2-furaldehyde (HMF) from d-fructose in the presence of DMSO. Biomass 13:67–74

24. Amarasekara AS, Ebede CC (2008) Mechanism of the dehydration of d-fructose to 5-hydroxymethylfurfural in dimethyl sulfoxide at 150°C: an NMR study. Carbohydr Res 343:3021–3024

25. Torres AI, Daoutidis P, Tsapatsis M (2010) Continuous production of 5-hydroxymethylfurfural from fructose: a design case study. Energy Environ Sci 3:1560–1572

26. Rosatella AA, Simeonov SP, Frade RFM, Afonso CAM (2011) 5-Hydroxymethylfurfural (HMF) as a building block platform: biological properties, synthesis and synthetic applications. Green Chem 13:754–793

27. Praznik W, Beck R (1984) Determination of fructan oligomers of degree of polymerization 2–30 by high-performance liquid chromatography. J Chromatogr 303:417–421

28. Bohm A, Kaiser I, Trebstein A, Henle T (2005) Heat-induced degradation of inulin. Eur Food Res Technol 220:466–471

29. Kays SJ, Nottingham SF (2008) Biology and chemistry of jerusalem artichoke. CRC Press, Taylor & Francis Group, Boca Raton

30. Franck A (2002) Technological functionality of inulin and oligofructose. Br J Nutr 87:S287–S291

31. Roberfroid M (2005) Inulin-type of fructans. CRC Press, Boca Raton

32. Ricca E, Calabro V, Curcio S, Iorio G (2007) Fructose production by chicory inulin enzymatic hydrolysis: a kinetic study and reaction mechanism. Process Biochem 2007(44):466–470

33. Hu Z, Liu B, Zhang Z, Chen L (2013) Conversion of carbohydrates into 5-hydroxymethylfurfural catalyzed by acidic ionic liquids in dimethyl sulfoxide. Ind Crop Prod 50:264–269

34. Wu S, Fan H, Xie Y, Cheng Y, Wang Q, Zhang Z et al (2010) Effect of CO_2 on conversion of inulin to 5-hydroxymethylfurfural and propylene oxide to 1, 2-propanediol in water. Green Chem 12:1215–1219

35. Benvenuti F, Carlini C, Patrono P, Raspolli Galletti AM, Sbrana G, Massucci MA et al (2000) Heterogeneous zirconium and titanium catalysts for the selective synthesis of 5-hydroxymethyl-2-furaldehyde from carbohydrates. Appl Catal A Gen 193:147–153

36. Carlini C, Patrono P, Gallettia AMR, Sbrana G (2004) Heterogeneous catalysts based on vanadyl phosphate for fructose dehydration to 5-hydroxymethyl-2-furaldehyde. Appl Catal A Gen 2004(275):111–118

37. Wu Q, Yan Y, Zhang Q, Lu J, Yang Z, Zhang Y et al (2013) Catalytic dehydration of carbohydrates on in situ exfoliatable layered niobic acid in an aqueous system under microwave irradiation. Chem Sustain Chem 6:820–825

38. Carlini C, Giuttari M, Galletti AMR, Sbrana G, Armaroli T, Busca G (1999) Selective saccharides dehydration to 5-hydroxymethyl-2-furaldehyde by heterogeneous niobium catalysts. Appl Catal A Gen 183:295–302

39. Savage PE (1999) Organic chemical reactions in supercritical water. Chem Rev 99:603–621

40. Akiya N, Savage PE (2002) Roles of water for chemical reactions in high-temperature water. Chem Rev 102:2725–2750

41. Oomori T, Khajavi SH, Kimura Y, Adachi S, Matsuno R (2004) Hydrolysis of disaccharides containing glucose residue in subcritical water. Biochem Eng J 2004(18):143–147

42. Kruse A, Dinjus E (2007) Hot compressed water as reaction medium and reactant properties and synthesis reactions. J Supercrit Fluids 39:362–380

43. Hansen TS, Woodley JM, Riis A (2009) Efficient microwave-assisted synthesis of 5-hydroxymethylfurfural from concentrated aqueous fructose. Carbohydr Res 344:2568–2572

44. Qi X, Watanabe M, Aida TM, Smith RL Jr (2008) Catalytical conversion of fructose and glucose into 5-hydroxymethylfurfural in hot compressed water by microwave heating. Catal Commun 9:2244–2249

45. Ranoux A, Djanashvili K, Arends IWCE, Hanefeld U (2013) 5-Hydroxymethylfurfural synthesis from hexoses is autocatalytic. ACS Catal 3:760–763

46. Harvey DJ (1996) Matrix-assisted laser desorption/ionisation mass spectrometry of oligosaccharides and glycoconjugates. J Chromatogr A 720:429–446

47. Kazmaier T, Roth S, Zapp J, Harding M, Kuhn R (1998) Quantitative analysis of malto-oligosaccharides by MALDI-TOF mass spectrometry, capillary electrophoresis and anion exchange chromatography. Fresenius J Anal Chem 361:473–478

48. Anan'ina NA, Andreeva OA, Mycots LP (2009) Standarization of inulin extracted from dahlia single tubers and some physicochemical properties of inulin. Pharm Chem J 43:157–158

49. Khajavi SH, Kimura Y, Oomori T, Matsuno R, Adachi S (2005) Degradation kinetics of monosaccharides in subcritical water. J Food Eng 68:309–313

50. Aida TM, Tajima K, Watanabe M, Saito Y, Kuroda K, Nonaka T et al (2007) Reactions of d-fructose in water at temperatures up to 100 MPa. J Supercrit Fluids 42:110–119

51. Aida TM, Sato Y, Watanabe M, Tajima K, Nonaka T, Hattori H et al (2007) Dehydration of d-glucose in high temperature water at pressures up to 80 MPa. J Supercrit Fluids 40:381–388

52. Brunner G (2009) Near critical and supercritical water. Part I. Hydrolytic and hydrothermal processes. J Supercrit Fluids 47:373–381

53. Jin FM, Zhou ZY, Moriya T, Kishida H, Higashijima H, Enomoto H (2005) Controlling hydrothermal reaction pathways to improve acetic acid production from carbohydrate biomass. Environ Sci Technol 39:1893–1902

54. Asghari SF, Yoshida H (2006) Acid-catalyzed production of 5-hydroxymethyl furfural from d-fructose in subcritical water. Ind Eng Chem Res 45:2163–2173

55. Fachri BA, Abdilla RM, Rasrendra CB, Heeres HJ (2015) Experimental and modeling studies on the acid-catalysed conversion of inulin to 5-hydroxymethylfurfural in water. Chem Eng Res Des (submitted)

56. De S, Dutta S, Saha B (2011) Microwave assisted conversion of carbohydrates and biopolymers to 5-hydroxymethylfurfural with aluminium chloride catalyst in water. Green Chem 12:1215–1219

57. Li Y, Lu X, Yuan L, Liu X (2009) Fructose decomposition kinetics in organic acids-enriched high temperature liquid water. Biomass Bioenergy 33:1182–1187

58. http://www.grandviewresearch.com/press-release/global-inulin-market. Accessed April 26, 2015

59. http://www.eblex.org.uk/wp/wp-content/uploads/2013/12/BRPplusChicory-and-Plantain031213.pdf. Accessed April 26, 2015

60. Rapp K (1988) Process for preparing pure 5-hydroxymethylfuraldehyde. US Patent No. 4740605

Chapter 9

BIOCATALYSIS FOR BIOMASS VALORIZATION

Joyeeta Mukherjee[1] and Munishwar Nath Gupta[2]

[1]Department of Chemistry, Indian Institute of Technology Delhi, Hauz Khas, New Delhi 110016, India

[2]Department of Biochemical Engineering and Biotechnology, Indian Institute of Technology Delhi, Hauz Khas, New Delhi 110016, India.

ABSTRACT

Biocatalysis is a more sustainable alternative to chemical catalysis. While trying to obtain value added products from a biomass, looking beyond lignocelluloses (e.g.: marine polysaccharides, plant gums) may pay dividend. Many waste materials can be looked upon as a source of oil. Oil can be a rich source of diverse types of chemicals apart from biodiesel. Similarly, glycerol, the common by-product during biodiesel production, irrespective of the oil used, itself is the starting material of many products. Biocatalysis already is a viable option for the biodiesel formation step and may have a greater potential scope in glycerol chemistry. The ways to obtain the necessary biocatalysts and tailoring a given biocatalyst for a particular activity in the context of biomass valorization are discussed.

Biocatalysis, in general, is considered to be a more suitable alternative to chemical catalysis [1, 2]. Hence valorization of biomass using biocatalysts enhances the sustainability quotient of such processes. This commentary article presents an overview of the use of biocatalysis as applied to the conversion of biomass into value added products.

Types of/Sources of Biomass

Looking up an excellent book published in 1995 [3], we find biomass defined as "all organic matter that grows by the photosynthetic conversion of solar energy". The net carbon production is highest in forests ($\sim33 \times 10^9$ tons/year) but is followed by another impressive figure of 25×10^9 tons/year corresponding to marine sources. The lignocellulosic material present in the plants has attracted sufficient attention from the beginning; interest in the marine sources has also intensified in the last several decades. Still, enzymology of cellulases is more extensively studied than chitinases. Much less work has been done on agar/ alginate or carrageenan degrading enzymes [4]. At present, polysaccharide degrading biocatalysts lead the list of biocatalysts used for the conversion of biomass into fuels and fine chemicals. It is not just degradation, modification of polysaccharides is equally important. For example, modification of xanthan which is used in food processing, textile, paper and paint industry is a valuable transformation [3]. Acylation of xanthan alters its functional properties in its applications. There is a scope for biocatalysts there. Some green shoots in this approach are beginning to appear [5].

Within plants, oil has become a starting material for the synthesis of numerous valuable chemicals. Oil itself is not generally treated as a biomass. However, let us base our discussion on a broader view of biomass: a renewable resource especially if it is not exploited yet and/or treated as a waste material. Metzger and Eissen pointed out that rapeseed oil has the lowest gross energy requirement (GER) of all the base chemicals [6]. (Base chemicals are chemicals produced > million ton per year worldwide and are starting materials for a large number of other products). The review also points out that "approximately 51% of the renewable raw materials used at present in Germany are fats and oils, carbohydrates are included in the other 6%!" [6]. Apart from biodiesel, obtaining biosurfactants, engineered lipids (for nutritional purposes) and new materials from fats/oils are all industrial level processes. It may be pertinent to point out that while it is still not a practical method, aqueous enzyme oil extraction (again the role of biocatalysts!) remain a very desirable technique to develop [7, 8]. More important, the concept of biorefinary is tied up with deriving platform molecules from rest of the plant material after the seeds are harvested [9].

It has been pointed out in recent years that enough oil is present in sources not tapped so far. This is not just a reference to the inedible oils like from jatropha and castor which have been fairly well studied for conversion to biodiesel and biolubricants [10–12]. The oils present in spent coffee grounds, bird feather and waste from tobacco industry constitute renewable resources [13–15]. With oil from spent coffee grounds, facile conversion to biodiesel has been reported [16]. Japan's experience has shown that oil from kitchen waste is a viable source for conversion to biodiesel [17]. Glycerol is the common by-product in all conversions. So, these renewable resources can also be tapped for all the excellent products which have been obtained from glycerol so far [18–20] and that list is growing [21].

Glycerol is a part of a rather small list of "green solvents". Given its high viscosity, it has not become popular as a reaction medium. However, there are reports which suggest that as a reaction medium for biocatalysis in low water media, glycerol does offer significant potential [18, 19].

Recent interest in microalgae as a source of oil for conversion to biodiesel [22–25] shows that search for new renewable resources is a continuous exercise. It is interesting to recall that National Renewable Energy Laboratory had published a "close out report" on biodiesel from algae in July 1998 on a program funded from 1978 to 1996. To quote from that report: "The cost analyses for large-scale microalgae production evolved from rather superficial analyses in the 1970s to the much more detailed and sophisticated studies conducted during the 1980s. A major conclusion from these analyses is that there is little prospect for any alternatives to the open pond designs, given the low cost requirements associated with fuel production. The factors that most influence cost are biological, and not engineering-related. These analyses point to the need for highly productive organisms capable of near-theoretical levels of conversion of sunlight to biomass. Even with aggressive assumptions about biological productivity, we project costs for biodiesel which are two times higher than current petroleum diesel fuel costs" [26].

Sometimes, we have to wait before the time for a particular idea arrives! May be the valorization of food waste is one such idea [27, 28]. Seaweed lipids constitute another renewable source of diverse compounds of nutritional value or with very interesting physiological effects [29].

Choosing/Tailoring Biocatalysts

While currently available information tools have facilitated cross-fertilization of ideas from different areas, it also has created what biochemists call feedback inhibition. Too much information of varied reliability still makes it difficult to identify the right alternative. Here is a quick list of choices available while deciding about the use of a biocatalyst for valorization of biomass (Table 1).

While earlier, one was limited by enzymes readily available from microbial/plant/animal sources, that is no longer the constraint. Cloning a gene in a suitable expression system is now a fairly well established route. *E. coli* remains the first choice as the host expression system. In case of the enzyme ending up in inclusion bodies (IBs), many refolding strategies are available [30, 31]. Lately, IBs are no longer looked upon as completely inactive proteins. Many turn out to have significant biological activity and IB formation is being exploited as carrier free immobilization [31]. In some cases it is necessary to obtain the enzyme in a post-translationally modified form. In such cases other host expression systems have to be used.

Table 1: Selecting a biocatalyst for biomass valorization

	Comments
(A) Source of biocatalyst	
Whole cell or free enzyme	Whole cells can also be used in permeabilized form [52]
Wild type or mutant enzyme produced by rDNA technology	Some organisms are frequently used as the source for industrial enzymes [53]
Refolded from IBs or use of IBs itself	Choice of the host is important. For enzymes where activity depends upon post trans-lational modification, higher organisms are used as host expression systems [54]
(B) Free or immobilized form	
Soluble conjugates	For insoluble substrates (which is often the case with biomass), soluble conjugates [55] or enzymes in membrane reactors are preferred [56]
Carrier free or insoluble support	Enzyme aggregates like CLEA [57] or CLEC [58] have high volumetric activity
(C) Operational stability	This may be different from storage stability [36]
(D) Reaction medium	The use of co-solvents is under exploited [59]
(E) Normal or promiscuous activity	As more enzymes, engineered for better promiscuous activity, become available, this application may increase [60, 61]

In many cases, whole cells as biocatalysts turn out to be a more economical option when mass transfer constraints are either not significant or can be minimized by the use of permeabilized cells. Many redox enzymes require co-enzymes, the latter either has to be regenerated [32] or whole cells become an attractive option. Unfortunately, in such cases permeabilization of the cells is ruled out as low molecular weight co-enzymes would diffuse out. Both enzymes and whole cells can be either used in free form or immobilized form [33, 34].

Given the complex nature of the biomass, the stability of the biocatalyst is a crucial parameter. It is necessary to be clear about the stability data available from the literature or the vendor. The thermal stability measured in aqueous buffers is not necessarily a reliable parameter. Measurement of the operational stability with a substrate dissolved/suspended in a mixture as close to the biomass composition may be a good starting point. It is not a good idea to base this choice on the basis of the optimum temperature as that depends upon both the assay composition and time period of assay measurements [35, 36].

Today, one can carry out biocatalytic processes in a wide range of non aqueous media. This is especially useful with biomass as a substrate since pre-incubation of the biomass with the reaction medium sometime can serve as a pre-treatment step. For biocatalytic processes in dry organic solvents, choosing the right water activity for the reaction medium and maintaining it throughout the process is important [37, 38].

It is now known that classical microbiological techniques for searching for suitable microbial sources do not tap a vast amount of microbial diversity. Metagenomics (initially called environmental DNA technology) is a powerful tool to search out for a suitable biocatalyst [39].

Directed evolution compliments protein engineering in tailoring enzymes for a particular purpose. Stability and specificity both can be altered [40–42]. Contrary to earlier belief, new enzymes continue to evolve [43].

The range of substrates on which enzymes can work has become considerably enlarged with the discovery of catalytic promiscuity [44–47]. In such cases, very unrelated substrates bind to the same active site. The binding and catalysis generally involves qualitatively or even quantitatively different active site residues. For example, till few years back, if one wanted a biocatalyst for a redox reaction, one would naturally choose a suitable redox enzyme. That still may be the best option but not necessarily the only option. Enzymes can catalyse reactions which are not expected according to their EC classification [46]. For example, recently we showed that a simple lipase can carry out Cannizaro reaction [47]. Not only that, starting from an aldehyde, formation of alcohol and/or acid could be controlled by a suitable choice of lipase and/or a reaction medium. At present, such accidental catalytic promiscuity generally is not of significant level for biomass valorization. However, there are enough results which show that both protein engineering and directed evolution can create a biocatalyst with a significant level of promiscuous activity [44, 45].

It is believed that "accidental catalytic promiscuity" results inherently from how new enzymes evolve [48]. Dellus-Gur et al. [49] have discussed how functional innovation (during evolution) depends upon protein structure. Conformational plasticity (which results in various conformations of a protein co-existing) correlates positively with new catalytic functions emerging. More interesting is the hypothesis that stability (opposite of plasticity) can also promote innovability/evolvability. According to these authors "Stability promotes evolvability only if stability is an additive, global parameter, whereby stabilizing mutations in one region (e.g., a protein's scaffold) readily compensate for the destabilizing effects of mutations in other locations (e.g., in the active site region). While this is the prevailing model, can it be taken for granted? In some proteins, higher stability is mediated by mutations in residues that mediate function, suggesting that stability and function do trade off" [49]. The above illustrates how our understanding of protein evolution can further help in engineering biocatalysts for an application such as biomass valorization.

CONCLUSION

Clearly, valorization of biomass is at the heart of developing sustainable chemical processes. The concept of biorefinary is based upon that. It is interesting to note that the area of process intensification is bringing chemists, engineers and biologists together [50]. Process intensification initially focussed upon reduction in the size of the process equipment but has evolved into a more inclusive version. The aim is to further develop multifunctional reactors. Microwaves, ultrasound and solar energy are emphasized as alternative energy sources. In the context of the present commentary, carrier free immobilization reduces reactor volume, biocatalyst formats like combi-CLEAs facilitate the design of multifunctional reactors [51].

AUTHORS' CONTRIBUTIONS

JM searched the literature, took part in the discussions during drafting and helped with the formatting. MNG shaped the final draft and proof read the article before submission. Both authors read and approved the final manuscript.

ACKNOWLEDGEMENTS

We acknowledge financial support from the Department of Science and Technology (DST), Govt. of India [Grant No.: SR/SO/BB-68/2010]. JM thanks the Council of Scientific and Industrial Research [Govt. of India] for the Senior Research Fellowship. We acknowledge interesting discussions on biomass valorization with Prof. P. Mishra (Dept. of Biochem. Engg. and Biotechnol., IIT Delhi, India).

REFERENCES

1. Schmid A, Dordick JS, Hauer B, Kiener A, Wubbolts M, Witholt B (2001) Industrial biocatalysis: today and tomorrow. Nature 409:258–268

2. Malhotra D, Mukherjee J, Gupta MN (2015) Sustainability of biocatalytic processes. In: Letcher TM, Scott JL, Patterson DA (eds) Chemical process technology for a sustainable future. Royal Society of Chemistry, Cambridge, pp 390–423

3. Glazer AN, Nikaido H (1995) Microbial biotechnology: fundamentals of applied microbiology. Freeman and Co., New York, p 364

4. Wood WA, Kellogg ST (1988) Biomass, methods in enzymology, vol 161. Academic press Inc., New York

5. Bajpai P (2012) Biotechnology for paper and pulp processing. Springer Science + Business Media, LLC, New York, pp 317–326

6. Metzger JO, Eissen M (2004) Concepts on the contribution of chemistry to a sustainable development. Renewable raw materials. C R Chemie 7:569–581

7. Rosenthal A, Pyle DL, Niranjan K (1996) Aqueous enzymatic processes for edible oil extraction. Enzyme Microb Technol 19:402–420

8. Sharma A, Khare SK, Gupta MN (2002) Enzyme-assisted aqueous extraction of peanut oil. J Am Oil Chem Soc 79:215–218

9. Wettstein SG, Alonso DM, Gurbuz EI, Dumesic JA (2012) A roadmap for conversion of lignocellulosics biomass to chemicals and fuels. Curr Opin Chem Eng 1:218–224

10. Shah S, Gupta MN (2007) Lipase catalyzed preparation of biodiesel from Jatropha oil in a solvent free system. Process Biochem 42:409–414

11. Koh MY, Ghazi ITM (2011) A review of biodiesel production from *Jatropha curcas* L oil. Renew Sust Energy Rev 15:2240–2251

12. Malhotra D, Mukherjee J, Gupta MN (2015) Lipase catalyzed transesterification of castor oil by straight chain higher alcohols. J Biosci Bioeng 119:280–283

13. Veljković VB, Lakićević SH, Stamenković OH, Todorović ZB, Lazic ML (2006) Biodiesel production from tobacco (*Nicotiana tabacum* L.) seed oil with a high content of free fatty acids. Fuel 85:2671–2675

14. Kondamudi N, Strull N, Misra M, Mohaptra S (2009) A green process for biodiesel from feather meal. J Agric Food Chem 57:6163–6166

15. Encinar JM, Snchez N, Martnez G, Garcia L (2011) Study of biodiesel production from animal fats with high free fatty acid content. Bioresour Technol 102:10907–10914

16. Banerjee A, Singh V, Solanki K, Mukherjee J, Gupta MN (2013) Combi-protein coated microcrystals of lipases for production of biodiesel from oil from spent coffee grounds. Sust Chem Processes 1:14

17. Shimada Y, Watanabe Y, Sugihara A, Tominaga Y (2002) Enzymatic alcoholysis for biodiesel fuel production and application of the reaction to oil processing. J Mol Catal B Enzym 17:133–142

18. Torres S, Castro GR (2004) Non-aqueous biocatalysis in homogeneous solvent systems. Food Technol Biotechnol 42:271–277

19. Diaz-Alvarez AE, Francos J, Lastra-Barreira B, Crochet P, Cadierno V (2011) Glycerol and derived solvents: new sustainable reaction media for organic synthesis. Chem Commun 47:6208–6227

20. Len C, Luque R (2014) Continuous flow transformations of glycerol to valuable products: an overview. Sust Chem Processes 2:1

21. Canabarro N, Soares JF, Anchieta CG, Kelling CS, Mazutti MA (2013) Thermochemical processes for biofuels production from biomass. Sust Chem Processes 1:22

22. Hu Q, Sommerfield M, Jarvis E, Girardi M, Posewitz M, Siebert M et al (2008) Microalgal triacylglycerols as feedstocks for biofuel production: perspectives and advances. Plant J 54:621–639

23. Mata TM, Martins AA, Caetano NS (2010) Microalgae for biodiesel production and other applications: a review. Renew Sust Energy Rev 14:217–232

24. Chuck CJ, Wagner JL, Jenkins RW (2015) Biofuels from microalgae. In: Letcher TM, Scott JL, Patterson DA (eds) Chemical process technology for a sustainable future. Royal Society of Chemistry, Cambridge, pp 425–442

25. Alabi AO, Tampier M, Bibeau E (2009) Microalgae technologies and processes for biofuels/bioenergy production in british Columbia: current technology, suitability and barriers to implementation. British Columbia Innovation Council, Victoria

26. Sheehan J, Dunahay T, Benemann J, Roessler P (1998) A look back at the US department of energy's aquatic species program-biodiesel from algae. National Renewable Energy Laboratory, Colorado

27. Luque R, Clark JH (2013) Valorisation of food residues: waste to wealth using green chemical technologies. Sust Chem Processes 1:10

28. Karmee S, Lin C (2014) Valorisation of food waste to biofuel: current trends and technological challenges. Sust Chem Processes 2:22

29. Narayana B, Kumar CS, Sashima T, Maeda H, Hasokawa M, Miyashita K (2008) Composition, functionality and potential applications of seaweed lipids. In: Hou CT, Shaw JF (eds) Biocatalysis and bioenergy. Wiley and Sons Inc., New Jersey, pp 463–490

30. Clark EB (2001) Protein refolding for industrial processes. Curr Opin Biotechnol 12:202–207

31. Mukherjee J, Gupta MN (2015) Paradigm shifts in our views on inclusion bodies. Curr Biochem Eng 2:1–9

32. Torrelo G, Hanefeld U, Hollmann F (2015) Biocatalysis. Catal Lett 145:309–345

33. Hanefeld U, Gardossi L, Magner E (2009) Understanding enzyme immobilization. Chem Soc Rev 38:453–468

34. Leak DJ, Feng X, Emanuelsson EAC (2015) Enzyme biotransformations and reactors. In: Letcher TM, Scott JL, Patterson DA (eds) Chemical process technology for a sustainable future. Royal Society of Chemistry, Cambridge, pp 320–346

35. Purich DL (2010) Enzyme kinetics: catalysis and control. Academic, London

36. Halling PJ, Gupta MN (2014) Measurement and reporting of data in applied biocatalysis. Perspect Sci 1:98–109

37. Halling PJ (1992) Salt hydrates for water activity control with biocatalysts in organic media. Biotechnol Tech 6:271–276

38. Vulfson EN, Halling PJ, Holland HL (2001) Enzymes in nonaqueous solvents: methods and protocol. Humana Press, New Jersey

39. Tringe SG, Rubin EM (2005) Metagenomics: DNA sequencing of environmental samples. Nat Rev Genet 6:805–814

40. Arnold FH, Georgiou G (2003) Directed evolution screening and selection methods. Humana Press, Towota

41. Goldsmith M, Tawfik DS (2012) Directed enzyme evolution: beyond the low-hanging fruit. Curr Opin Struc Biol 22:406–412

42. Porter JL, Boon PL, Murray TP, Huber T, Collyer CA, Ollis DL (2015) Directed evolution of new and improved enzyme functions using an evolutionary intermediate and multidirectional search. ACS Chem Biol 10:611–621

43. Janssen DB, Dinkla IJT, Poerlands GJ, Terpstra P (2005) Bacterial degradation of xenobiotic compounds: evolution and distribution of novel enzyme activities. Environ Microbiol 7:1868–1882

44. Kapoor M, Gupta MN (2012) Lipase promiscuity and its biochemical applications. Process Biochem 47:555–569

45. Busto E, Gotor-Fernandez V, Gotor V (2010) Hydrolases: catalytically promiscuous enzymes for non-conventional reactions in organic synthesis. Chem Soc Rev 39:4504–4523

46. Arora B, Mukherjee J, Gupta MN (2014) Enzyme promiscuity: using the dark side of enzyme specificity in white biotechnology. Sust Chem Processes 2:25

47. Arora B, Pandey PS, Gupta MN (2014) Lipase catalyzed Cannizzaro-type reaction with substituted benzaldehydes in water. Tetrahedron Lett 55:3920–3922

48. Khersonsky O, Roodveldt C, Tawfik DS (2006) Enzyme promiscuity: evolutionary and mechanistic aspects. Curr Opin Chem Biol 10:498–508

49. Dellus-Gur E, Toth-Petroczy A, Elias M, Tawfik DS (2013) What makes a protein fold amenable to functional innovation? Fold parity and stability trade-offs. J Mol Biol 425:2609–2621

50. Boodhoo K, Harvey A (eds) (2013) Process intensification for green chemistry; engineering solution for sustainable chemical processing. Wiley and Sons Ltd., London

51. Dalal S, Kapoor M, Gupta MN (2007) Preparation and characterization of combi-CLEAs catalyzing multiple non-cascade reactions. J Mol Catal B Enzym 44:128–132

52. Raghava S, Gupta MN (2009) Tuning permeabilization of microbial cells by three-phase partitioning. Anal Biochem 385:20–25

53. Wubbolts MG, Bucke C, Bielecki S (2000) How to get the biocatalyst. In: Straathof AJJ, Adlercreutz A (eds) Applied biocatalysis. Harwood Academic Publishers, Amsterdam, pp 157–158

54. Greene JJ (2004) Host cell compatibility in protein expression. Methods Mol Biol 267:3–14

55. Roy I, Sharma S, Gupta MN (2004) Smart biocatalysts: design and applications. Adv Biochem Eng Biotechnol 86:159–189

56. Pederson S, Christensen MW (2000) Immobilized biocatalysts. In: Straathof AJJ, Adlercreutz A (eds) Applied biocatalysis. Harwood Academic Publishers, Amsterdam, pp 213–228

57. Minteer SM (ed) (2011) Enzyme stabilization and immobilization: methods and protocols. Humana Press, Totowa

58. Montanez-Clemente I, Alvira E, Macias M, Ferrer A, Fonceca M, Rodriguez J et al (2002) Enzyme activation in organic solvents: co-lyophilization of subtilisin Carlsberg with methyl-betacyclodextrin renders an enzyme catalyst more active than the crosslinked enzyme crystals. Biotechnol Bioeng 78:53–59

59. Roy I, Mukherjee J, Gupta MN (2013) High activity preparations of lipases and proteases for catalysis in low water containing organic solvents and ionic liquids. In: Guisan JM (ed) Immobilization of enzymes and cells, 3rd edn. Humana Press, New York, pp 275–284

60. Hult K, Berglund P (2007) Enzyme promiscuity: mechanism and applications. Trends Biotechnol 25:231–238

61. Kourist R, Bartsch S, Fransson L, Hult K, Bornscheuer UT (2008) Understanding promiscuous amidase activity of an esterase from *Bacillus subtilis*. Chembiochem 9:67–69

Chapter 10

BUTADIENE SULFONE AS 'VOLATILE', RECYCLABLE DIPOLAR, APROTIC SOLVENT FOR CONDUCTING SUBSTITUTION AND CYCLOADDITION REACTIONS

Yong Huang[1,3], Esteban E. UreñaBenavides[1,3], Afrah J. Boigny[1] , Zachary S. Campbell[1] , Fiaz S. Mohammed[1,3], Jason S. Fisk[4] , Bruce Holden[4] , Charles A. Eckert[1,2,3], Pamela Pollet[2,3] and Charles L. Liotta[1,2,3]

[1] School of Chemical and Biomolecular Engineering, Georgia Institute of Tech nology, Atlanta, GA 30332, USA

[2] School of Chemistry and Biochemistry, Geor gia Institute of Technology, Atlanta, GA 30332, USA

[3] Specialty Separations Center, Georgia Institute of Technology, Atlanta, GA 30332, USA

[4] The Dow Chemical Company, Midland, MI 48674, USA.

ABSTRACT

Butadiene sulfone has been employed as a "volatile", recyclable dipolar, aprotic solvent in the reaction of benzyl halide with metal azides to form benzyl azide (1) and the subsequent reaction of benzyl azide with p-toluenesulfonyl cyanide (3) to produce 1-benzyl-5-(p-toluenesulfonyl)tetrazole (2). Comparisons are made with the solvent DMSO and an analogous sulfolene solvent—piperylene sulfone. In addition, recycling protocols for butadiene sulfone and piperylene sulfone are also presented.

BACKGROUND

Dimethylsulfoxide (DMSO) is an outstanding solvent for conducting a wide variety of organic reactions. Its specific dipolar, aprotic properties allow for the dissolution of a range of organic molecules and ionic species. Unfortunately, the isolation of reaction products and the recyclability of the solvent is often times difficult as well as economically expensive. Recently a couple of sulfolene solvents have been proposed as possible recyclable substitutes for DMSO [1–6]. Piperylene sulfone is a liquid at room temperature and butadiene sulfone is a liquid at 64 °C. Both possess similar properties to that of DMSO. Unlike DMSO, however, each of these solvents can undergo a thermally promoted reversible retro-cheletropic process to form SO_2 and the respective diene (Fig. 1). This reversible characteristic provides a strategy for both solvent removal from the products of reaction as well as solvent recovery and reuse. Piperylene sulfone undergoes a smooth reversal at 110 °C while butadiene sulfone requires temperatures in the 135–140 °C range. In both cases the gaseous diene and SO_2 can be captured by condensing at low temperatures (−76 °C) and reacting at room temperature to reform the original sulfolene solvent.

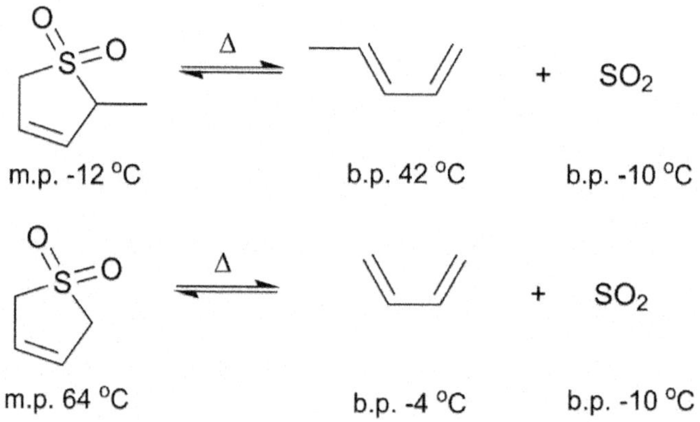

m.p. -12 °C   b.p. 42 °C   b.p. -10 °C

m.p. 64 °C   b.p. -4 °C   b.p. -10 °C

Figure. 1: Thermally induced retro-cheletropic switch enabling the recyclability of piperylene sulfone (*top*) and butadiene sulfone (*bottom*). Piperylene sulfone melts at −12 °C, making it a liquid at room temperature while butadiene sulfone is a stable solid that melts at 64 °C

Several reports have employed sulfolenes as primary solvents for conducting various organic reactions along with the subsequent recycling of the solvent. Vinci et al. [5] reported the substitution reactions and associated rates of a wide variety of nucleophiles with benzyl chloride in both DMSO and in piperylene sulfone solvent. In general the reactions conducted in DMSO proceeded at faster rates than those in piperylene sulfone. It was discovered, however, that the addition of trace quantities of water (1–3 %) added to piperylene sulfone increased the rates of the nucleophilic substitution reactions. Furthermore, the reaction of benzyl chloride with thiocyanate ion in piperylene sulfone resulted in a 96 % isolated yield of benzyl thiocyanate upon reversal of piperylene sulfone to gaseous SO_2 and piperylene. The reformation and recovery of piperylene sulfone solvent was also demonstrated with 87 %

efficiency [5]; a clear demonstration of the sulfolene's advantage over its DMSO counterpart. Ragauskas et al. [2] reported the TEMPO oxidation of substituted benzyl alcohols to benzaldehydes in piperylene sulfone. Not only were the product yields as high as the reactions conducted in DMSO but, in addition, the turn-over frequencies (TOF) were greater.

Herein is reported a reaction sequence which exemplifies the potential superiority of sulfolene solvents over DMSO. Specifically, the synthesis of 1-benzyl-5-(p-toluenesulfonyl) tetrazole (2) in piperylene sulfone (PS) and butadiene sulfone (BS) via a two-step process is reported (Scheme 1). The first step involves the reaction of azide with benzyl chloride or benzyl bromide to form the corresponding benzyl azide. The second step involves the cycloaddition reaction of benzyl azide with p-toluenesulfonyl cyanide (TsCN, 2). Each of these reactions was investigated individually and in tandem in both DMSO and a sulfolene solvent. In addition, a detailed protocol is presented for the recycling of the sulfolene solvents (piperylene and butadiene sulfones).

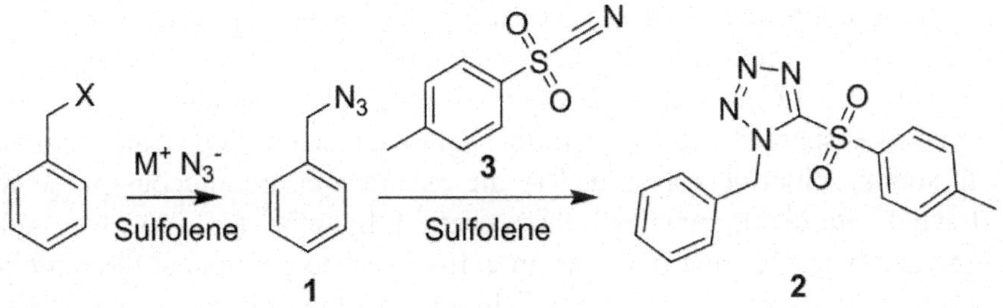

Scheme 1: Overall reaction sequence in the synthesis of 1-benzyl-5-(4-toluenesulfonyl)tetrazole (2)

EXPERIMENTAL

Materials

Piperylene (*cis*- and *trans*- mixutres) (97 %) was purchased from TCI America (Portland, OR, USA). Sulfur dioxide (>99.9 %) was purchased from Airgas (Kennesaw, GA, USA), Sigma-Aldrich (St. Louis, MO, USA) and Matheson (Montgomeryville, PA, USA). p-Toluenesulfonyl cyanide (>95 %) was purchased from AK Scientific, Inc (Union City, CA, USA) and Accel Pharmtech, LLC (East Brunswick, NJ, USA). Benzyl bromide (98 %), benzyl chloride (98 %), cesium azide (99.99 %), sodium azide (99.5 %) and dimethyl sulfoxide (99.9 %) were obtained from Sigma-Aldrich (St. Louis, MO, USA). Butadiene sulfone (98 %) and all other chemicals were purchased from VWR International (Suwanee, GA, USA). All compounds were used as received. Authentic samples of benzyl azide were prepared in lab batches (see Additional file 1).

Synthesis of Piperylene Sulfone (PS)

Piperylene sulfone was synthesized in large quantities (200–500 mL) from piperylene (*cis-* and *trans-* mixture) (200–630 mL) and sulfur dioxide (12 eq.) using 8-anilino-1-naphthalenesulfonic acid hemi-magnesium salt (0.012 eq.) as a polymerization inhibitor [3, 7]. The inhibitor was weighed and added to an Ace-Glass 5 L glass reactor. The experimental apparatus was then purged with N_2. The reactor was filled with 2 atm of vapour SO_2 and purged to remove N_2, this process was repeated three times. Liquid SO_2 was allowed to flow into the reactor while keeping the temperature at −30 °C or less. Once, all the desired amount of SO_2 was introduced, the piperylene was added into the reactor using an air tight syringe. The reactor was then sealed and allowed to warm up to room temperature around 21 °C.

The reaction was carried for at least 15 h after which the excess SO_2 was vented and collected in a bubbler containing 2.2 L of saturated potassium carbonate (K_2CO_3) solution, yielding an orange slurry product mixture. The mixture was sparged with N_2 to further remove residual SO_2. Water saturated with sodium chloride was added to the reactor and the aqueous phase was extracted with dichloromethane three times. Ethyl ether (1/3, v/v) was added to the combined organic phase as an anti-solvent to precipitate the inhibitor. The resulting liquid was dried over $MgSO_4$, and then filtered. A clear yellow liquid was obtained after evaporating the ethyl ether and dichloromethane under reduced pressure, affording 78 % yield of PS based on the trans isomer content. The resulting piperylene sulfone was characterized by ^{1}H and ^{13}C NMR to verify nearly pure production.

Recycling of Sulfolenes

The recycle of sulfolene solvents was demonstrated by beginning with a certain quantity of sulfolene, thermally decomposing it, and subsequently reforming it. The difference between the initial and final weights was designated as the percent recovery. The experiments were conducted in the prototype apparatus described in Fig. 2. Reactors R-1 and R-2 are Ace-Glass pressure tubes which could be easily removed and reattached to the setup. Both reactors were weighted before recycling. A desired amount of sulfolene was first added to R-1 (1 % by weight hydroquinone with regard to BS was added to R-2 for recycling of BS); the system was then purged with N_2. Liquid SO_2 was introduced into R-2, which was kept cold using a dry ice/isopropanol bath. When the desired amount of liquid SO_2 was introduced, the extra SO_2 was released through the base bath B-1. The decomposition flask was heated with an oil bath to the desired temperature, 120 °C for PS and 135 °C for BS. During the decomposition process, the reformation flask was kept between −76 and −55 °C (Table 1) to trap the volatile dienes and SO_2. N_2 was allowed to enter from V-5 towards B-1 to maintain a constant near atmospheric pressure and drive the decomposition products from R-1 to R-2. It should be noted that the line connecting R-1 to R-2 was heated during the decomposition process. In the case of PS, the tube was kept above 42 °C to prevent condensation of piperylene and polymerization in the lines; for BS, the tube was maintained at 70 °C to prevent clogging from solid BS reforming inside the tube. When reactor R-1 was completely empty, 100 mL/min of N_2 were allowed to enter from V-1 to further transport volatile products to R-2; this final wash was performed for at least 30 min. Reactor R-2 was then sealed and allowed to warm to room temperature.

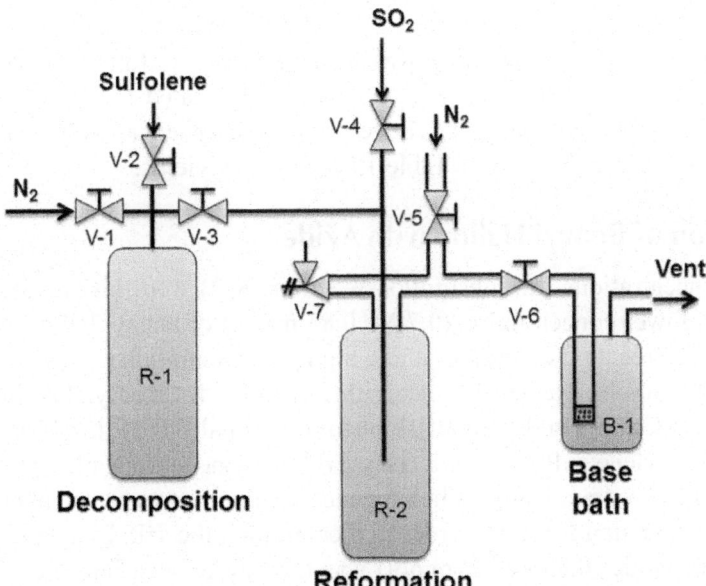

Figure. 2: Diagram showing the experimental setup for the recycle of sulfolene solvents. *V-1* to *V-6* are needle valves, *V-7* is pressure relief valve, *R-1* is the decomposition reactor, *R-2* is the reformation reactor and *B-1* is a bubbler submerged in a base bath used to neutralize excess SO_2.

Table 1: Recycling results of sulfolene solvents

Entry	Solvent	Scale (mL)	SO_2/diene molar ratio	Cold bath (°C)	Recovered solvent (%)
1	PS	5	8	−55	89 ± 2
2	PS	20	6	−60	98 ± 0
3	BS	20	6	−76	95 ± 1[a]

[a]Two replicates, one of them is added with inhibitor

The reformation reaction was carried for at least 40 h. During that time a pressure relief valve (V-7) ensured that the entire system was held under the pressure rating for the glass reactors. Upon conclusion, excess SO_2 was vented through B-1 until no bubbling was observed. The sulfolenes were sparged with N_2 to remove residual SO_2; BS had to be heated to 70 °C to prevent crystallization during sparging. The reformed sulfolene was weighted to obtain a recovery measurement. See Table 1 for recovery yields.

Reaction of Benzyl Halide with Azide

Two concentrations for substitution reaction (S_N2) were investigated in this study. A lower concentration (0.7 M, 0.85 mmol) of benzyl halides was used to optimize reaction conditions and a higher concentration (2.5 M, 3.5 mmol) of benzyl halides was used to determine accurate isolated yields. Sodium or cesium azide (0.93 or 3.8 mmol), biphenyl (internal standard, 0.37 mmol) and liquid piperylene sulfone (1 mL) or solid butadiene sulfone (1.3 g) were first added to a two-dram vial. The heterogeneous mixture was heated at 60 °C with vigorous agitation for 2 h in order to precondition the salt. Benzyl chloride or benzyl bromide (0.85 or 4.2 mmol) was then introduced into the mixture to initiate reaction. Samples of 0.05–0.1 mL were taken hourly for the first 3 h. The samples were diluted with 1 mL of benzene and filtered with a syringe filter before GC-FID analysis. The GC oven temperature was ramped from 90.0 to 300 °C at a heating rate of 15 °C/min. The GC injector was held at a constant temperature of 300 °C. For NMR analysis, the same amount of sample was diluted with 0.5 mL of DMSO-D$_6$, and then filtered with a syringe filter. Reactions for synthesis of benzyl azide were all run in duplicate, and Table 2 includes the reaction yields.

Table 2: Nucleophillic substitution reaction of benzyl halides with inorganic azide salts in sulfolenes and DMSO

Entry	Solvent	Reactants	Reaction time (h)	Yield[a] (%)
1	DMSO [8]	BnBr, NaAz	1	98
2	PS	BnCl, NaAz	3	6 ± 0
3	BS	BnCl, NaAz	1	49 ± 5
			3	86 ± 4
4	BS	BnCl, 1.5 NaAz	1	65 ± 2
			3	93 ± 1
5	DMSO (1 % H_2O)[b]	BnCl, NaAz	1	97 ± 0
6	BS (1 % H_2O)	BnCl, NaAz	1	93 ± 1
			3	95 ± 1
7	BS (1 % H_2O)	BnCl, CsAz	1	96 ± 0
8	BS	BnBr, NaAz	1	100[c]

[a]GC yield

[b]1 % water by volume

[c]NMR yield, BnBr (2.5 M)

Reaction of Benzyl Azide (1) with *p*-Toluenesulfonyl Cyanide (3)

Benzyl azide (1.09 g, 8.19 mmol), TsCN (1.63 g, 9.00 mmol) and BS (3.05 g, 25.81 mmol) were all added to a three neck round bottom flask and heated to the designated reaction temperature. At the end of the predefined reaction time, the reaction was cooled and the contents in the flask were dissolved in acetone. Samples for NMR analysis were taken from the acetone solution. A known amount of dimethyl sulfone was added to the samples as an internal standard for NMR quantitation.

Product isolation was carried out for selected high yield reactions. The post reaction mixture containing product was heated to 135 °C for approximately 2 h in order to promote the decomposition of the sulfolene; after approximately

2 h the bubbling had ceased. The product residue was a brown liquid which solidified upon cooling. Methanol (4 mL) and ethyl ether (1 mL) were added and stirred overnight to extract impurities from the solid. The mixture was cooled in an ice/water bath and filtered. The solid was then recovered and washed once more with the methanol-ether mixture. After filtration and drying, a cream colored powder (compound 3) was obtained (see Table 3 for yields and conversions). DSC: m.p. = 136 °C, ΔH = 106.0 J/g. ^{1}H NMR (DMSO-d$_6$, ppm): δ = 2.40 (s, 3H), 6.00 (s, 2H), 7.24–7.27 (m, 2H), 7.36–7.39 (m, 3H), 7.46 (d, J = 8.1 Hz, 2H), 7.83 (d, J = 8.4 Hz, 2H). ^{13}C NMR (DMSO-d$_6$, ppm): δ = 21.23, 52.67, 127.97, 128.67, 128.69, 128.91, 130.53, 133.56, 133.87, 147.46, 154.29. ESI–MS (m/z): 315 [M + H]$^+$, 332 [M + H$_2$O]$^+$.

Table 3: Synthesis of compound 3 through [2 + 3] cycloaddition of p-toluenesulfonyl cyanide (3) and benzyl azide (1)

Entry	Solvent	Temp (°C)	Time (days)	Yield (%)	Conv. 1 (%)	Conv. 2 (%)
1	DMSO	50	4	30 ± 1	100 ± 0	47 ± 1
2	BS	70	1	70 ± 2	99.2 ± 0.1	89 ± 1
3	BS	50	1	50 ± 1	57 ± 2	51 ± 2
4	BS	50	4	77 ± 1	95 ± 1	91 ± 4
5	BS	50–60	4^a	82 ± 2^b	99 ± 1	93 ± 2

Yields and conversion were measured by NMR
^{a}Reaction was run 2 days at 50 °C, then 2 days at 60 °C
^{b}Isolated yield was 71 ± 2 %

Tandem, two-step synthesis of 1-benzyl-5-(p-toluenesulfonyl) tetrazole (2)

Benzyl bromide (0.641 g, 3.75 mmol), sodium azide (0.269 g, 4.14 mmol), BS (1.315 g, 11.13 mmol) and biphenyl (0.06 g, GC internal standard) were added to a 2 dram vial. The reaction was allowed to react for 3 h at 60 °C. The reaction mixture was filtered hot using a syringe filter and added to a second vial containing TsCN (0.750 g, 4.14 mmol). The cycloaddition reaction was allowed to run for 2 days at 50 °C and 2 days at 60 °C. The reaction mixture was then cooled at which point the product solidified; dimethyl sulfone was added as an internal standard for NMR. The contents of the vial were dissolved in acetone and DMSO-d_6 and samples were analysed by NMR, and a yield of 72 ± 5 % 1-benzyl-5-(p-toluenesulfonyl) tetrazole (2) was obtained.

RESULTS AND DISCUSSION

Synthesis of Benzyl Azide (1)

Literature contains many examples of alkyl azide syntheses in a wide variety of solvents, using sodium azide and an alkyl halide [8, 9]. Alvarez et al. [8] reported high yields when the displacement reaction was conducted in DMSO. In particular, an isolated yield of 98 % was obtained in the reaction of benzyl bromide with sodium azide at ambient temperature. Nevertheless, while this reported yield is excellent, solvent recycle was not addressed. Indeed, the post-reaction mixture in DMSO was quenched with water. The product was subsequently extracted with ethyl ether, followed by several washes with brine, drying, and finally ether solvent evaporation. Thus, while the isolated product was obtained in excellent yield, the DMSO solvent was no longer usable. This is almost always the case when DMSO is employed as the reaction medium. Since PS and BS have similar properties to those of DMSO and since they are recyclable, these solvents could represent a more sustainable approach to the production of alkyl azides.

Scheme 2 shows the reaction of benzyl halide with inorganic azide salts in either PS or BS and Table 2 summarizes the results in these solvents. For comparison, the result from Alverez et al. for this same reaction in DMSO is included (Table 2, entry 1). When PS (Table 2, entry 2) was used as reaction medium, the substitution reaction afforded a 6 ± 0 % yield of benzyl azide in 3 h. However, employing the more thermally stable BS, a yield of 49 ± 5 % and 86 ± 4 % at 1 and 3 h, respectively, was obtained (Table 2, entry 3). The low yield in PS is attributed to the much lower solubility of azide salt in this solvent as compared to BS. It is postulated that the presence of the methyl group in the 2-position of PS sterically interferes with its ability to solvate the cationic portion of the salt and, as a consequence, results in reduced solubility. This steric factor is absent in BS. Several attempts were made to reduce the reaction time and increase the yield of benzyl azide in BS. For instance, the concentration of sodium azide was increased from 1.1 to 1.5 equivalents resulting in an increase of yield at 3 h reaction time from 86 ± 4 % to 93 ± 1 % (Table 2, entries 3, 4). Although the yield was marginally increased, the excess sodium azide presented concerns from both safety and atom-economy points of view. In addition, the excess azide might also interfere with the subsequent cycloaddition reaction step in the synthesis of 1-benzyl-5-(p-toluenesulfonyl) tetrazole (2) [10]. As mentioned previously it was discovered that trace amounts of water added to PS improved the rates of several nucleophilic substitution reactions [5]. Addition of 1 % water to DMSO did not appear to have any noticeable effect on the rate or the yield of benzyl azide (Table 2, entry 5). In contrast, addition of 1 % water to BS resulted in a 93 % yield of benzyl azide in a 1 h time period (from 49 % in anhydrous conditions). Extending further the reaction time to 3 h resulted in only a marginal increase in yield (Table 2, entry 6). The use of the more expensive cesium azide in place of the sodium salt also resulted in excellent yields (Table 2, entry 7). While these increased yields looked good, it was recognized that the addition of water to BS could potentially form small quantities of sulphurous acid which can subsequently react with azide to produce the extremely explosive and toxic hydrazoic acid. As a consequence, *the addition of water was avoided in further experiments.* Finally, by replacing benzyl chloride with its bromide counterpart a quantitative yield of benzyl azide was achieved in 1 h in BS in the absence of added water (Table 2, entry 8). This latter protocol provides a relatively less expensive and much safer procedure.

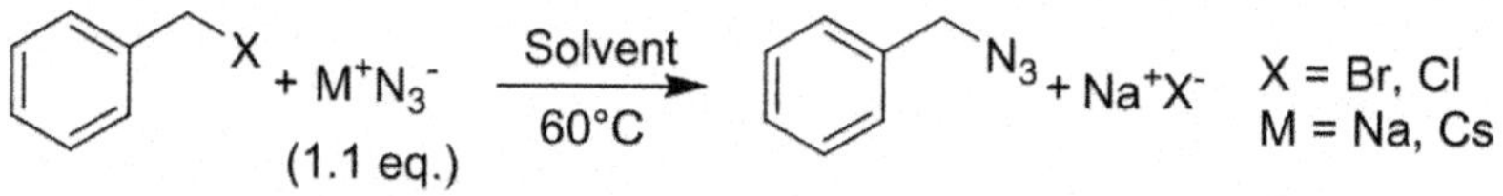

Scheme 2: General reaction of benzyl halide with metal azide at 60 °C, to form the corresponding benzyl azide (1).

next step in the reaction sequence involved the reaction of benzyl azide (BnAz, 1) with p-toluenesulfonyl cyanide (TsCN, 3) to form 1-benzyl-5-p-toluenesulfonyl tetrazole (2). Tetrazoles have a broad range of applications. They are found in number of pharmaceutical compounds, they can be surrogates for peptides [11] and carboxylic acids [12], and they have been used to tag drug receptor proteins [13]. In addition, tetrazole ligands have also been used for fabricating coordination polymers [14, 15]. Moreover, the synthesis of tetrazoles with labile groups like the toluene sulfonyl substituent can enable their use as building blocks for further functionalization. As such, compounds like 1-benzyl-5-p-toluenesulfonyl tetrazole (2) is of especial interest [10]. Demko and Sharpless synthesized 1-benzyl-5-p-toluenesulfonyl tetrazole (2) in the absence of solvent from BnAz and TsCN with a near quantitative yield of product being reported [10]. The solid product, however, had to be chipped off the reactor. Although feasible on a laboratory scale, from an industrial standpoint, a scalable protocol that facilitates post-reaction processing and simultaneously minimizes waste is more desirable.

Synthesis 1-Benzyl-5-(*p*-Toluenesulfonyl)Tetrazole (2)

Since excellent yields of benzyl azide were obtained in BS it was decided to conduct the second step in the synthetic sequence in the same solvent (Scheme 3). The results of the [2 + 3] cycloaddition of TsCN with benzyl azide (1) to form 1-benzyl-5-*p*-toluenesulfonyl tetrazole (2) in both DMSO and BS are summarized in Table 3. The reaction in DMSO (Table 3, entry 1) at 50 °C for a period of 4 days produced a modest yield of 30 ± 1 %. In contrast, the reaction conducted in BS (Table 3, entry 3 and 4) resulted in yields of 50 % and 77 ± 1 % for reaction times of 1 and 4 h, respectively, under the same conditions. In both cases nearly all the TsCN cyanide is consumed at the end of 4 days. However, it is interesting to note that when DMSO is used, the conversion of benzyl azide is only 47 ± 1 % in contrast to a 91 ± 4 % conversion in BS. A control reaction was conducted: TsCN was added to DMSO at 50 °C; no benzyl azide was present. After 4 days most of the TsCN disappeared with the formation of a major product, and ^{1}H NMR analysis of the control reaction at 2 and 4 days showed that the major product is a salt of *p*-toluenesulfonic acid (see ^{1}H NMR spectra in Additional file 1: Fig. S1). The major product was precipitated from the DMSO solution upon the addition of water, and its exact mass analysis was consistent with the salt of *p*-toluenesulfonic acid [ion trap/ orbitrap tandem mass spectrometer (*m/z*): calcd. for $C_7H_7O_3S$ 171.0110, found 171.0119 [M]$^-$]. It is clear that, in addition to the reaction of TsCN with benzyl azide to form the desired tetrazole (2), a competing reaction of TsCN with the solvent is taking place. As a consequence, *DMSO is not an appropriate solvent for this pericyclic process*. BS, in contrast, does not react with TsCN. In this particular case, therefore, BS is a dipolar, aprotic solvent alternative to DMSO.

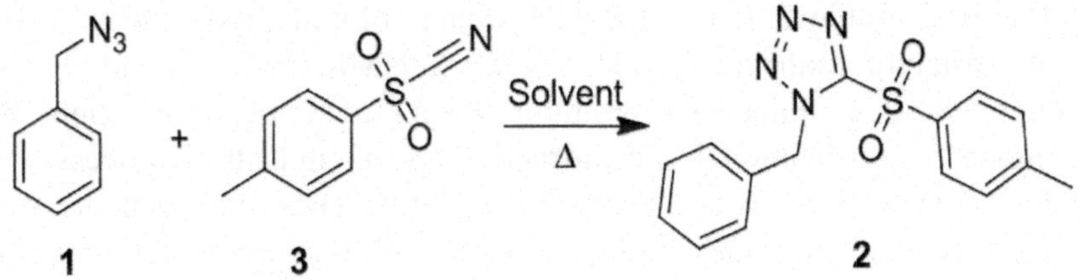

Scheme 3: *p*-Toluenesulfonyl cyanide (3) reacting with benzyl azide (1) to form 1-benzyl-5-(*p*-toluenesulfonyl) tetrazole (2).

In order to improve the yield of the tetrazole and reduce the reaction time, experiments were performed at slightly elevated temperatures. Since both the TsCN and the benzyl azide are thermally labile, the reaction temperatures employed had to be carefully adjusted. In addition, care had to be taken to avoid the retrochelotropic reaction of the solvent. Fortunately, BS undergoes negligible decomposition up to 100 °C. The effect of heating from 50 to 70 °C was studied in BS (Table 3 entries 2 and 5). It was observed that after 1 day the yield at the higher temperature was 70 ± 2 %, while at the lower temperature it was 50 ± 1 %. Nevertheless, the conversion of TsCN at 50 °C was only 57 ± 2 %, and at 70 °C nearly all TsCN reacted within 1 day. The reaction at 50 °C proceeds slower than at 70 °C, but it can ultimately reach a higher yield since side reactions are not as competitive at the lower temperature. Temperature is also important on the phase behaviour of the reaction mixtures. Even though pure BS melts at 64 °C, the reaction mixture becomes a homogenous clear liquid at 45 °C. When the reaction is carried at 50 °C, some product precipitates with time and at the end of 4 days the mixture takes the appearance of a thick paste. However, if the temperature is raised to 60 °C after 2 days of reacting at 50 °C, the reaction mixture ends as a fluid slurry that can be easily poured out of the reaction flask. Table 3 shows that using entries 4 and 5 temperature scheme, the product yield is slightly increased from 77 ± 1 to 82 ± 2 %. It is postulated that the lower viscosity obtained by increasing the temperature favours the bimolecular cycloaddition reaction over the decomposition of the starting materials.

One of the greatest advantages of sulfolene solvents is the simplicity of product isolation. Figure 3 shows the scheme employed for the isolation of 1-benzyl-5-*p*-toluenesulfonyl tetrazole (2). Heat was used to decompose BS, leaving a liquid mixture containing the product and some unreacted starting materials. The residual impurities were extracted from the product residue with a 4/1 v/v mixture of methanol/ethyl ether (Fig. 3). This simple purification scheme gave an isolated yield of 71 ± 2 % for the highest yielding reaction conditions in Table 3 (entry 5). The cream-colored product was analysed and its structure confirmed by ^{1}H NMR, ^{13}C NMR, ESI–MS and DSC. The product was also subjected to column chromatographic purification. While this mode of purification is not attractive for an industrial process, it was conducted in order to see if there are any differences between the "cream colored" product and the product derived from column chromatography. Figure 4shows a comparison of the ^{1}H NMR spectra of the product prior to and after the column chromatographic procedure. No appreciable differences were observed suggesting that comparable purities can be obtained using the simple isolation scheme described above without the need for subsequent column chromatography.

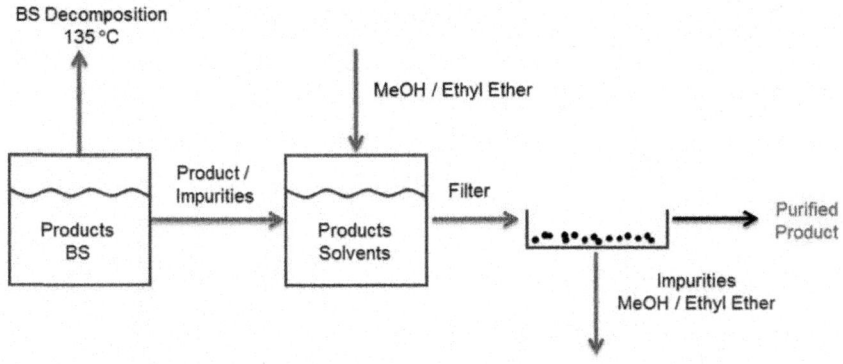

Figure. 3: Purification scheme for the isolation of 1-benzyl-5-*p*-toluenesulfonyl tetrazole (2). The crude product mixture obtained after the cheletropic removal of the butadiene sulfone is washed with a 4:1 v:v mixture of methanol:ethyl ether and then filtered. The product residue is then dried to afford a cream-colored product.

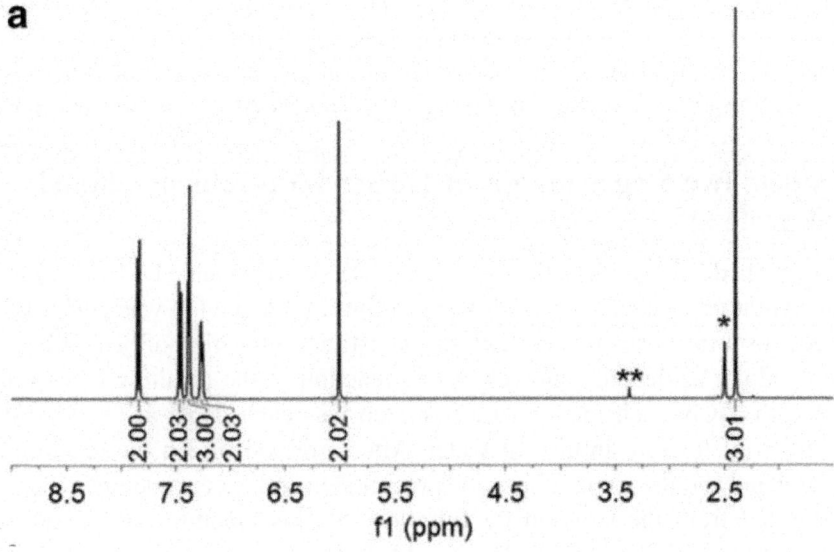

b

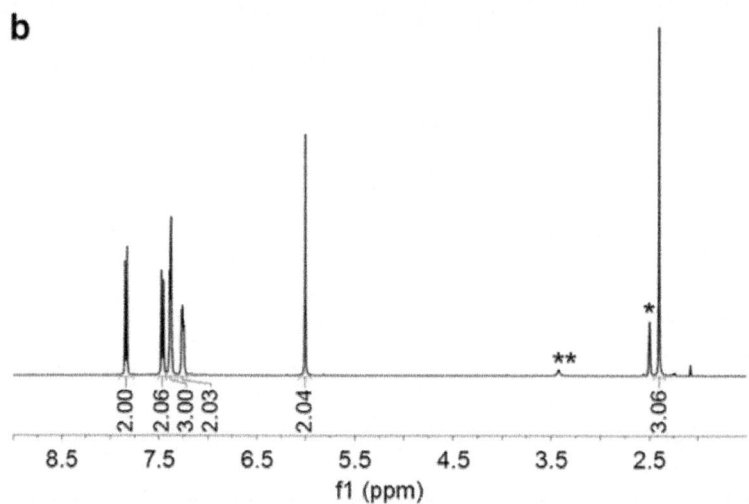

Figure. 4: ¹H NMR of 1-benzyl-5-*p*-toluenesulfonyl tetrazole (2). The purity of the isolated tetrazole obtained using the purification scheme outlined in Fig. 3 is compared to the isolated tetrazole product obtained via traditional purification techniques. **a** cheletropic switch, followed by extraction and **b** column chromatography. Peak labelled *asterisk* corresponds to d-DMSO and peak labelled *doubleasterisk* corresponds to water.

Tandem Two Step Synthesis of 1-Benzyl-5-p-Toluenesulfonyl Tetrazole (2)

In order to demonstrate the broad utility of BS as a recyclable DMSO substitute, the synthesis of 1-benzyl-5-*p*-toluenesulfonyl tetrazole (2) was performed in tandem starting from the nucleophilic substitution reaction of benzyl bromide and sodium azide and followed by the reaction of the resulting benzyl azide (1) with TsCN (Scheme 1). The first step was carried out at 60 °C without addition of trace quantities of water. After a period of 3 h the reaction was completed and the sodium bromide precipitated and excess sodium azide were separated from the solution by filtration. At this juncture the benzyl azide product was not isolated. TsCN was added to the filtered reaction solution and then diluted to match the concentration used for the cycloaddition experiments depicted in Table 3. The reaction solution was then heated to 50 °C for 2 days and subsequently to 60 °C for an additional 2 days. At the end of the tandem process NMR analyses showed that all the benzyl bromide was consumed and only traces of (3 ± 2 %) of benzyl azide remained. The conversion of TsCN (3) was 93 ± 5 %. Compound 2 (based on the initial moles of benzyl bromide) was obtained in 72 ± 5 %: a yield comparable to that obtained in the single step process (Table 3entry 5, 82 ± 2 %).

Recycling of Sulfolenes

There are a number of reports dealing with the reaction of conjugated dienes with SO_2 as well as the reverse process. Two simple addition products have been reported for the reaction of piperylene with SO_2 (Scheme 4). One is the product of a hetero-Diels–Alder process which produces a 6-membered ring sultine. This product is the result of a [4 + 2] cycloaddition. The other product is the expected 5-membered ring sulfone- the result of a [4 + 1] cycloaddition. Deguin and Vogel reported that at temperatures ranging from −80 to −60 °C the hetero-Diels–Alder sultine product can be obtained in the reaction of a 0.3 M solution of (E)-piperylene in CD_2Cl_2/SO_2 in the presence of an acid catalyst (0.2 M of CF_3COOH). In contrast, however, the reaction of butadiene or (Z)-piperylene did not produce the sultine products under the same conditions [16]. The latter results are not surprising since the necessary syn-conformation of the Z-piperylene is relatively high in energy due to steric factors. More recent studies revealed that at the same temperatures the reaction of dimethylidenecycloalkanes with variable ratios of CD_2Cl_2/SO_2 produce sultine products without employing acid catalysis. It was also reported that the sultine products isomerized into the corresponding sulfolenes at temperatures above −40 °C [17, 18].

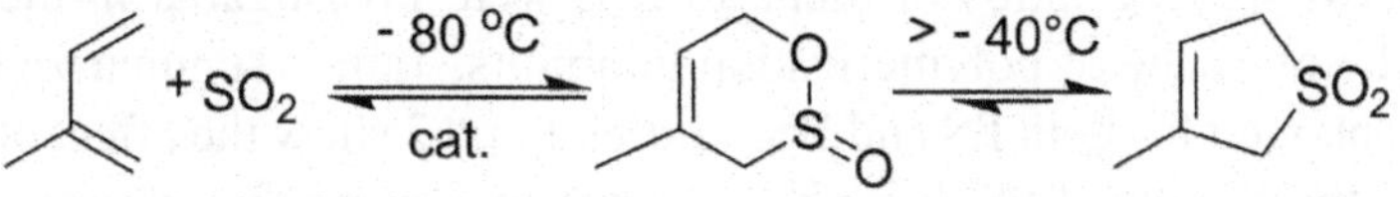

Scheme 4: Heat controlled hetero-Diels–Alder addition and retrocheletropic addition of isoprene [16, 18].

Cheletropic reactions of SO_2 and a diene are often carried in the presence of radical inhibitors in order to avoid undesired polymerization of the dienes [19]. Morris and Finch proposed that organic peroxides, often present in dienes feedstock, are the major cause for polymerization. They claimed that a diene free of organic peroxides allows the cheletropic reaction to occur in the absence of polymerization inhibitors [20]. Staudinger et al. reported that reaction between SO_2 and butadiene at room temperature produced an amorphous product in 11 % yield and the crystalline cyclic sulfone in 89 % yield. The amorphous solid was identified as a linear polysulfone [21]. Finally, it has been reported that the rate of cheletropic and retrocheletropic reactions is affected by polarity of solvents. Polar solvents, like methanol, slow down the decomposition process, but accelerate the reformation reaction; the opposite occurs with non-polar solvents [22]. In this study, it is the pure sulfolene solvent which is thermally decomposed while the reformation process (the reaction of the conjugated diene with SO_2) takes place in liquid SO_2.

Sulfolene solvents have been proposed as recyclable substitutes for DMSO. The reversible reaction between a conjugated diene and SO_2 (cheletropic reaction) shown in Fig. 1 is the basis for the recyclability of these solvents. Initial experiments concerning the decomposition process of piperylene sulfone, trapping the volatile compounds, and reforming the solvent has previously been reported. As mentioned before, an 87 % recovery yield was obtained [5]. The loss of 13 % of the solvent was attributed to the small scale (5 mL) of the recycle process and the accompanying material loss due to surface adhesion to the tubing in the recycling apparatus. It was appropriate therefore to demonstrate the efficiency of the recycle process on a scale and in equipment which would minimize material losses. The results for the recycling of PS and BS reported here were conducted in the apparatus described in the "Experimental" section (Fig. 2). Each of the pure sulfolenes was allowed to undergo a thermal retro-cheletropic process at a specific temperature for a specified length of time. PS and BS undergo decomposition at a reasonable rate at 120 and 135 °C, respectively [23]. The pertinent processes were conducted on a 20 mL scale and compared to results conducted on a 5 mL scale. The products of the decomposition (the conjugated diene and SO_2) were captured and allowed to react to reform the original sulfolene. The overall process was meant to demonstrate the recyclability of these solvents. In this latter part of the process specific ratios of diene to SO_2 were investigated in the absence and in the presence of polymerization inhibitors. Table 1 summarizes the final results obtained for both PS and BS. Entries 1 and 2 show the effect of reaction scale for the recycle of PS. It is observed that increasing the amount of starting PS from 5 to 20 mL increased the recovery from an acceptable 89 ± 2 % to a near quantitative 98.3 ± 0.3 %. In addition, it is interesting to note that even though the molar ratio of SO_2 to piperylene in the reforming step was reduced from 8 to 6, the yield of PS was still excellent. Vinci's result of 87 % recovery was performed at a 5 mL scale [5]. The results reported herein are consistent with his data.

First, the recycling process was investigated in the absence of any polymerization inhibitor. For recycling of PS, when the SO_2/diene molar ratio was 6 or higher, minimal or no polymerization was detected. Minimal polymerization could be observed in the tubing connecting two reaction vessels; however it did not affect the recovery yields due to negligible volume of connecting tubing. Lower SO_2/diene molar ratios yielded significant amounts of polymers which had to be removed by antisolvent precipitation using a 3/1 mixture of dichloromethane and ethyl ether. For recycling of BS, the same procedure was performed six times in the absence of a polymerization inhibitor. Only one of these experiments was successful—a 94 % recovery yield was obtained. The other five experiments resulted in the formation of the white amorphous polysulfone polymer [24]. However, with the addition of 1 % of the polymerization inhibitor hydroquinone (by weight with respect to butadiene sulfone) to reformation flask, 96 % yield of butadiene sulfone was obtained.

In our recycle experiments for both PS and BS, the dienes/SO_2 mixture is kept at temperature between −55 and −76 °C for at least 2–3 h. The temperatures at which the hetero-Diels–Alder products in Fig. 2 [16, 18] were observed are the same used here to trap the products of the retrocheletropic decomposition. The kinetic product (hetero-Diels–Alder) may be favoured at low temperature in this case; but at higher temperatures, the more thermodynamically stable sulfolene is formed. It is hypothesized that if kinetic products were formed at low temperatures, the undesired polymerization of dienes would be significantly reduced upon warming up to room temperature.

Figure 5 shows a schematic diagram of the desired process. The reactants for nucleophilic reaction or cycloaddition reaction are introduced to a reactor in which the synthesis is performed, followed by the retrocheletropic decomposition of BS, and leaving the desired product (and side products) behind. The volatile SO_2 and butadiene are trapped in the presence of excess SO_2. The neat reformation (cheletropic) reaction is performed at room temperature; after which SO_2 is vented. Any residual SO_2 can be separated by bubbling N_2 through the sulfolene. Since SO_2 has a very low vapour pressure compared to N_2, it can be easily condensed and recycled. In this study, only small scale two step synthesis (1 mL) was performed due to safety concern, and recycling of sulfolene solvents in larger scale (20 mL) were conducted separately.

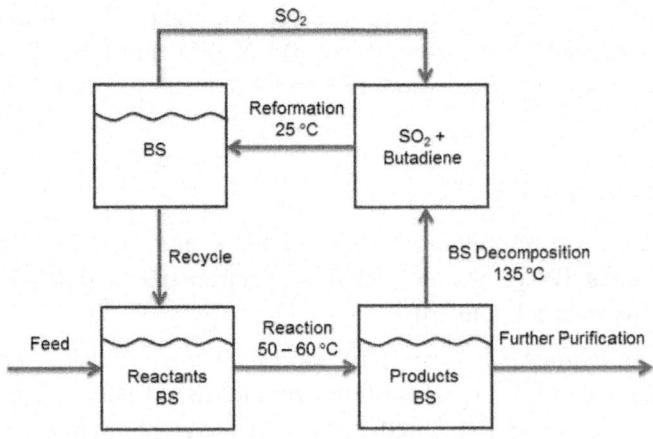

Figure. 5: Process diagram for the synthesis of tetrazoles using butadiene sulfone. The reactants are introduced to a reactor in which the synthesis is performed. Post reaction,

the retrocheletropic decomposition of Butadiene sulfone leaves the desired product (and side products) behind. These can then further purified using techniques described in Fig. 3. The volatile SO_2 and butadiene are trapped in the presence of excess SO_2. The neat reformation (cheletropic) reaction is performed at room temperature; after which SO_2 is vented. Any residual SO_2 can be separated by bubbling N_2 through the sulfolene. Since SO_2 has a very low vapour pressure compared to N_2, it can be easily condensed and recycled.

CONCLUSIONS

In conclusion, piperylene sulfone and butadiene sulfone have been shown to be recyclable solvents as a consequence of the reversible reactions between SO_2 and the respective diene. They are dipolar, aprotic solvents and serve as potential substitutes for DMSO. This is especially true for BS in the synthesis of organic azides by nucleophilic substitution and, tetrazoles by the reaction of organic azides with p-toluenesulfonyl cyanide (3). Both reactions using sulfolene solvent have noticeable advantages: operational simplicity, low cost and environmental safety.

ADDITIONAL FILE

Preparation of Benzyl Azide (1)

The benzyl azide used for the cycloaddition reactions was synthesized and isolated first in acetonitrile water. Benzyl chloride (6.2 g) and sodium azide (4.4 g) were added to a three neck round bottom flask, followed by addition of 20 mL of water and 80 mL of acetonitrile. The reaction mixture was heated to 60 °C and allowed to react overnight. Benzyl azide (1) was extracted with dichloromethane, dried over $MgSO_4$ and filtered. Acetonitrile and dichloromethane were evaporated under reduced pressure to yield a clear yellow liquid.

Characterizations

Nuclear magnetic resonance (NMR): ^{1}H-NMR and ^{13}C-NMR spectra were measured on a Bruker Avance III 400 spectrometer, and NMR yield was quantified by internal standard.

Gas chromatography/flame ionization detector (GC-FID): GC-FID was using a Shimadzu GC2010 gas chromatograph fitted with a Supelco PTA-5 (30m x 0.32 mm x 1.00 µm, length x inside diameter x film thickness) capillary column. The injector temperature was held constantly at 300 °C, and column oven was increased from 90 °C to 300 °C at a ramp rate of 15 °C/min. GC-FID detector temperature was held at 320 °C. The used calibration curves are listed as below in **Fig.S6**.

Triple quadrupole tandem mass spectrometer with ionization via ESI (ESI-MS): experiments were run on a Quattro LC, made by Micromass, which is now part of Waters. The capillary voltage was 3.5kV, and the cone voltage was 20V. The instrument was scanned from 150-1500Da in 3 seconds. Nitrogen was used as both the nebulizing gas, at a flow of 100L/hr, and the desolvation gas, at 600L/hr.

Ion trap/orbitrap tandem mass spectrometer (LTQ Orbitrap XL): made by Thermo Instruments. The source voltage was 5kV, the capillary voltage was 35V, and the tube lens voltage was 110V. The sheath gas was nitrogen, at a flow of 10 arbitrary units, and the auxiliary gas was nitrogen, at a flow of 5 arbitrary units. The mass resolution was 30,000, and the instrument was scanned from 75-2000Da.

Differential scanning calorimeter (DSC): DSC was carried out on a TA DSC Q20, under nitrogen flow, at a scanning rate of 10 °C min^{-1}.

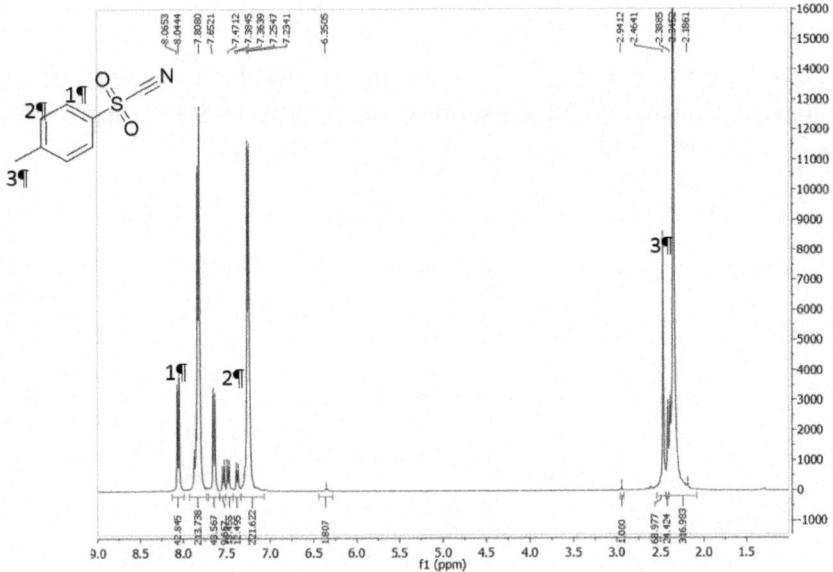

Figure.S1.a ^{1}H-NMR of *p*-Toluenesulfonyl Cyanide (3) in DMSO-d6 for 2 days at 50 °C.

p-Toluenesulfonyl Cyanide ^{1}H NMR (DMSO-d$_6$, ppm): δ = 2.46 (s, 3H), 7.64 (d, *J*= 8.6 Hz, 2H), 8.05 (d, *J*= 8.4 Hz, 2H). The same reaction solution (TsCN and DMSO-d$_6$) was then spiked with toluenesulfonic acid, and the ^{1}H-NMR result was shown in **Fig.S1.b.**

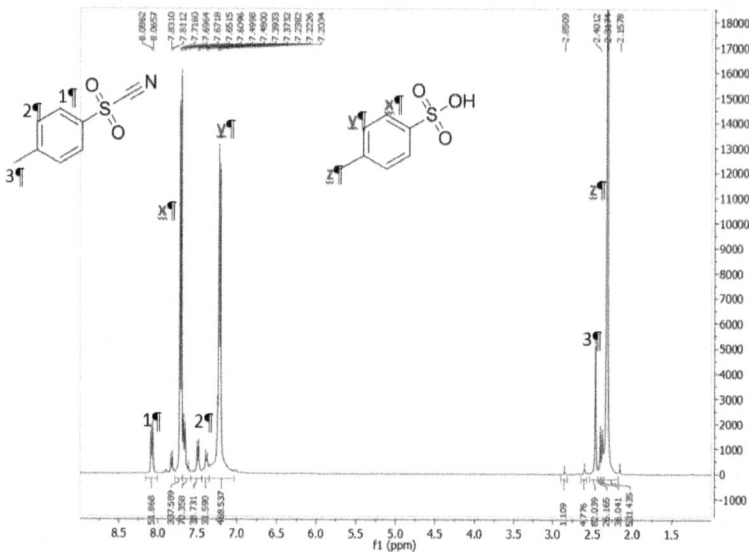

Figure.S1.b ¹H-NMR of *p*-Toluenesulfonyl Cyanide in DMSO-d6 for 2 days at 50 °C, spiked with toluenesulfonic acid.

The increased peaks x, y, z correspond to the increased amount of p-toluenesulfonate. *p*-Toluenesulfonate ¹H NMR (DMSO-d$_6$, ppm): δ = 2.30 (s, 3H), 7.19 (d, *J*=7.7 Hz, 2H), 7.69 (d, *J*=7.8 Hz, 2H).

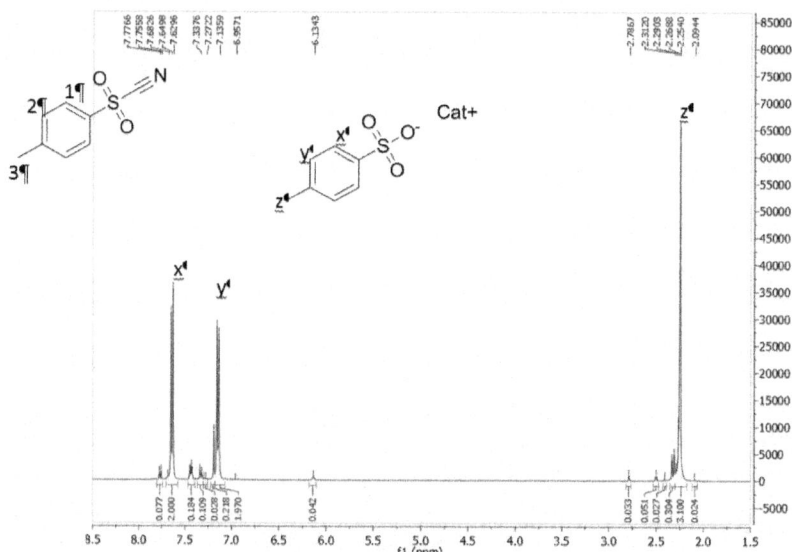

Figure.S1.c ¹H-NMR of *p*-Toluenesulfonyl Cyanide in DMSO-d6 for 4 days at 50 °C.

Comparing Fig.S1.C (reaction at 4 days) with Fig.S1.a (reaction at 2 days), *p*-Toluenesulfonyl Cyanide was further consumed and its peaks became much smaller.

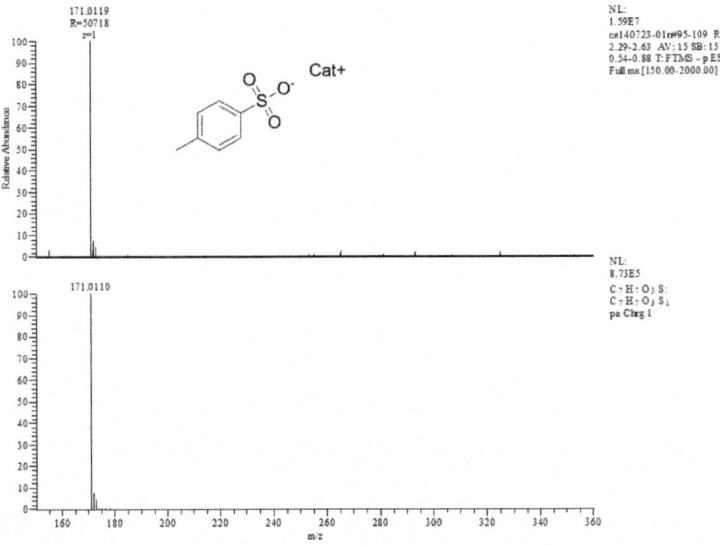

Figure.S2 Negative Ion trap/orbitrap tandem MS of major product obtained upon reaction of DMSO with p-toluenesulfonyl cyanide.

Exact mass analysis was consistent with the *p*-toluenesulfonate (ESI-FTMS (m/z): calcd. for $C_7H_7O_3S$ 171.0110, found 171.0119 [M]$^-$).

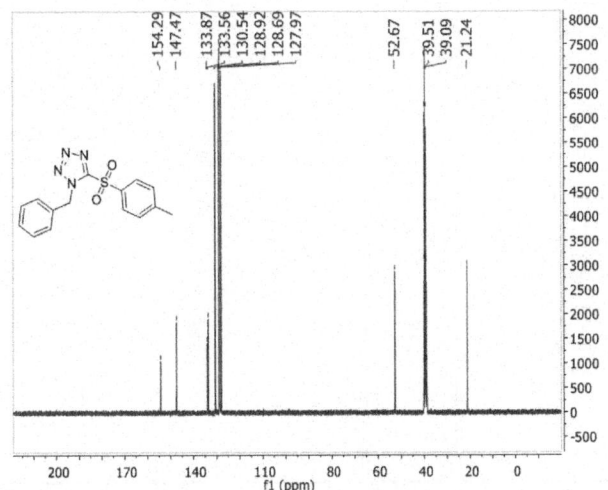

Figure.S3.a ^{13}C NMR of 1-benzyl-5-*p*-toluenesulfonyl tetrazole (2) using cheletropic switch and followed by solvent extraction prior to column chromatographic procedure.

No appreciable differences were observed between **Fig.S2.a** and **Fig.S2.b**, which suggests that comparable purities can be obtained without the need for subsequent column chromatography.

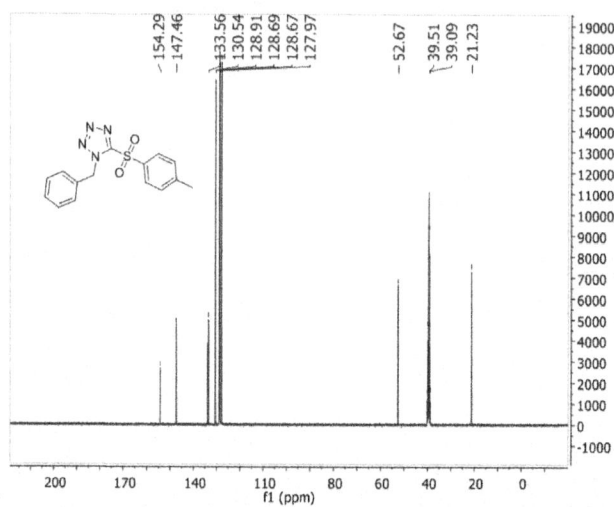

Figure.S3.b ^{13}C NMR of 1-benzyl-5-*p*-toluenesulfonyl tetrazole (2) using cheletropic switch and followed by solvent extraction after the column chromatographic procedure.

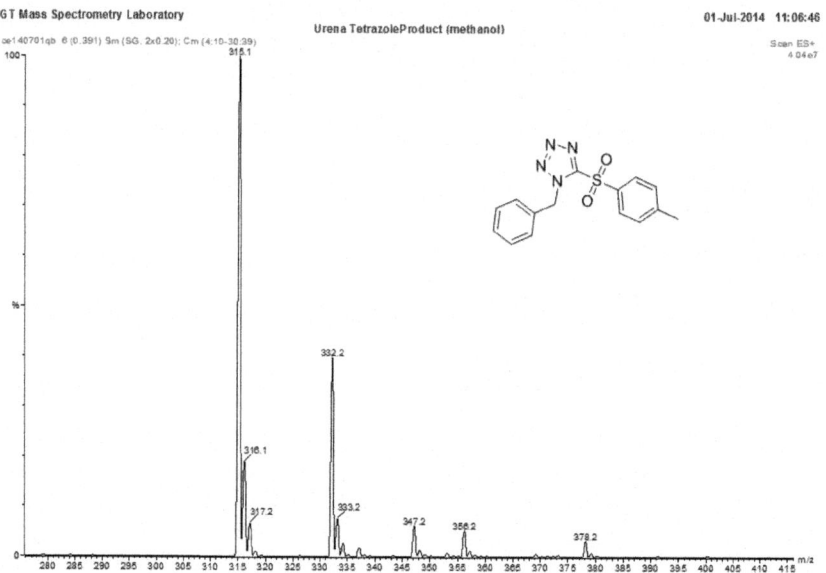

Figure.S4 ESI-MS spectra of 1-Benzyl-5-(p-toluenesulfonyl)tetrazole (2) from cycloaddition chemistry using BS as solvent: 315 [M+H]$^+$, 332 [M+H$_2$O]$^+$

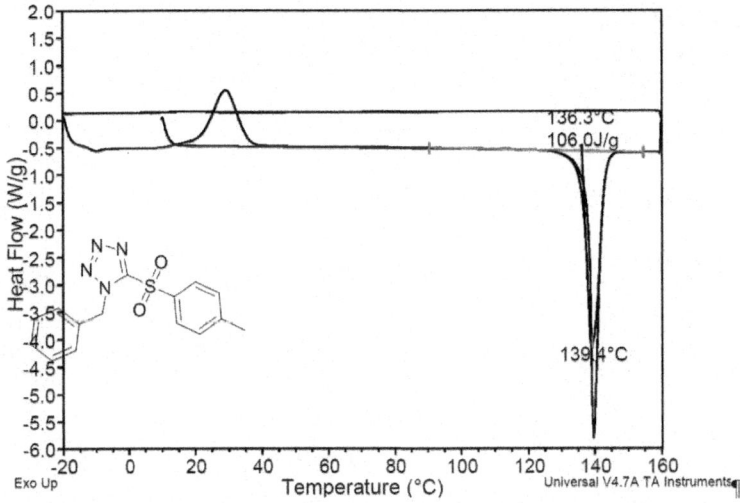

Figure.S5 DSC scan of 1-Benzyl-5-(4-methylphenylsulfonyl)tetrazole (2) product from cycloaddition chemistry reaction using BS as solvent.

DSC scan started at 25 °C, and then temperature was increased to 160 °C at a ramp rate of 10 °C/min. The initial heating process helped to eliminate thermal history of sample. Next, DSC was cooled down to -20 °C for sample to crystallize, and then was heated to 160 °C to determine its melting point and enthalpy of fusion.

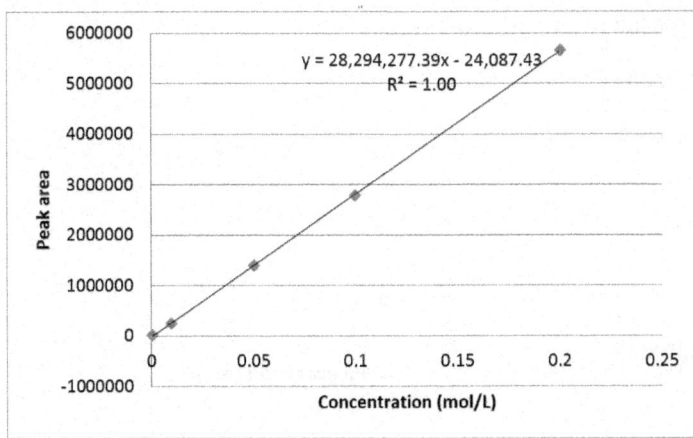

Figure.S6.a GC-FID calibration curve for internal standard biphenyl.

This curve was used to calculate the concentration of biphenyl in reaction solution, which was then used to determine the concentration of BnCl before reaction.

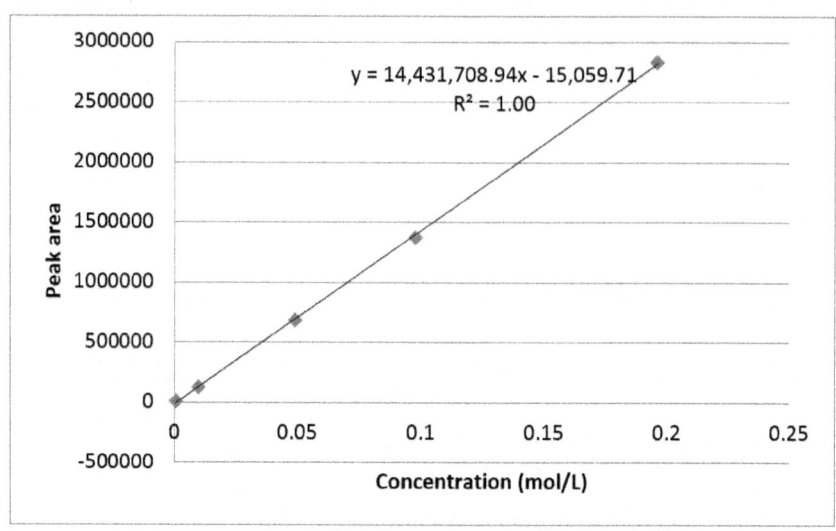

Figure.S6.b GC-FID calibration curve for BnAz (1).

This curve kept track of concentration of BnAz in reaction solution. GC yield of BnAz was able to be calculated.

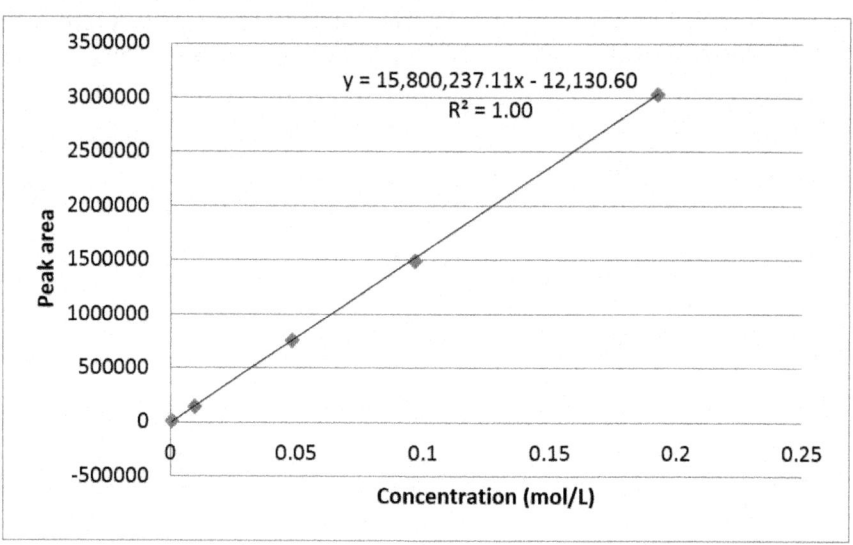

Figure.S6.c GC-FID calibration curve for BnCl.

This curve was used to calculate the concentration of BnCl in reaction solution. Conversion of BnCl at different reaction time was able to be determined.

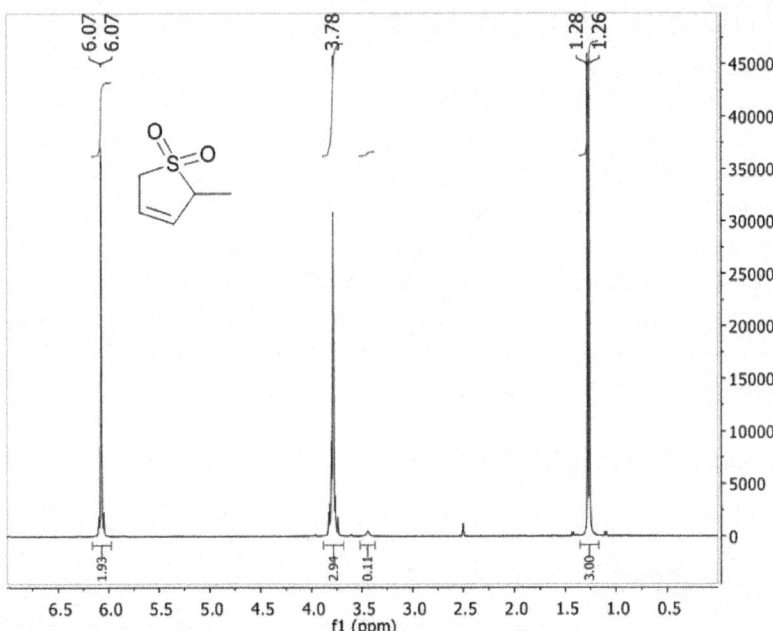

Figure.S7.a Piperylene sulfone ¹H NMR (DMSO-d$_6$, ppm): δ = 2.27 (s, *J*=7.2 Hz, 3H), 3.78 (m, 3H), 6.07 (m, 2H).

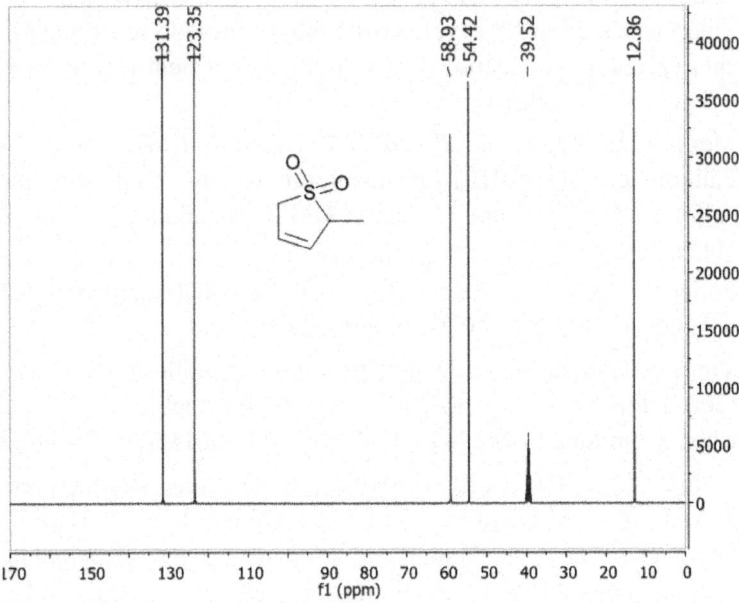

Figure.S7.b Piperylene sulfone 13C NMR (DMSO-d6, ppm): δ = 12.86, 54.42, 58.93, 123.35, 131.39.

AUTHORS' CONTRIBUTIONS

The reported work is a collaboration between researchers at Dow Chemical Company and the Research Teams of CLL, CAE and PP. YH and EU contributed equally: Experimentally determined, conducted and interpreted the bulk of the data on the synthesis of benzyl azide, the synthesis of the tetrazole and the drafting of this manuscript. FM aided in the synthesis of the piperylene sulfone, provided technically input and revised the manuscript. AB and ZC contributed experimentally to all aspects of the work while JF, BH and CAE contributed to the design of experiments. PP and CLL oversaw the entire research study and coordinated the redaction of the manuscript. All authors read and approved the final manuscript.

ACKNOWLEDGEMENTS

We are grateful for financial support from The Dow Chemical Company.

REFERENCES

1.   Donaldson ME, Mestre VL, Vinci D, Liotta CL, Eckert CA (2009) Switchable solvents for in-situ acid-catalyzed hydrolysis of beta-pinene. Ind Eng Chem Res 48(5):2542–2547. doi:10.1021/ie801149z

2.   Jiang N, Vinci D, Liotta CL, Eckert CA, Ragauskas AJ (2008) Piperylene sulfone: a recyclable dimethyl sulfoxide substitute for copper-catalyzed aerobic alcohol oxidation. Ind Eng Chem Res 47(3):627–631. doi:10.1021/ie070616y

3.   Marus GA, Vyhmeister E, Pollet P, Donaldson ME, Llopis-Mestre V, Faltermeier S et al (2010) Sustainable and scalable synthesis of piperylene sulfone: a "volatile" and recyclable DMSO substitute. Ind Eng Chem Res 50(1):23–27

4.   Pollet P, Eckert CA, Liotta CL (2011) Switchable solvents. Chem Sci 2(4):609–614. doi:10.1039/c0sc00568a

5.   Vinci D, Donaldson M, Hallett JP, John EA, Pollet P, Thomas CA et al (2007) Piperylene sulfone: a labile and recyclable DMSO substitute. Chem Commun 14:1427–1429. doi:10.1039/b616806j

6.   de Frias JA, Feng H (2013) Switchable butadiene sulfone pretreatment of Miscanthus in the presence of water. Green Chem 15(4):1067–1078

7.   Marus GA (2011) The application of green chemistry and engineering to novel sustainable solvents and processes [Dissertation]. Georgia Institute of Technology, Atlanta

8. Alvarez SG, Alvarez MT (1997) A practical procedure for the synthesis of alkyl azides at ambient temperature in dimethyl sulfoxide in high purity and yield. Synthesis 1997(04):413–414. doi:10.1055/s-1997-1206

9. Zeng HY, Tian Q, Shao HW (2011) PEG 400 promoted nucleophilic substitution reaction of halides into organic azides under mild conditions. Green Chem Lett Rev 4(3):281–287. doi:10.1080/17518253.2011.571717

10. Demko ZP, Sharpless KB (2002) A click chemistry approach to tetrazoles by Huisgen 1,3-dipolar cycloaddition: synthesis of 5-sulfonyl tetrazoles from azides and sulfonyl cyanides. Angew Chem Int Ed 41(12):2110–2113. doi:10.1002/1521-3773(20020617)41:12<2110:AID-ANIE2110>3.0.CO;2-7

11. Tymtsunik AV, Bilenko VA, Kokhan SO, Grygorenko OO, Volochnyuk DM, Komarov IV (2011) 1-Alkyl-5-((di)alkylamino) tetrazoles: building blocks for peptide surrogates. J Org Chem 77(2):1174–1180. doi:10.1021/jo2022235

12. Singh H, Singh Chawla A, Kapoor VK, Paul D, Malhotra RK (1980) 4 Medicinal chemistry of tetrazoles. In: Ellis GP, West GB (eds) Progress in medicinal chemistry. Elsevier, Amsterdam, pp 151–183

13. Otsuki S, Nishimura S, Takabatake H, Nakajima K, Takasu Y, Yagura T et al (2013) Chemical tagging of a drug target using 5-sulfonyl tetrazole. Bioorg Med Chem Lett 23(6):1608–1611

14. Sharifzadeh Z, Abedi S, Morsali A (2014) Fabrication of novel multi-morphological tetrazole-based infinite coordination polymers; transformation studies and their calcination to mineral zinc oxide nano- and microarchitectures. J Mater Chem A 2(13):4803–4810. doi:10.1039/C3TA14904H

15. Wang D-Z (2012) Zinc(II) and cadmium(II) coordination polymers with bis(tetrazole) ligands: syntheses, structures and luminescent properties. Polyhedron 35(1):142–148. doi:10.1016/j.poly.2012.01.012

16. Deguin B, Vogel P (1992) Hetero-diels-alder addition of sulfur dioxide to 1,3-dienes. Suprafaciality, regioselectivity, and stereoselectivity. J Am Chem Soc 114(23):9210–9211. doi:10.1021/ja00049a076

17. Fernandez T, Sordo JA, Monnat F, Deguin B, Vogel P (1998) Sulfur dioxide promotes its hetero-Diels–Alder and cheletropic additions to 1,2-dimethylidenecyclohexane. J Am Chem Soc 120(50):13276–13277. doi:10.1021/ja982565p

18. Monnat F, Vogel P, Sordo JA (2002) Hetero-Diels–Alder and cheletropic additions of sulfur dioxide to 1,2-dimethylidenecycloalkanes.

Determination of thermochemical and kinetics parameters for reactions in solution and comparison with estimates from quantum calculations. Helv Chim Acta 85(3):712–732. doi:10.1002/1522-2675(200203)85:3<712:aid-hlca712>3.0.co;2-5

19. Turk SD, Cobb RL (1967) Formation of five-membered cyclic sulfones. In: Hamer J (ed) 1,4-Cycloaddition reactions: the Diels–Alder reaction in heterocyclic syntheses. Organic chemistry: a series of monographs, vol 8. Academic Press, New York, pp 13–45

20. Morris RC, Finch HDV (1947) Production of cyclic mono-sulfones. Google Patents

21. Staudinger H, Ritzenthaler B (1935) Ber. 68B:445

22. Desimoni G, Faita G, Garau S, Righetti P (1996) Solvent effect in pericyclic reactions. X. The cheletropic reaction. Tetrahedron 52(17):6241–6248

23. Drake L, Stowe S, Partansky A (1946) Kinetics of the diene sulfur dioxide reaction1. J Am Chem Soc 68(12):2521–2524

24. Minoura Y, Nakajima S (1966) Polymerization of butadiene sulfone. J Polym Sci Part A-1 Polym Chem 4(12):2929–2944

Chapter 11

ENHANCED PRODUCTION OF FRUCTOSE ESTER BY BIOCATALYZED CONTINUOUS FLOW PROCESS

Felipe K Sutili[1] , Halliny S Ruela[1] , Daniel De O Nogueira[1,2], Ivana CR Leal[2] , Leandro SM Miranda[1] and Rodrigo OMA De Souza[1]

[1]Biocatalysis and Organic Synthesis Group, Chemistry Institute, Federal University of Rio de Janeiro, Rio de Janeiro CEP 22941 909, Brazil

[2] Faculdade de Farmácia, Federal University of Rio de Janeiro, Rio de Janeiro CEP22941909, Brazil.

ABSTRACT

Background

Fatty acid sugar esters are non-toxic, odorless, non-irritanting surfactants. They can be synthesized by renewable resources and are completely biodegradable in aerobic and anaerobic conditions. Their application has been expanded in innumerous areas including pharmaceuticals, cosmetics, detergents and food industry. Lipase-catalyzed esterification have been investigated as a potential substitute to the traditional chemical, demanding milder reaction conditions, allowing better reaction control and providing higher-quality products. So, the lipase catalyzed sugar ester synthesis becomes an interesting strategy for producing biodegradable, non- ionic surfactants. The main disadvantage of this protocol is the poor solubility of substrates and long reaction time required for performed the esterification reaction with moderated to good yields.

Results

Here in, we report the enzymatic esterification of steric hindered fructose derivative with free fatty acids derived from palm oil refining process (RePO) under continuous flow conditions at concentrations up to 0.5 M, increasing the productivity up to 100 mg. min $^{-1}$.g immob. enzyme $^{-1}$.

Conclusions

The immobilized commercial enzyme from *Rhizomucor miehei* generated the best performance between the catalysts tested.

Fructose Ketal   Free Fatty Acid   Fructose Ketal Ester yield 20-98%

Syringe pump   *Rhizomucor miehei* Immobilized enzyme   Residence Time 8-24 minutes

BACKGROUND

Sugar esters derived from fatty acids are a very interesting class of surfactants with wide application in industry [1,2]. These esters can be applied in innumerous areas including pharmaceuticals, cosmetics, detergents and food industry [3], because of their wide range of hydrophilic/lipophilic balance (HLB) depending on the length of the fatty acid chain and the nature of the sugar [4].

Fructose based surfactants possess good interfacial tension values when compared with commercial sucrose esters [5]. Despite all importance related to this class of molecules, the synthesis of fructose esters is still a challenging task. Most of the chemical process used to the synthesis of these molecules uses mineral or heterogeneous solid acids, under mild conditions in order to preserve the sugar moiety, leading to moderate conversions and colored products; often a mixture of mono-, di-ester is produced, after long reaction times [6]. Lipase-catalyzed protocols have already been used to overcome the drawbacks of chemical synthesis of sugar esters with good conversions/ selectivity and reaction time ranging from 24 to 72 hours [7-11].

While enzymatic catalyzed reactions present important advantages over chemical catalysis, especially concerning green chemistry demands, some developments are still necessary [12]. One important characteristic of many enzyme-catalyzed reaction that prevents its use in large scale industrial process are the high dilutions needed in order to avoid product inhibition, leading to an economically inefficient processes. Despite all progress made on the biocatalyzed synthesis of sugar esters, in most cases the concentration of the starting material solution is still to dilute (mM range) to encourage an industrial application of this process [13].

But, one of the most important issues related to the synthesis of sugar esters is finding a solvent in which both reactants: carbohydrate and fatty acid can have a good solubility and the enzyme is still active. Several authors have tried to overcome the solubility issue by the use of protecting group strategies [14-16]. Recently, we reported the use of sugar ketal derivatives, on biocatalyzed reactions, as an alternative strategy in order to improve solubility of starting material [17].

Here in we report our efforts on the enhanced production of fructose ester by a biocatalyzed continuous flow approach. Aiming to add value to a residue from palm oil industry we have performed esterification reactions catalyzed by lipase between fructose ketal and free fatty acids derived from palm oil refining process (fatty acid residue) under continuous flow conditions at concentrations up to 0.5 M.

RESULTS AND DISCUSSION

In our continuous work on developing bio-catalyzed reactions under continuous-flow conditions, we decided to carry out the esterification between 2,3:4,5-O-D-diisopropylidene frutopyranose (FK) and acidic residue of palm oil refine process (RePO) under continuous flow conditions at increasing concentrations (0.1, 0.3 and 0.5 M), using the commercial *Rhizomucor miehei* immobilized lipase, which is recognized in the literature by having remarkable esterification activity [18] (Scheme 1).

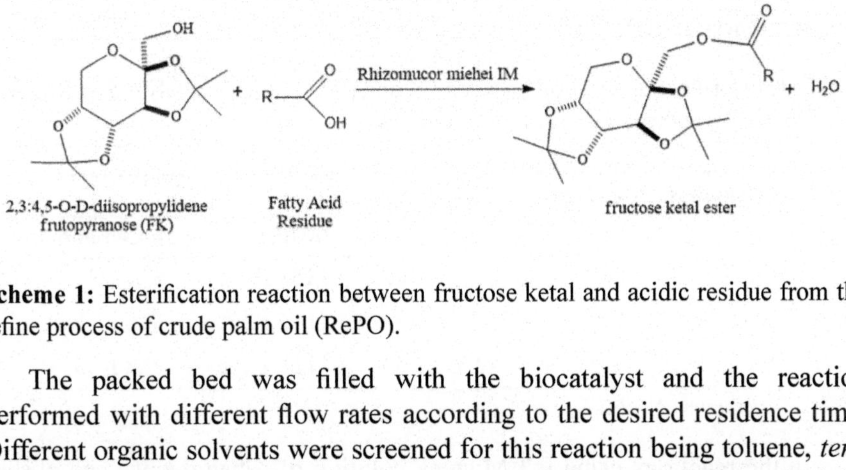

Scheme 1: Esterification reaction between fructose ketal and acidic residue from the refine process of crude palm oil (RePO).

The packed bed was filled with the biocatalyst and the reaction performed with different flow rates according to the desired residence time. Different organic solvents were screened for this reaction being toluene, *tert*-butylmethyl ether and *p*-cymene the most promising ones because had the best solubilization capacities, thereby used for further investigations. First, we investigated the conversions by using 0.1 M solution of starting materials.

As presented in Table 1, all solvents tested showed high conversions at high residence times (24 minutes). However the results obtained for the reactions carried out in *p*-cymene presented lower conversions (87.03%±3.5) when compared with the other solvents and a dramatic decrease on conversion is observed when reducing the residence time from 24 minutes to 8 and 5.4 minutes. The decrease on conversion is less pronounced in Toluene at a residence time of 8 minutes. But in a reaction time of 5.4 minutes the conversion decreased considerably (13.5%). MTBE seems to be the best solvent choice, since even at high flow rates (short residence times, i.e. 5.4 minutes) high conversions are still observed (95.7±2.8), leading to a productivity of 34.3 mg. min $^{-1}$.g immob. enzyme $^{-1}$.

Table 1: Esterification reaction between FK and RePO (0.1 M) catalyzed by Rhizomucor miehei IM under continuous flow conditions

Residence time (min)	Conversion (%)[a]		
	Toluene	MTBE	p-Cymene
24	98.2 ± 1.0	98.6 ± 0.5	87.0 ± 3.5
8	87.6 ± 3.0	98.1 ± 0.5	18.5 ± 2.2
5.4	13.5 ± 2.8	95.7 ± 2.8	2.6 ± 4.3

[a]based on Lowry-Tinsley analysis.

In order to enhance the production of protected fructose ester we moved then for a more concentrated starting material solution (0.3 M) and the results are summarized in Table 2.

Table 2: Esterification reaction between FK and RePO (0.3 M) catalyzed by Rhizomucor miehei IM under continuous flow conditions

Residence time (min)	Conversion (%)[a]		
	Toluene	MTBE	p-Cymene
24	99.91 ± 0.08	93.89 ± 3.84	57.7 ± 5.1
8	99.69 ± 0.12	21.55 ± 2.50	30.8 ± 3.7
5.4	93.03 ± 2.60	11.33 ± 3.66	33.3 ± 2.5

[a]based on Lowry-Tinsley analysis.

In the reactions using 0.3 M stock solution of substrates we can observe a different behavior of solvent effect. Excellent conversions were obtained in the reaction performed in toluene with very short residence times. In contrary to the observed on 0.1 M solutions, MTBE has shown a significant decrease on conversion for reactions performed under residence times of 8 and 5.4 minutes (21.55±2.50 and 11.33±3.66, respectively). The use of 0.3 M stock solution with p-cymene as solvent lead to moderate results at high residence times. Under these conditions, toluene presented the best result leading to a productivity of 100 mg. min $^{-1}$.g immob. enzyme $^{-1}$.

Due the poor solubility of FK in *p*-Cymene at 0.5 M was not possible performed a test with this solvent under continuous flow conditions at this concentration. At this time, only Toluene and MTBE were tested and results are presented on Table 3.

Table 3: Esterification reaction between FK and RePO (0.5 M) catalyzed by *Rhizomucor miehei* IM under continuous flow conditions

Residence time (min)	Conversion (%)[a]	
	Toluene	**MTBE**
24	99.76 ± 0.24	99.91 ± 0.01
8	13.84 ± 0.12	22.25 ± 3.07
5.4	12.71 ± 3.66	7.37 ± 4.11

[a]based on Lowry-Tinsley analysis.

At this substrate concentration (0.5 M) the best results were obtained with high residence times for both solvents (24 minutes). A dramatic decrease on conversion is observed when reducing the residence time from 24 minutes to 8 minutes for both solvents, which can be explained as a substrate inhibition at such high concentration. Under these conditions the productivity obtained for the production of the protected fructose ester derived from the fatty acidic residue under continuous flow conditions is 40.3 mg of ester.h^{-1}.g of immob. enzyme^{-1}and 40.2 mg of ester.h^{-1}.g of immob. enzyme^{-1} for the reactions performed with MTBE and toluene, respectively. To the best of our knowledge such values have no precedent in the synthesis of sugars fatty acids esters. The recyclability of the enzyme source was also evaluated and was observed that the immobilized commercial enzyme from *Rhizomucor miehei* can be recycled for 6 times without change on reaction conversions.

The higher substrate concentration in the esterification reaction was also evaluated for other enzymes sources and the results are presented in Table 4.

Table 4: Esterification reaction between FK and acid residue from palm oil refining process (0.5 M) catalyzed by different immobilized enzymes under continuous flow conditions

Immob. Enz.	Residence Time (min)	Conversion (%) [a]	
		Toluene	MTBE
Novozyme 435	24	52.39±0.87	76.65±0.98
	8	44.96±3.77	60.32±2.34
	5.4	29.21±3.21	25.27±3.34
Lipozyme TL IM	24	90.18±0.87	86.6±0.9
	8	28.8±3.24	22.9±0.82
	5.4	17.9±5.43	17.4±2.44
PS Amano IM	24	46.75±3.4	62±4.3
	8	25.0±1.21	34±2.19
	5.4	7.98±0.99	12±1.36

[a]based on Lowry-Tinsley analysis.

We can see in the Table 3 the best results for reactions catalyzed by TL IM enzyme in 24 min of residence time performed with substrate solubilized in toluene (90.1%). When compared the results of esterification catalyzed by RM IM lipase (Table 2) can be observed the greater conversion this enzyme (99.7%) thus demonstrating that the best catalyst for the reaction. The protected fructose ester can be easily cleaved by several protocols already reported over literature [6].

CONCLUSIONS

In conclusion, we performed enzymatic synthesis of a secondary ester of protected fructose achieved with high substrate concentration and high percentage of conversion, by a simple method that add value to an acid residue of oil palm refining process. Continuous flow conditions allowed high product productivity in short reaction times.

METHODS

Chemicals and Materials

All reagents and immobilized enzymes were purchased from Sigma–Aldrich and used without further purification.

The acidic residue from the refine process of crude palm oil (RePO) was kindly provided by Agropalma Industry. The composition of the residue was determined through gas chromatography. The approximate composition is palmitic acid (44%), oleic acid (42%), and stearic acid (14%).

Synthesis of 2,3:4,5-*O*-diisopropylidene-D-frutopyranose (FK) (2)

In a 2000 mL reactor was added 30 g (44.8 mmol) of sucrose and 400 mL of acetone being vigorously mechanically stirred at 5°C for 15 min. Then, 16 mL of concentrate sulfuric acid (H_2SO_4) was slowly added to the reaction mixture. The solution was kept under stirring for 150 min. Subsequently, the reaction mixture was cooled (0–10°C) in ice bath and neutralized with 50% NaOH (w/v). The pH was adjusted with saturated sodium carbonate. The final mixture was filtered to remove the solids and subsequently, the solvent was evaporated under reduced pressure. The solid crude ketal was diluted with 400 mL of dichloromethane. A 0.5 M H_2SO_4 solution was added and stirred vigorously for 120 min. The organic phase was separated and washed consecutively with sodium bicarbonate ($NaHCO_3$) and water and dried with anhydrous sodium sulfate (Na_2SO_4). The solvent was evaporated under reduced pressure until obtaining a white solid, which was crystallized in hexane with 30% final yield after filtration through activated charcoal [19].

Continuous Flow Reaction Procedure

An equimolar stock solution (*tert*-butylmethyl ether (MTBE), toluene or *p*-cymene) of 2,3:4,5-*O*-D-diisopropylidene frutopyranose (FK) and the RePO was prepared (the molarity of the residue was expressed in palmitic acid). The starting mixture was stirred for 5 min while the instrument Asia Flow Reactor was equipped with Omnifit column (2.4 mL) containing the immobilized lipase from *R. miehei* (600 mg). The reaction parameters were selected on the flow reactor, and processing was started, whereby only pure solvent was pumped through the system until the instrument had achieved the desired reaction parameters and stable processing was assured. At this point, the inlet pipe of the flask was switched to HPLC bottle containing the prepared reaction mixture. After processing through the flow reactor, the inlet tube was dipped back into the flask containing respective pure solvent and processed in order to wash the system of any remaining reactant.

Lowry–Tinsley Method

The conversion of the esterification reactions was measured by the adapted Lowry and Tinsley method [20]. In an eppendorf tube was placed 0.3 mL of reaction medium, 1 mL heptane, 0.3 ml of 5% copper acetate–pyridine solution (pH 6) and stirred vigorously for 30 seconds. The supernatant was measured by spectrophotometer UV/715 nm visible wavelength. Each reaction was measured in triplicate and conversion calculations were based on the analytical curve with the fatty acid used. This methodology for determination of residual acid is aligned with the analytical chromatographic methods, being widely used.

Statistical Analysis

A comparison of the data was performed using the software Statistica 6.0 (Statsoft, Inc., USA), through ANOVA followed by Tukey test, considering statistically significant when $p < 0.05$.

ACKNOWLEDGEMENTS

We thank CAPES (Coordenação de Aperfeiçoamento de Pessoal de Nível Superior) FAPERJ (Fundação de Apoio a Pesquisa do Estado do Rio de Janeiro) and CNPq (Conselho Nacional de Desenvolvimento Científico e Tecnológico).

AUTHORS' CONTRIBUTIONS

All authors contributed equally in this work. All authors read and approved the final manuscript.

REFERENCES

1.   Chang SW, Shaw JF. Biocatalysis for the production of carbohydrate esters. New Biotechnol. 2009;26:109–16.

2.   Shi Y, Li J, Chu Y-H. Enzyme-catalyzed regioselective synthesis of sucrose-based esters. Chem Technol Biotechnol. 2011;86:1457–68.

3.   Flores MV, Engasser JM, Halling PJ. Dissolution kinetics of crystalline -glucose in 2-methyl 2-butanol. Biochem Eng J. 2005;22:245–52.

4.   Baker IJA, Matthews B, Suares H, Krodkiewska I, Furlong DN, Grieser F, et al. Sugar fatty acid ester surfactants: structure and ultimate aerobic biodegradability. J Surfact Deterg. 2000;3:1–11.

5.   Soultan S, Ognier S, Engasser J-M, Ghoul M. Comparative study of some surface active properties of fructose esters and commercial sucrose esters. Colloids Surf A: Physicochem. 2003;227:35–44.

6.   Yan YC, Bornscheuer UT, Stadler G, Lutz-Wahl S, Otto RT, Reuss M, et al. Production of sugar fatty acid esters by enzymatic esterification in a stirred-tank membrane reactor: optimization of parameters by response surface methodology. J Am Oil Chem Soc. 2001;78:147–52.

7.   Neta NDAS, Dos Santos JCS, Sancho SDO, Rodrigues S, Gonçalves LRB, Rodrigues LR, et al. Enzymatic synthesis of sugar esters and their potential as surface-active stabilizers of coconut milk emulsions. Food Hydrocoll. 2012;27:324–31.

8.   Therisod M, Klibanov AM. Facile enzymatic preparation of monoacylated sugars in pyridine. J Am Chem Soc. 1986;108:5638–40.

9.   Therisod M, Klibanov AM. Regioselective acylation of secondary hydroxyl-groups in sugars catalyzed by lipases in organic-solvents. J Am Chem Soc. 1987;109:3977–81.

10. Riva S, Therisod J, Kieboom APG, Klibanov AM. Protease-catalyzed regioselective esterification of sugars and related-compounds in anhydrous dimethylformamide. J Am Chem Soc. 1988;110:584–9.

11. Ikeda I, Klibanov AM. Lipase-catalyzed acylation of sugars solubilized in hydrophobic solvents by complexation. Biotechnol Bioeng. 1993;42:788–91.

12. Rousseau D, Marangoni AG. The effects of interesterification on the physical properties of fats. In: Narine SS, Marangoni AG, editors. Physical properties of lipids. New York: Marcel Dekker; 2002. p. 479–564.

13. Pauly G, Engasser J, Ghoul M. Method for enzymatic synthesis of sucrose esters. U. S. Patent: Laboratoires Serobiologiques (Societe Anonyme), France. US 6,355,455, 2002 March 12.

14. Ward OP, Fang JW, Li ZY. Lipase-catalyzed synthes, is of a sugar ester containing arachidonic acid. Enzyme Microb Technol. 1997;20:52–6.

15. Fregapane G, Sarney DB, Vulfson EN. Enzymic solvent free synthesis of sugar acetal fatty acid esters. Enzyme Microb Technol. 1991;13:796–800.

16. Gao CL, Whitcombe MJ, Vulfson EN. Enzymatic synthesis of dimeric and trimeric sugar-fatty acid esters. Enzyme Microb Technol. 1999;25:264–70.

17. Sutili FK, Ruela HS, Leite SGF, Miranda LSM, Leal ICR, De Souza ROMA. Lipase-catalyzed esterification of steric hindered fructose derivative by continuous flow and batch conditions. J Mol Catal B Enzym. 2013;85–86:37–42.

18. Rodrigues RC, Fernandez-Lafuente R. Lipase from Rhizomucor miehei as an industrial biocatalyst in chemical process. J Mol Catal B Enzym. 2010;64:1–22.

19. Ferreira VF, Perrone CC. Sacarose no laboratório de química orgânica de graduação. Quim Nova. 2001;24:905–7.

20. Lowry RR, Tinsley IJ. Rapid colorimetric determination of free fatty acids. J Am Oil Chem Soc. 1976;53:470–2.

Chapter 12

HIGH PERFORMANCE GREEN BARRIERS BASED ON NANOCELLULOSE

Sandeep S Nair[1] , JY Zhu[2] , Yulin Deng[3] and Arthur J Ragauskas[4]

[1]School of Chemistry and Biochemistry, Georgia Institute of Technology, 500 10th Street, N.W, Atlanta, GA 30332, USA

[2] USDA Forest Service, Forest Products Laboratory, One Gifford Pinchot Drive, Madison, WI 53726, USA

[3] School of Chemical and Biomolecular Engineering, Georgia Institute of Technology 500 10th Street, N.W, Atlanta, GA 30332, USA

[4] Department of Chemical and Biomolecular Engineering, Department of Forestry, Wildlife, and Fisheries, Center for Renewable Carbon, University of Tennessee, Knoxville, TN 37996-2200, USA.

ABSTRACT

With the increasing environmental concerns such as sustainability and end-of-life disposal challenges, materials derived from renewable resources such as nanocellulose have been strongly advocated as potential replacements for packaging materials. Nanocellulose can be extracted from various plant resources through mechanical and chemical ways. Nanocellulose with its nanoscale dimensions, high crystalline nature, and the ability to form hydrogen bonds resulting in strong network makes it very hard for the molecules to pass through, suggesting excellent barrier properties associated with films made from these material. This review paper aim to summarize the recent developments in various barrier films based on nanocellulose with special focus on oxygen and water vapor barrier properties.

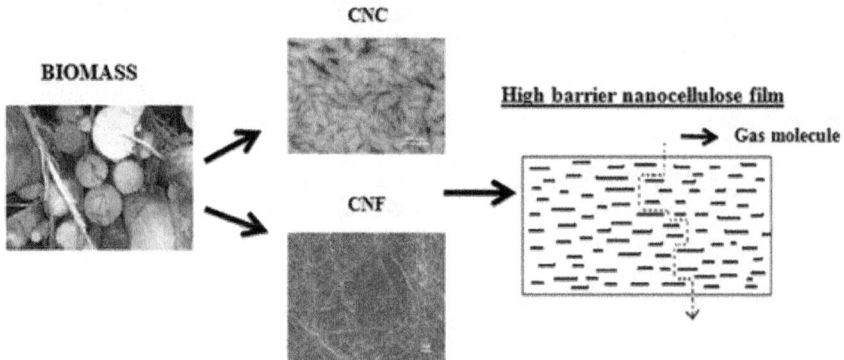

INTRODUCTION

Packaging materials are widely used to prevent food and drink, healthcare, cosmetics and other consumer goods against physical, biochemical, and microbiological deterioration. They should provide sufficient barrier against oxygen, water vapor, grease, and microorganisms. Currently, the packaging materials are largely based on glass, aluminum and tin, and fossil derived synthetic plastics. These materials possess high strength and barrier properties. However, they are unsustainable, some are fragile such as glass, and their weight adds to energy costs for shipping [1-6]. The global consumer packaging demand is valued at approximately US$400b-$500b and is one of the faster-growing markets, forecasted to grow at ~4% per year until 2015 [7].

With the increased environmental concerns over sustainability and end-of-life disposal challenges, materials derived from renewable resources have been strongly advocated as potential replacements [8]. Cellulose is the most abundant polymer in nature and accounts for approximately 40% of lignocellulosic biomass. Cellulose paper-based packaging is lightweight, low-cost, and most important, sustainable. Unfortunately, common paper made from lignocelluloses does not provide sufficient barrier for water, oxygen or oil. Currently, paper based packages are made with unsustainable coatings of wax, plastics, or aluminum. Cellophane is the only cellulose based material (not modified or coated) currently used for barrier packaging due to its high gas barrier. However, the production of cellophane is via a viscose route which produces byproducts and uses reagents (CS_2 and H_2S) that are harmful to the environment [9].

The production of cellulose nanomaterial such as cellulose nanofibrils (CNFs) and cellulose nanocrystals (CNCs) have opened vast possibilities of utilizing cellulose based materials for packaging. Cellulose nanomaterial has diameter in the range of 2–50 nm with large surface area [10-12]. The ability to form hydrogen bonds resulting in strong network makes it very hard for the molecules to pass through, excellent for barrier applications [13]. This review paper aim to summarize the recent developments in various barrier films based on nanocellulose with special focus on oxygen and water vapor barrier properties.

Nanocellulose and its Preparation

Cellulose nanofibrils (CNFs) or microfibrils have diameter in the range of 2–50 nm and lengths up to several micrometers depending on their origin [10-12]. CNFs have exceptional optical and mechanical properties, and therefore can be used as a building block for a variety of high-performance materials [14-16]. Intensive mechanical treatment is required to disintegrate the cellulose fiber to nanofibrils [17]. Several methods of mechanical fibrillation have been used for the production of CNFs such as homogenizers [18,19], microfluidizers [20], and grinders [21,22]. Cellulose nanocrystals (CNCs) are often prepared by treating cellulose fiber with sulfuric acid or hydrochloric acid. Strong acidic condition leads to aggressive hydrolysis to attack the noncrystalline fractions within the cellulose fiber which results in the formation of low aspect cellulose fibril aggregates known as CNCs [23-25].

Mechanical fibrillation of cellulose fibers to CNFs are very energy intensive with reported values ranging from 4500–10000 kWh/tonne [21,22]. Chemical pretreatments and enzymatic pretreatments before mechanical fibrillation have been used to reduce this energy consumption. Cellulose fibers were oxidized by a 2,2,6,6-tetramethylpiperidine-1-oxyl radical (TEMPO)-mediated system. These treated fibers were further mechanically fibrillated to CNFs while reducing the energy costs [26]. Aulin et al. [27] carboxymethylated softwood pulp and then fibers were mechanically fibrillated using a high-pressure fluidizer to produce CNFs of diameter of 5–10 nm. Enzyme pretreatment of biomass followed by mechanical homogenization has been used as an environmentally friendly alternative to chemical pretreatment for nanocellulose production [28-30]. Figure 1 shows the morphological difference between the CNCs and CNFs.

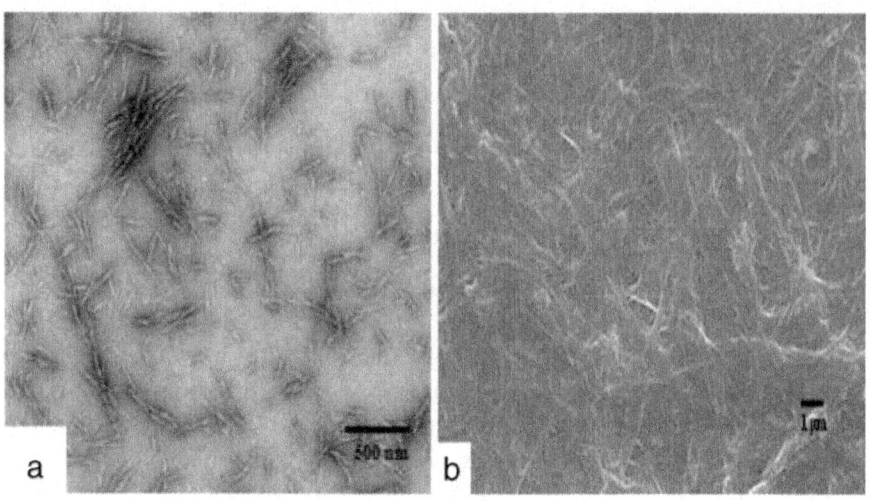

Figure 1: Images of (a) CNCs [24] , and (b) CNFs [18] .

Migration Process of Molecules through Nanocellulose Film

Migration of molecules between two adjacent volumes separated by a thin film of solid polymer or membrane occurs in three basic steps. In the first step, the diffusing molecule is adsorbed on to the sample surface. In the second step, the molecule then diffuses through the film or membrane. Finally, the diffusing molecule exits the sample by desorbing from the surface. The gas permeability through a nanocellulose film mainly depends on the dissolution of gas and its rate of diffusion in the film. Thus, permeability of gas molecules through nanocellulose film can be expressed as:

$$P = DS \qquad (1)$$

Where P is the permeability, D is the diffusion coefficient, and S is the solubility coefficient.

The permeability coefficient P is obtained from the application of Henry's law of solubility to Fick's law of diffusion,

$$P = DS = ql/At\Delta p \qquad (2)$$

Where q is the amount of material passing through the film, l is the thickness, A is the cross sectional area, t is time, and Δp is the pressure difference between the two sides of film.

The gas molecules should be first dissolved in the membrane or film before diffusing. Even though the surface of films influences the permeating gas molecules, the most dominant factor in molecular migration is bulk flow, i.e., rate of molecule diffusion in the membrane or film [31]. The good oxygen barrier properties of nanocellulose can be attributed to the dense network formed by nanofibrils with smaller and more uniform dimensions. Considering this fact, the pores within the films serve as the major path for permeating oxygen molecules. The dense nanofibrils forms more complex and smaller pores compared to pure cellulose fibers which are in micro scale. This complex dense network increases the tortuosity within the film and thereby decreases the permeability within the films [13]. Figure 2 shows a schematic representation of increased diffusion path within the nanocellulose films. Also, the high crystalline structure within the nanofibrils or whiskers contributes to the gas barrier properties [23,31]. Cellulose is composed of both crystalline and disordered regions. High crystallinity ranging from 40-90% has been reported for the nanocellulose, with CNCs showing higher crystallinity than the CNFs because strong acid hydrolyzes disordered cellulose to result in highly crystalline CNCs [21,32]. Even though the CNCs have higher crystallinity than CNFs, mechanically fibrillated CNF films were found to have much lesser oxygen permeability than CNCs. Both showed similar solubility, but the oxygen molecules penetrated more slowly though the CNF films. This is mainly due to the structural organization within the films. The CNF films have higher entanglements within the film which increased the tortuosity factor or increase the diffusion path [33]. Even though nanocellulose provides a high barrier for oxygen, the water vapor barrier properties are low. This is mainly due to high affinity between water and the nanocellulose film. Nanocellulose is much better water vapor barrier than cellulose fiber. Nanocellulose has a strong reducing effect on water vapor diffusion due to its size, and swelling constraints formed due to rigid network within the films. However, at a high relative humidity these structural organizations can be disrupted due to high swelling and can lose barrier properties for both oxygen and water vapor [34].

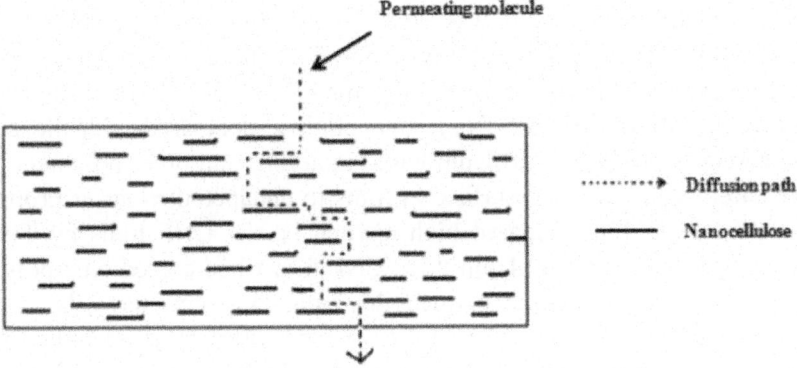

Figure 2: Shows schematic representation of increased diffusion path within the nano-cellulose films.

CNFs for Barrier Application

CNFs is a strong gas barrier material. Compared to CNCs, CNFs consists of crystalline and disordered regions. In most of the cases, crystallinity ranging from 40-75% has been reported for the CNFs obtained from softwoods and hardwoods [11,12,21,22]. Saito and Isogai (2004) showed that the degree of crystallinity varied from 78-91% for CNFs produced from TEMPO oxidation of cotton linter [35]. Films made purely of mechanically fibrillated CNFs have very high air and oxygen barrier property. The oxygen transmission rates (OTR) of CNF films with thickness of 21 μm were as low as 17 ± 1 ml m^{-2} day^{-1}. These values are competitive with other best synthetic polymers such as ethylene vinyl alcohol (EVOH) (3–5 ml m^{-2} day$^{-1)}$ and polyvinylidene chloride (PVdC) coated polyester films (9–15 ml m^{-2} day^{-1}) of approximately same thickness with respect to OTR [13]. Recently, Osterberg et al. [36] demonstrated a rapid method of making robust CNF films with high oxygen barrier property. The CNF solutions were first filtered followed by hot pressing at high pressure followed by air drying. At a relative humidity below 65%, the oxygen permeability of these films was below 0.6 cm^3 μm m^{-2} d^{-1} kPa^{-1}. However, oxygen permeability of CNF films increases with the increase in relative humidity. This is mainly due to the plasticizing and swelling of nanofibrils through the adsorption of water molecules at high relative humidities.

CNFs also have tunable barrier properties. Rodionova et al. [5] showed that both pure and partially acetylated CNF films can be used for modified atmosphere packaging (MAP) with permeability below 10–20 mL m^{-2} day^{-1}. Similarly, carboxymethylated CNF films with thicknesses of 2.54 and 3.19 μm have very low oxygen permeability of 0.009 and 0.0006 cm^3 μm m^{-2} day^{-1} kPa^{-1}, respectively [27]. High oxygen performance barrier films with permeability of 0.007 cm^3 μm m^{-2} day^{-1} kPa^{-1} can be made by controlled thermal treatment at different temperatures. It was demonstrated that after thermal treatment at 175°C, the oxygen permeability of the CNF films was reduced by 96% compared to the films without heat treatment [37]. Table 1 shows the oxygen permeability of films made from nanocellulose compared with those from commercially available petroleum based materials and other polymers. Also, CNFs are used to coat the surface of films to enhance the barrier properties. Fukuzumi et al. [38] prepared high oxygen barrier CNF films by TEMPO mediated oxidation. The 0.1 μm thick CNF film when coated on a polylactic acid film, reduced the oxygen permeability from 746 ml m^{-2} day^{-1} Pa^{-1} for pure PLA film to 1 ml m^{-2} day^{-1} Pa^{-1} for PLA film with CNF layer. The oxygen barrier of PET films was raised by more than 350 times through a 1 μm thick layer of the same TEMPO-oxidized CNFs [39]. Similarly, Fujisawa et al. [40] found that the use of TEMPO-oxidized CNFs with sodium carboxylate groups improved the oxygen barrier of PET films more than when coated with TEMPO-oxidized CNFs with carboxyl groups. Hult et al. [41] showed that the oxygen transmission rate of paper substrates were decreased by several logarithmic units by the deposition of CNFs and shellac coating layers. CNFs have also been used as filler material to obtain a nanocomposite material. Plackett et al. [42] showed that the addition of 15 wt% of CNFs substantially increased the oxygen barrier of amylopectin films. The use of CNFs in xylan films showed very low oxygen permeability of 0.19 cm^3 μm m^{-2} day^{-1} kPa, which is comparable to previously reported values for 100% CNF films [43]. Several studies have been conducted on making composites using CNF with nanoclay [44], and talc [45]. Oxygen barrier properties of these composites were highly competitive and much better than various synthetic polymers currently used in the packaging industry.

Table 1: Oxygen permeability of nanocellulose film compared to those made form commercially available petroleum based materials and other polymers

Material	Oxygen permeability (cc.μm/m^2.day.kPa)	References, conditions
CNF	0.6	[35], at 65% RH
CNF (car-boxymethyl-ated)	0.0006	[27], at 0% RH
CNF (car-boxymethyl-ated)	0.85	[27], at 50% RH
Cellophane	0.41	[46], at 0%RH
PVdC	0.1-3	[47], at 50% RH
Polyvinyl alcohol (PVOH)	0.20	[47], at 0% RH
EVOH	0.01-0.1	[47], at 0% RH

Even though the oxygen barrier properties of CNF films are competitive with current commercial films made from synthetic polymers, their water vapor barrier remains very low or the water vapor transfer rate remain very high. Table 2 shows the water vapor transmission rate (WVTR) of nanocellulose films than those of commercially available films from petroleum based materials and other polymers. This is mainly due to the strong hydrophilic nature of the cellulose nanofibers. However, the water vapor transmission rate

(0.02 g/m² day/m) was significantly lower compared with films made from bleached softwood fibers (0.11 g/m² day/m). The CNF film structure is more compact, therefore less water can penetrate through the film compared to the films made of cellulose fibers. The water contact angle was substantially increased for the CNF film compared to films made from bleached softwood film. The hydrophilic nature is not the only important criteria for water vapor barrier property. The films with high lignin content created more hydrophobic surface with higher water contact angles but showed the least water vapor barrier enhancement. However, the CNF films without lignin produced a more compact structure. The presence of lignin hindered the hydrogen bonding and created more hydrophobic pores which aided in more water vapor transmission [34]. Recently, researchers have modified the properties of CNFs using various pretreatments to enhance the water vapor barrier property. Sharma et al. [37] showed that the water vapor permeability was reduced by 50% upon heat treatment at 175°C for 3 hours as compared to the untreated CNF films. They attributed that the enhancement in barrier property was mainly due to increase in hydrophobicity as evidenced by the increase in water contact angle and reduced porosity by heat treatment. Rodionova et al. [5] showed that the WVTR of pure CNF films (234 g m⁻² day⁻¹) was reduced to 167 g m⁻² day⁻¹ by 30 mins of acetylation treatment. The acetylation created a much higher hydrophobic film surface indicates by higher water contact angles. Minelli et al. [48] showed that the use of carboxymethylated pretreated CNFs produced a more homogenous and less porous film than enzyme pretreated CNFs and showed a much enhancement in water vapor barrier property below 80% relative humidity. Recently, the water vapor permeability of the paper substrates was substantially decreased with a multilayer coating of nanocellulose and alkyd resins. The WVTR reached very low and reached values considered as high barriers in packaging applications [49].

Table 2: WVTR of nanocellulose compared to commercially available petroleum based materials and other polymers

Material	Water vapor transmission rate (WVTR) (g/m² day)	Average thickness of the film (μm)	References, conditions
CNF	234	42	[5], 50% RH
CNF (acetylated for 0.5 h)	167	46	[5], 50% RH
PVdC	3.07	12.7	[50], 100% RH

Polyethylene (PE)	16.8	18.3	[50], 100% RH
Plasticized polyvinyl chloride (PVC)	118.56	12.7	[50], 100% RH

CNCs for Barrier Application

Contrary to CNFs, very few studies have been directed toward study of 100% pure CNC film or treated CNC films. Belbekhouche et al. [33] compared the gas barrier properties between CNF and CNC films. They found that the films made of CNCs were more permeable to oxygen than those made of CNFs. The oxygen molecules penetrated much more slowly within CNF film due to the higher fibril entanglements within the film which increased the tortuosity factor. CNCs, which have crystallinity greater than 60% combined with their ability to form a dense hydrogen bonded network can increase gas barrier property. Bacterial cellulose nanocrystals (BNCs) films present excellent oxygen barrier at low relative humidity, but their high moisture sensitivity results in dramatically decreased barrier when the relative humidity is higher than 70%. The oxygen permeability of 6.99×10^{-22} m^3 m/m^2 s Pa at 0% humidity increased to 5.97×10^{-18} m^3 m/m^2 s Pa at 80% humidity. However, this permeability was reduced by 97% and 74% when BNC films were coated with annealed PLA electro spun nanostructured fibers and 3-aminopropyl) trimethoxysilane (APTS), respectively [51]. Herrara et al. [52] studied thin spin coated films made from CNCs prepared with sulfuric acid and hydrochloric acid. The hydrochloric acid made CNCs resulted in films with low permeability for oxygen, while the sulfuric acid made CNCS resulted in films with higher permeability.

CNCs have been studied as filler for various natural polymers for enhancing the barrier properties. Saxena et al. [23] produced nanocomposite film with low oxygen permeability by casting an aqueous solution containing xylan, sorbitol and nanocrystalline cellulose. Oxygen permeability of films prepared from xylan, sorbitol and 50% by weight of sulfonated CNC exhibited a significantly reduced oxygen permeability of 0.1799 $cm^3.\mu m/$ $m^2.d.kPa$ compared with films prepared solely from xylan and sorbitol with an oxygen permeability of 189.1665 $cm^3.\mu m/m^2.d.kPa$. Poly lactic acid (PLA) nano-biocomposites containing 5 wt% of nanocrystals exhibited the highest oxygen barrier. The OTR for PLA nanocomposites with 5% w/w of unmodified CNCs was 17.4 ± 1.4 cm^3 mm m^{-2} day^{-1}, while that for modified CNCs with an acid phosphate ester of ethoxylated nonylphenol in a 1/4 (wt/wt) ratio was 15.8 ± 0.6 cm^3 mm m^{-2} day^{-1} [53]. Addition of 1 wt% of silver nanoparticles to these modified CNC- PLA composites further decreased the

OTR to 12.6 ± 0.1 cm^3 mm m^{-2} day^{-1} [54]. The OTR values of ternary systems consisting of PLA, PHB (poly hydroxybutyrate) and 5 wt% unmodified CNCs was 15.3 cm^3 mm m^{-2} day^{-1}, while that for modified CNCs with an acid phosphate ester of ethoxylated nonylphenol in a 1/1 (wt/wt) ratio was 13 cm^3 mm m^{-2} day^{-1}. Water contact angle measurements showed that the ternary system had high hydrophobicity and the presence of sulphate groups with low polarity on the surface of CNCs increased the surface hydrophobicity of the final composite material [55].

CNCs were used as fillers in polyvinyl alcohol (PVOH) matrix. The addition of 5 wt% CNCs decreased the WVP of pure PVOH films from 0.61 ± 0.04 g. mm/kPa.h.m^2 to 0.44 ± 0.01 g.mm/kPa.h.m^2 [56]. The reinforcement of natural biopolymers with CNCs was found to reduce WVTR of the resulting nanocomposites. The films prepared using xylan as reinforcement polymer with 10% sulfonated CNCs exhibited a 74% reduction in specific water transmission properties compared with the film without CNCs and a 362% improvement compared with xylan films reinforced by 10% softwood kraft fibers. The xylan/sulphonated CNC nanocomposites showed a WVTR of 174 g/ hm^2 [57]. They also compared xylan films reinforced with CNC made from hydrochloric acid with those reinforced with sulphonated CNC. Even though, films showed a significant reduction in water transmission, the reduction was not as significant as those using sulfonated CNCs. The xylan/ hydrochloric acid made CNC films showed a WVTR of 281 g/hm^2 [58]. Khan et al. [59] showed that the values of water vapor permeability (WVP) decreased sharply as the content of CNCs increased in the methyl cellulose based films. The WVP of control films (without CNCs) was 6.3 g.mm/m^2.day.kPa, while those films in cooperated with 1 wt% CNC showed a permeability of 4.7 g.mm/m^2.day.kPa.

CONCLUSIONS

Nanocellulose such as CNFs and CNCs have opened vast possibilities of utilizing cellulose based materials. The use of CNFs in films, composites, and coatings has found to substantially reduce the oxygen permeability within these materials. The oxygen barrier efficiency of pure CNF films is highly competitive and even be comparable with commercial synthetic polymers. The improvement of oxygen barrier properties by CNFs can be attributed to the dense network formed by nanofibrils with smaller and more uniform dimensions. Even though CNCs have higher crystallinity than CNFs, mechanically fibrillated CNF films were found have much lesser oxygen permeability than CNCs. The CNF films have higher entanglements within the film which increases the diffusion path for gas molecules. Also, nanocellulose has a strong reducing effect on water vapor diffusion due to its size, and swelling constraints formed due to rigid network within the films compared to cellulose fibers. The use of CNFs and CNCs in various natural polymer based composites has found to substantially reduce the gas permeability within these composites.

ACKNOWLEDGEMENTS

This work was partially supported by the USDA Forest Service R & D special funding on Cellulose Nano-Materials (2012).

AUTHORS' CONTRIBUTIONS

All the authors have contributed to the literature review and manuscript writing. All authors read and approved the final manuscript.

REFERENCES

1. Bayer IS, Fragouli D, Attanasio A, Sorce B, Bertoni G, Brescia R, Di Corato R, Pellegrino T, Kalyva M, Sabella S, Pompa PP, Cingolani R, Athanassiou A: Water-repellent cellulose fiber networks with multifunctional properties. *ACS Appl Mater Interfaces* 2011, 3:4024–4031.

2. Hansen NML, Plackett D: Sustainable films and coatings from hemicelluloses: a review. *Biomacromolecules* 2008, 9:1493–1505.

3. Priolo MA, Gamboa D, Holder KM, Grunlan JC: Super gas barrier of transparent polymer−clay multilayer ultrathin films. *Nano Lett* 2010, 10:4970–4974.

4. Reis AB, Yoshida CMP, Reis APC, Franco TT: Application of chitosan emulsion as a coating on Kraft paper. *Polym Int* 2011, 60:963–969.

5. Rodionova G, Lenes M, Eriksen O, Gregersen O: Surface chemical modification of microfibrillated cellulose: improvement of barrier properties for packaging applications. *Cellulose* 2011, 18:127–134.

6. Spence KL, Venditti RA, Rojas OJ, Pawlak JJ, Hubbe MA: Water vapor barrier properties of coated and filled microfibrillated cellulose composite films. *BioResources* 2011, 6:4370–4388.

7. Neil-Boss N, Brooks K: Unwrapping the packaging industry: seven factors for success. 2013. http://www.ey.com/Publication/vwLUAssets/ Unwrapping_the_packaging_industry_%E2%80%93_seven_factors_ for_success/$FILE/EY_Unwrapping_the_packaging_industry_-_seven_ success_factors.pdf.

8.   Nair SS, Wang SQ, Hurley DC: Nanoscale characterization of natural fibers and their composites using contact-resonance force microscopy. *Compos Part A* 2010, 41:624–631.

9.   Hyden WL: Manufacture and properties of regenerated cellulose films. *Ind Eng Chem* 1929, 21:405–410.

10. Stelte W, Sanadi AR: Preparation and characterization of cellulose nanofibers from two commercial hardwood and softwood pulps. *Ind Eng Chem Res* 2009, 48:11211–11219.

11. Nair SS, Zhu JY, Deng Y, Ragauskas AJ: Hydrogels prepared from cross-linked nanofibrillated cellulose. *ACS Sustainable Chem Eng* 2014, 2:772–780.

12. Hoeger IC, Nair SS, Ragauskas AJ, Deng Y, Rojas OJ, Zhu JY: Mechanical deconstruction of lignocellulose cell walls and their enzymatic saccharification. *Cellulose* 2013, 20:807–818.

13. Syverud K, Stenius P: Strength and barrier properties of MFC films. *Cellulose* 2009, 16:75–85.

14. Abraham E, Deepa B, Pothan LA, Jacob M, Thomas S, Cvelbar U, Anandjiwala R: Extraction of nanocellulose fibrils from lignocellulosic fibers: a novel approach. *Carbohydr Polym* 2011, 86:1468–1475.

15. Kaushik A, Singh M: Isolation and characterization of cellulose nanofibrils from wheat straw using steam explosion coupled with high shear homogenization. *Carbohydr Res* 2011, 346:76–85.

16. He L, Li X, Li W, Yuan J, Zhou H: A method for determining reactive hydroxyl groups in natural fibers:application to ramie fiber and its modification. *Carbohydr Res* 2012, 348:95–98.

17. Uetani K, Yano H: Nanofibrillation of wood pulp using a high-speed blender. *Biomacromolecules* 2011, 12:348–353.

18. Chinga-Carrasco G, Syverud K: On the structure and oxygen transmission rate of biodegradable cellulose nanobarriers. *Nanoscale Res Lett* 2012, 7:192.

19. Chinga-Carrasco G, Kuznetsova N, Garaeva M, Leirset I, Galiullina G, Kostochko A, Syverud K: Bleached and unbleached MFC nanobarriers: properties and hydrophobisation with hexamethyldisilazane. *J Nanopart Res* 2012, 14:1280.

20. Henriksson M, Berglund LA, Isaksson P, Lindstrom T, Nishino T: Cellulose nanopaper structures of high toughness. *Biomacromolecules* 2008, 9:1579–1585.

21. Nair SS, Zhu JY, Deng Y, Ragauskas AJ: Charaterization of cellulose nanofibrillation by micro grinding. *J Nanopart Res* 2014,16:2349.

22. Wang QQ, Zhu JY, Gleisner R, Kuster TA, Baxa U, McNeil SE: Morphological development of cellulose fibrils of a bleached eucalyptus pulp by mechanical fibrillation. *Cellulose* 2012, 19:1631–1643.

23. Saxena A, Elder TJ, Kenvin J, Ragauskas AJ: High oxygen nanocomposite barrier films based on xylan and nanocrystalline cellulose. *Nano-Micro Lett* 2010, 2:235–241.

24. Wang QQ, Zhu JY, Reiner RS, Verril SP, Baxa U, Mc Neil SE: Approaching zero cellulose loss in cellulose nanocrystal (CNC) production: recovery and characterization of cellulosic solid residues (CSR) and CNC. *Cellulose* 2012, 19:2033–2047.

25. Beck-Candanedo S, Roman M, Gray DG: Effect of reaction conditions on the properties and behavior of wood cellulose nanocrystal suspensions. *Biomacromolecules* 2006, 6:1048–1054.

26. Saito T, Kimura S, Nishiyama Y, Isogai A: Cellulose nanofibers prepared by TEMPO-mediated oxidation of native cellulose. *Biomacromolecules* 2007, 8:2485–2491.

27. Aulin C, Gallstedt M, Lindstrom T: Oxygen and oil barrier properties of microfibrillated cellulose films and coatings. *Cellulose* 2010,17:559–574.

28. Henriksson M, Henriksson G, Berglund LA, Lindström T: An environmentally friendly method for enzyme-assisted preparation of microfibrillated cellulose (MFC) nanofibers. *Eur Polym J* 2007, 43:3434–3441.

29. Hayashi N, Kondo T, Ishihara M: Enzymatically produced nano-ordered short elements containing cellulose I-beta crystalline domains. *Carbohydr Polym* 2005, 61:191–197.

30. Zhu JY, Sabo R, Luo XL: Integrated production of nano-fibrillated cellulose and cellulosic biofuel (ethanol) by enzymatic fractionation of wood fibers. *Green Chem* 2011, 13:1339–1344.

31. Lagaron JM, Catala R, Gavara R: Structural characteristics defining high barrier properties in polymeric materials. *Mater Sci Technol* 2004, 20:1–7.

32. Guo J, Catchmark JM: Surface area and porosity of acid hydrolyzed cellulose nanowhiskers and cellulose produced by Gluconacetobacter xylinus. *Carbohydr Polym* 2012, 87:1026–1037.

33. Belbekhouche S, Bras J, Siqueira G, Chappey C, Lebrun L, Khelifi B, Marais S, Dufresne A: Water sorption behaviour and gas barrier properties of cellulose whiskers and microfibrils films. *Carbohydr Polym* 2011, 83:1740–1748.

34. Spence KL, Venditti RA, Rojas OJ, Habibi Y, Pawlak JJ: The effect of chemical composition on microfibrillar cellulose films from wood pulps: water interactions and physical properties for packaging applications. *Cellulose* 2010, 17:835–848.

35. Saito T, Isogai A: TEMPO-mediated oxidation of native cellulose. The effect of oxidation conditions on chemical and crystal structures of the water- insoluble fractions. *Biomacromolecules* 2004, 5:1983–1989.

36. Osterberg M, Vartiainen J, Lucenius J, Hippi U, Seppala J, Serimaa R, Laine J: A fast method to produce strong NFC films as a platform for barrier and functional materials. *ACS Appl Mater Interfaces* 2013, 5:4640–4647.

37. Sharma S, Zhang X, Nair SS, Ragauskas AJ, Zhu JY, Deng Y: Thermally enhanced high performance cellulose nano fibril barrier membranes. *RSC Adv* 2014, 4:45136–45142.

38. Fukuzumi H, Saito T, Wata T, Kumamoto Y, Isogai A: Transparent and high gas barrier films of cellulose nanofibers prepared by TEMPO-mediated oxidation. *Biomacromolecules* 2009, 10:162–165.

39. Rodionova G, Saito T, Lenes M, Eriksen O, Gregersen O, Fukuzumi H, Isogai A: Mechanical and oxygen barrier properties of films prepared from fibrillated dispersions of TEMPO-oxidized Norway spruce and Eucalyptus pulps. *Cellulose* 2012, 19:705–711.

40. Fujisawa S, Okita Y, Fukuzumi H, Saito T, Isogai A: Preparation and characterization of TEMPO-oxidized cellulose nanofibrils films with free carboxyl groups. *Carbohydr Polym* 2011, 84:579–583.

41. Hult EL, Lotti M, Lenes M: Efficient approach to high barrier packaging using microfibrillar cellulose and shellac. *Cellulose* 2010, 17:575–586.

42. Plackett D, Anturi H, Hedenqvist M, Ankerfors M, Gallstedt M, Lindstrom T, Siro I: Physical properties and morphology of films prepared from microfibrillated cellulose and microfibrillated cellulose in combination with amylopectin. *J Appl Polym Sci* 2010, 117:3601–3609.

43. Hansen NML, Blomfeldt TOJ, Hedenqvist MS, Plackett DV: Properties of plasticized composite films prepared from nanofibrillated cellulose and birch wood xylan. *Cellulose* 2012, 19:2015–2031.

44. Aulin C, Salazar-Alvarez G, Lindstrom T: High strength, flexible and transparent nanofibrillated cellulose – nanoclay biohybrid films with

tunable oxygen and water vapor permeability. *Nanoscale* 2012, 4:6622–6628.

45. Liimatainen H, Ezekiel N, Sliz R, Ohenoja K, Sirvio JA, Berglund L, Hormi O, Niinimaki J: High-strength nanocellulose – talc hybrid barrier films. *ACS Appl Mater Interfaces* 2013, 5:13412–13418.

46. Wu J, Yuan Q: Gas permeability of a novel cellulose membrane. *J Membr Sci* 2002, 204:185–194.

47. Lange J, Wyser Y: Recent innovations in barrier technologies for plastic packaging—a review. *Packag Technol Sci* 2003, 16:149–158.

48. Minelli M, Baschetti MG, Doghieri F, Ankerfors M, Lindstrom T, Siro I, Plackett D: Investigation of mass transport properties of microfibrillated cellulose (MFC) films. *J Membr Sci* 2010, 358:67–75.

49. Aulin C, Strom G: Multilayered alkyd resin/nanocellulose coatings for use in renewable packaging solutions with a high level of moisture resistance. *Ind Eng Chem Res* 2013, 52:2582–2589.

50. Steven MD, Hotchkiss JH: Comparison of flat film to total package water vapor transmission rates for several commercial food wraps. *Packag Technol Sci* 2002, 15:17–27.

51. Martinez-Sanz M, Lopez-Rubio A, Lagaron JM: High-barrier coated bacterial cellulose nanowhiskers with reduced moisture sensitivity. *Carbohydr Polym* 2013, 98:1072–1082.

52. Herrera MA, Mathew AP, Oksman K: Gas permeability and selectivity of cellulose nanocrystals films (layers) deposited by spin coating. *Carbohydr Polym* 2014, 112:494–501.

53. Fortunati E, Peltzer M, Armentano I, Torre L, Jimenez A, Kenny JM: Effects of modified cellulose nanocrystals on the barrier and migration of PLA nano-composites. *Carbohydr Polym* 2012, 90:948–956.

54. Fortunati E, Peltzer M, Armentano I, Jimenez A, Kenny JM: Combined effects of cellulose nanocrystals and silver nanoparticles on the barrier and migration properties of PLA nano-biocomposites. *J Food Eng* 2013, 118:117–124.

55. Arrieta MP, Fortunati E, Dominici F, Rayon E, Lopez J, Kenny JM: PLA-PHB/cellulose based films: mechanical, barrier and disintegration properties. *Polym Degrad Stab* 2014, 107:139–149.

56. Pereira ALS, do Nascimento DM, Souza Filho MM, Morais JPS, Vasconcelos NF, Feitosa JPA, Brigida AIS, Rosa MF: Improvement of polyvinyl alcohol properties by adding nanocrystalline cellulose isolated from banana pseudostems. *Carbohydr Polym* 2014, 112:165–172.

57. Saxena A, Ragauskas AJ: Water transmission barrier properties of biodegradable films based on cellulosic whiskers and xylan.*Carbohydr Polym* 2009, 78:357–360.

58. Saxena A, Elder TJ, Ragauskas AJ: Moisture barrier properties of xylan composite films. *Carbohydr Polym* 2011, 84:1371–1377.

59. Khan RA, Salmieri S, Dussault D, Uribe-Calderon J, Kamal MR, Safrany A, Lacroix M: Production and properties of nanocellulose-reinforced methycellulose-based biodegradable films. *J Agric Food Chem* 2010, 58:7878–7885.

Chapter 13

CONTINUOUS-FLOW PROCESSES IN HETEROGENEOUSLY CATALYZED TRANSFORMATIONS OF BIOMASS DERIVATIVES INTO FUELS AND CHEMICALS

Juan Carlos Serrano-Ruiz, Rafael Luque, Juan Manual Campelo and Antonio A. Romero

Departamento de Química Orgánica, Universidad de Córdoba, Campus de Excelencia Internacional Agroalimentario CeiA3, Edificio Marie Curie (C-3), Ctra Nnal IV-A, Km 396, Córdoba E-14014, Spain

ABSTRACT

Continuous flow chemical processes offer several advantages as compared to batch chemistries. These are particularly relevant in the case of heterogeneously catalyzed transformations of biomass-derived platform molecules into valuable chemicals and fuels. This work is aimed to provide an overview of key continuous flow processes developed to date dealing with a series of transformations of platform chemicals including alcohols, furanics, organic acids and polyols using a wide range of heterogeneous catalysts based on supported metals, solid acids and bifunctional (metal + acidic) materials.

INTRODUCTION

Fossil derived fuels (*i.e.*, coal, natural gas and petroleum) currently supply most of the energy consumed on the planet. This high reliance on fossil resources represents a serious issue in terms of availability of resources, environmental pollution, and future development of society since these natural resources are highly contaminant, unevenly distributed around the world and they are in diminishing supply. These important concerns have stimulated the search for new well-distributed and non-contaminant renewable sources of energy including solar, wind, hydroelectric power, geothermal activity, and biomass. Biomass has attracted a great deal of interest in recent years as the only renewable source of organic carbon currently available on Earth.

Consequently, it is considered to be the perfect replacement for petroleum in the production of fuels and chemicals [1].

In any case, the progressive replacement of petroleum with biomass involves a number of challenges in terms of processing approaches since both resources are diametrically opposed chemically speaking (Scheme 1). Petroleum is chemically composed of a mixture of hydrocarbons (e.g., lineal, cyclic, aromatics, alkenes, *etc.*) as opposed to highly functionalized biomass feedstocks which typically comprise highly-oxygenated compounds embedded in some cases in complex polymeric structures. As a result, the highly-optimized catalytic approaches developed over the past 50 years in the petrochemical industry cannot be directly projected to process biomass feedstocks. For example, while in the petrochemical industry the production of hydrocarbon fuels (e.g., diesel, gasoline and jet fuels) is a relatively simple process involving separation of hydrocarbons (by fractional distillation) and subsequent catalytic upgrading (to control molecular weight and structure), the conversion of biomass into fuels is typically associated with deep chemical changes and multistep processing [2]. The particular composition of biomass has also a decisive effect on catalysts design. Typical petroleum catalysts designed to resist high temperatures and hydrophobic environments might not be effective upon the new conditions required to process biomass.

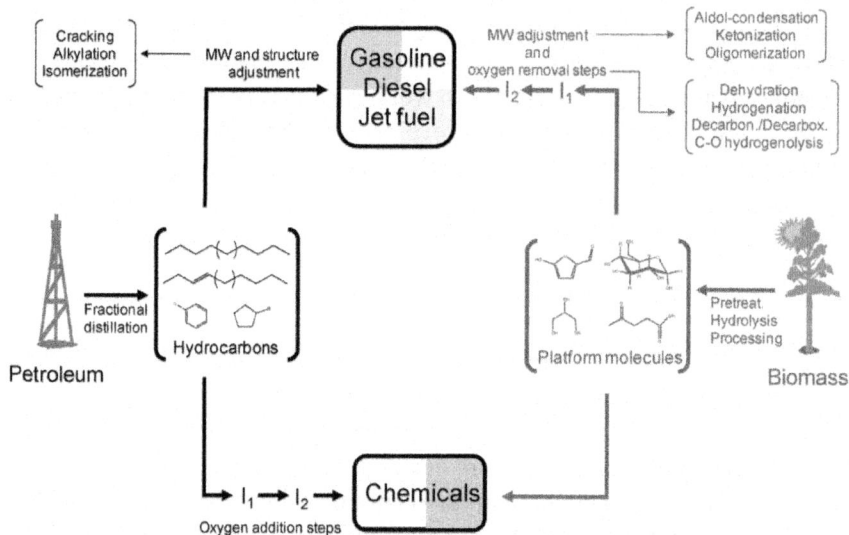

Scheme 1. Divergences between petroleum and biomass processing to chemicals and liquid hydrocarbon transportation fuels. Reproduced by permission of the Royal Society of Chemistry from [2].

The high level of structural and chemical complexity of biomass is an important issue when envisioning new processes. One potential solution to overcome biomass complexity involves its conversion into simpler fractions which are more easily handled in downstream processes. This approach, resembling that utilized in the petrochemical industry, appropriately facilitates the simultaneous production of fuels, chemicals, and energy in a single facility denoted as biorefinery. As an important part of this biorefinery concept, a small group of biomass-derived molecules have been recently identified to be especially relevant according to a series of indicators [3]. These important biomass derivatives (the so-called platform molecules or building blocks) are relatively simple compounds containing multiple functionalities in their structures which are suitable for a range of chemical transformations to valuable compounds. Platform molecules are carefully selected on the basis of a series of factors including the availability of commercial technologies for their production and their platform potential for the simultaneous production of fuels and chemicals in biorefineries. Relevant examples of biomass platform molecules include sugars (glucose, xylose), polyols (sorbitol, xylitol, glycerol), furans (hydroxymethylfurfural or HMF, furfural), acids (lactic acid, levulinic acid) and alcohols (ethanol).

Controlling chemical reactivity represents an important challenge when operating with biomass derivatives. Overfunctionalization of many of these compounds typically leads to high (and sometimes uncontrolled) reactivity. Additionally, many of these molecules have a natural tendency to decompose with temperature.

New processing approaches are thus required to control reactivity in biomass derivatives thereby directing the conversion to the desired products. In this regard, continuous flow processing has a number of significant and inherent advantages for biomass processing as compared to batch reactor technologies:

- Continuous flow processing allows a better of control of reaction conditions. This is advantageous when dealing with highly reactive feedstocks such as those derived from biomass. Furthermore, as will be shown in some of the examples detailed bellow, continuous approaches offer more flexibility to modify conditions through the course of a reaction thereby allowing an optimum control of intermediates in consecutive A → B → C type of reactions. Double-bed reactors, operating at different temperature conditions, could in principle be designed to achieve the desired and optimum control of reactivity in consecutive reactions involving several intermediates.

- Flow processing also facilitates scaling up which is an important point taking into consideration that many of the biomass processes are still in the lab scale. The development of flow technologies will thus contribute to the commercialization of biomass technologies in the near future.

- Since the chemical composition of biomass feedstocks is normally very different from that of the final products, multiple processing steps are typically required in such transformations, negatively affecting the economy of the process. The utilization of flow processing technologies allows intensification of the chemical processes, thereby significantly contributing to simplify technologies, as detailed in some of the examples of the present review.

- Unlike batch processing, fixed-bed flow technologies do not require catalyst separation after reaction and regeneration, if required, is readily performed over the same catalytic bed. In the case of biomass processing, easy regeneration is crucial since the high reactivity of biomass derivatives typically leads to overreactions generating carbonaceous deposits and tars that poison and deposit on the catalysts surface.

- Many of the biomass processes will require oxygen removal steps to produce the final product. Oxygen is generally removed from biomass molecules in the form of H_2O or CO_x (e.g., CO and CO_2). When operating under batch conditions, these gases build up in the reactor leading to increasing pressure and, potentially, new and uncontrolled processes. Flow operation allows continuous removal of these gases which may not interfere in the main catalytic process.

- The microwave-to-flow paradigm, recently highlighted by Kappe's group [4] is a smart approach to translating batch microwave chemistries to more scalable flow conditions upon mimicking the relatively high pressures and temperatures obtained in a microwave experiment in a continuous flow reactor equipped with a back pressure regulator [4]. The proposed methodology is envisaged to be particularly useful for biomass valorisation practises which could be in principle screened in a quick and efficient manner under microwave batch conditions and then translated to flow chemistry protocols after optimisation of reaction parameters, catalysts and conditions.

In this work, we aim to provide a general overview of the most relevant catalytic flow technologies developed (as well as under development) for the conversion of biomass platform molecules into fuels and chemicals. Important aspects such as the reactions conditions (pressure, temperature, space velocity) and the catalysts used are analyzed, with the platform potential for each biomass derivative discussed in the different sections. It is important to note that most of the processes to convert biomass into platform molecules are based on biological or chemical batch technologies. Consequently, many of the flow processes herein described require a previous batch conversion step. Even though an entire continuous process from biomass to the final product would be highly desirable in terms of scalability, the transformation of batch processes into continuous technologies would involve strong processing challenges, especially for those molecules produced by biological routes.

CONTINUOUS-FLOW TRANSFORMATIONS OF BIO-MASS DERIVATIVES INTO FUELS AND CHEMICALS

Ethanol

Ethanol is the predominant biomass-derived fuel at the present time representing 80% of the world biofuel production. Ethanol is added to conventional gasoline to improve the combustion of the mixture, decreasing the generation of pollutants such as CO, NO_x and SO_x. A very simple fermentation technology along with a partial compatibility with the existing infrastructure for gasoline have allowed ethanol industry to grow fast, and production reached 87 billion L in 2010 [5]. Ethanol is thus likely to become a cheap and abundant renewable resource taking into consideration that technologies for the conversion of lignocellulosic feedstocks (more abundant that currently used starches and edible sugars) are under extensive investigation and may become economically feasible in the near future [3]. However, ethanol possesses important limitations as a fuel namely, low energy density (23.4 MJ/L $vs.$ 34.4 MJ/L of gasoline, which negatively affects the gas mileage of vehicles), and serious compatibility issues with the gasoline infrastructure (only dilute blends are tolerated by current gasoline vehicles because of corrosiveness and water absorption issues of ethanol).

Apart from its use as a biofuel, ethanol possesses an enormous potential as a chemical feedstock. A number of valuable chemicals can be derived from ethanol by using continuous-flow catalytic technologies (Figure 1). Steam reforming is the most studied continuous catalytic process of ethanol which allows the gasification of aqueous solutions of ethanol into hydrogen-rich mixtures at high temperatures (typically 600–800 °C) and atmospheric pressure using Ni, Co and noble metals (e.g., Pt, Pd, and Rh) supported on stable oxides [6,7]. Hydrogen is co-produced along with CO_2 with a maximum yield of 6 moles of gas per mol of ethanol fed into the process, although this theoretical maximum is never reached since by-products including CH_4 (produced by hydrogenation of CO and CO_2 reforming products over metal catalysts) are typically produced at steam reforming conditions. The main challenge of these technologies lies in a careful control over the chemistry of the process, which is challenging at the high temperatures required to reform ethanol.

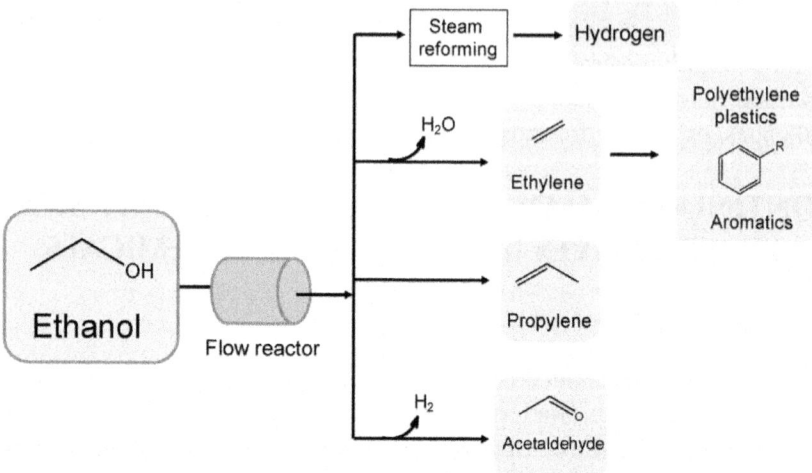

Figure 1. Production of fuels and high added value chemicals from ethanol by flow catalytic processing. Adapted from [2]. Reproduced by permission of the Royal Society of Chemistry.

A proper control of reaction temperature and space velocity conditions has been found to be crucial to minimize hydrogen-consuming reactions (e.g., methanation) which in turn leads to acceptable hydrogen yields. The catalysts formulation also plays an important role determining the outlet gas composition in steam reforming technologies. For example, metal-oxide supports able to activate water (ZnO, CeO_2, and La_2O_3) are typically employed as supports of metal catalysts to favor water-gas shift processes (WGS, CO + $H_2O \rightarrow CO_2 + H_2$) which lead to purified H_2 streams with very low levels of CO to be potentially relevant in fuel cell applications. These materials also possess basic characteristics avoiding coke-forming polymerization reactions (typically catalyzed on acidic sites) and improving the stability of the catalysts. Double-bed continuous approaches are utilized to enhance hydrogen yields in ethanol reforming processes. This proposed strategy allows the use of different catalytic beds operating at fixed optimum conditions which is advantageous to avoid undesirable processes and to increase stability. For example, dehydrogenation of acetaldehyde is favored with respect to ethylene dehydration employing a double-bed catalytic configuration with a first bed of a Cu-containing catalyst operating at low temperatures (e.g., 300–400 °C). The intermediate acetaldehyde species is subsequently passed through a second bed containing a reforming catalysts (e.g., Ni-based material) operating at higher temperatures [8]. In this way, subsequent polymerization of organic species to carbonaceous deposits (coke) is minimized. This concept enhances hydrogen yields while preserving the stability of the catalysts via minimization of coke-generating reactions.

There are a number of emerging processes for the continuous flow conversion of ethanol into valuable chemicals which in any case have been less investigated compared to reforming reactions (Figure 1). Bio-ethanol can be readily dehydrated into bio-ethylene with quantitative yields using acidic catalysts [9]. This process is of significant importance as ethylene is the most produced organic compound worldwide (annual production 120 million tons) and one of the seven primary building blocks of the petrochemical industry. With current growing oil prices, ethanol could potentially serve as a renewable and cheap feedstock for the production of polyethylene plastics, and commercial plants are already operative in Brazil with a capacity to produce 200,000 tons per year of green polyethylene from sugar cane ethanol [10].

Important aromatic compounds including benzene, toluene, and xylene (BTX, base of the petrochemical industry) could potentially be generated from as-produced bio-ethylene in the bioethanol dehydration process. However, the controlled oligomerization of ethylene is quite challenging and a good deal of research will be required to identify suitable catalysts (with optimum acid strength and porosity properties) and reaction conditions to yield BTX products in acceptable quantities. Alternatively, the generated bio-ethylene could be partially dimerized to butene, with the mixture subsequently undergoing methathesis to form propylene, the second most important starting product in the petrochemical industry after ethylene. Interestingly, the process has been recently patented [11]. The main challenge of this technology for the use of bioethanol lies in the extremely high purity ethylene required to perform the methathesis reaction which requires catalysts that are highly-sensitive to nitrogen and sulfur impurities (typically present in bioethanol feeds). This will require an extensive investigation on catalyst design and development under conditions and purities present in bioethanol feedstocks.

Apart from olefins, important C_2 commodity chemicals such as acetaldehyde can be also continuously derived from ethanol with 100% selectivity. The process involves the dehydrogenation of ethanol over inexpensive Cu catalysts at mild conditions (low temperatures and ambient pressure) and it represents an attractive route to co-produce a large-scale consumed chemical (acetaldehyde) as well as hydrogen in a single catalytic bed [12]. The process has not been commercially exploited so far, since stability issues associated with Cu sintering still have to be resolved.

Furans

Biomass-derived sugars, obtained from cellulose, hemicellulose or starches, can be dehydrated to form furan compounds including hydroxymethyl furfural (HMF) and furfural typically in aqueous mineral acids such as HCl or H_2SO_4 and/or solids acids which allow a facile separation and recycling of the catalysts during the dehydration process. The industrial production of furfural is a well established batch process that benefit from the easy depolymerization of hemicelluloses to C_5 sugars (mainly xylose). Furfural production reaches 200,000 tons per year being one of the most common industrial chemicals derived from biomass. Comparatively, HMF requires deconstruction of highly-recalcitrant and crystalline cellulose via an additional glucose isomerization step to fructose, which is more easily converted into the furanic intermediate.

Figure 2 summarizes the most relevant continuous flow catalytic processes for the conversion of furfural and HMF into fuels and chemicals. HMF can be converted into biofuel additives via hydrogenation of the carbonyl group and subsequent C−O hydrogenolysis of the hydroxyl intermediate dihydroxymethyl furan (DHMF) in a cascade-type continuous flow process reported by Dumesic and co-workers [13]. The process utilizes a Cu-based catalyst under hydrogen pressure to achieve removal of OH groups while preserving the C=C bonds in the furanic ring. The final product, dimethylfuran (DMF), possesses excellent energy-density and boiling point characteristics to be used as a transportation liquid fuel 100% compatible with current hydrocarbon-based infrastructure. Continuous flow conditions are finely adjusted to allow carbonyl hydrogenation and OH hydrogenolysis to take place in a single step. Interestingly, the utilization of continuous flow conditions avoids the potential poisoning of the Cu catalysts by the chlorides used to salt out HMF from the aqueous phase. Vapor phase hydrogenolysis can be carried out in a flow mode circumventing NaCl contact with the catalyst bed which would not be possible in the case of batch processing.

HMF derivatives such as 2,5-furandicarboxylic acid (FDCA) can also find markets in the production of biopolymers as replacement of petroleumderived terephtalic acid (Figure 2). FDCA can be synthesized from HMF by catalytic oxidation under pressurized air/oxygen at mild temperature conditions (100–150 °C) using metal catalysts. Even though the majority of studies used batchpressurized reactors to perform oxidation, this process has been recently scaled up to benchscale flow conditions by Gray and coworkers [14]. These authors found FDCA yields to be highly dependent on the employed oxidant (e.g., air, oxygen), the pH of the HMF aqueous solution used as feed, the nature of the metal support, and the pressure and temperature conditions in the reactor.

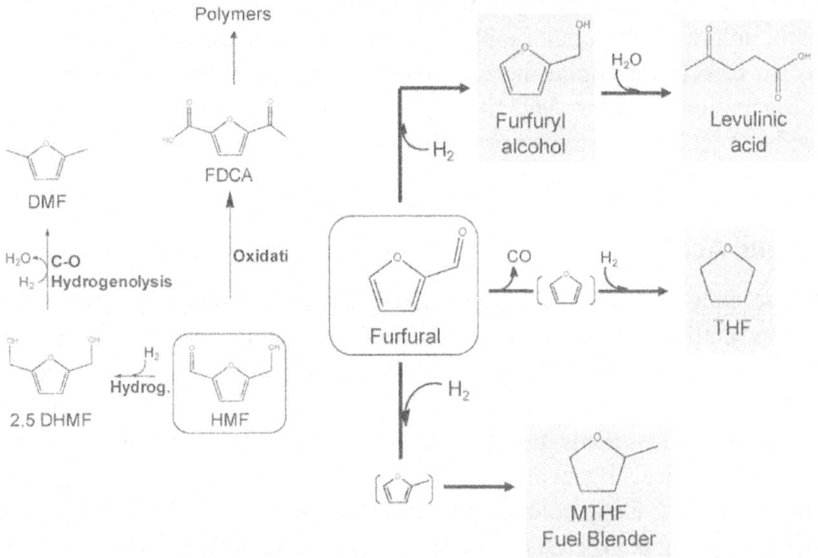

Figure 2. Hydroxymethylfurfural (HMF) and furfural as platform molecules for the production of high-added value chemicals and biofuels by flow catalytic processing. Adapted from [2]. Reproduced by permission of the Royal Society of Chemistry.

Similarly, furfural possesses a great platform potential for the simultaneous production of fuels and important chemicals in biorefineries (Figure 2). Hydrogenation of furfural is able to provide a range of valuable products by a number of technologies with the potential to be developed in a continuous operating mode. In particular, reduction at mild conditions (e.g., 120 °C, atmospheric pressure) using transition metal catalysts based on Cu yields furfuryl alcohol [15], a compound highly-demanded for the production of foundry resins. Furfuryl alcohol can additionally be hydrolyzed under acid conditions producing levulinic acid (4-oxopentanoic acid), an important biomass platform molecule that will be subsequently discussed. Decarbonylation was found to be prevalent over Pd-based catalysts (even at high space velocities) producing tetrahydrofuran (THF) after hydrogenation [16]. THF is a common organic solvent and precursor of polymers with an annual production that reaches 200,000 tons. This reaction was performed under microwave heating and formic acid as a hydrogen source [17], which is currently being translated into more scalable continuous flow processes by mimicking the moderate to high temperature achieved in the sealed microwave vessel using an X-Cube flow reactor [18].

More severe hydrogenation conditions (e.g., higher pressures, utilization of noble metal catalysts) favor a complete hydrogenation of the furanic ring to afford methyl tetrahydrofuran (MTHF). MTHF is a hydrophobic molecule which, unlike ethanol, can be blended with gasoline up to 60% (v/v) without adverse effects on engine performances or gas mileage. The most relevant hydrogenation routes to MTHF employing continuous flow reaction systems have been recently reviewed by Lange and coworkers [19], which explore the potential of furfural for the production of transportation biofuels.

Organic Acids

Organic acids are common biomass intermediates and some of the most relevant biomass platform molecules [20]. Carboxylic acids are present in certain quantities in biomass liquids (e.g., bio-oils) and many known biological routes are able to convert biomass sugars into carboxylic acids. Among the large number of biomass acids, this section will summarize the potential to produce fuels and chemicals by means of continuous flow catalytic technologies of two of them, namely, lactic acid (2hydroxypropanoic acid, LA) and levulinic acid (4-oxopentanoic acid, LVA). These two acids have been selected on the basis of their potential to be exploited as feedstocks under flow conditions.

Lactic acid is the most widely occurring carboxylic acid in nature. It is mainly produced by bacterial fermentation of biomass sugars in batch reactors. Traditional uses of lactic acid in the food industry (as additives) have been complemented by new recently developed applications in the fields of specialty chemicals and polymers for this platform molecule [21]. The improvement of existing fermentation and separation technologies, as well as the development of non-biological routes for the conversion of aqueous sugars into lactic acid, and the aforementioned novel applications of bio-lactic acid have considerably increased the platform potential of this molecule in recent years [22].

The special chemical composition of LA (containing two functional –OH and –COOH groups in a small molecule of only 3 carbon atoms) provides rich and versatile chemistry that allows a variety of flow catalytic transformations to useful products (Figure 3). LA is readily decarbonylated to acetaldehyde using metal catalysts under flow conditions, with the process being reported to accelerate in the presence of acidic catalysts [23,24]. This tendency of LA to acid-catalysed decomposition into acetaldehyde can be utilized to design novel routes to produce advanced biofuels (e.g., higher alcohols). LA readily dehydrates to produce acrylic acid (monomer employed to manufacture acrylic fibers) when strong dehydration conditions are applied [25]. The acidic strength of the catalyst as well as the residence time of LA in the fixed catalytic bed have been reported to be key aspects determining the acrylic/acetaldehyde molar ratio in flow experiments. This continuous flow route allows the production of propanoic acid when operating under hydrogen conditions and in the presence of bifunctional (metal and acid site) catalysts [26]. Propanoic acid is typically used as a preservative in foodstuffs [27].

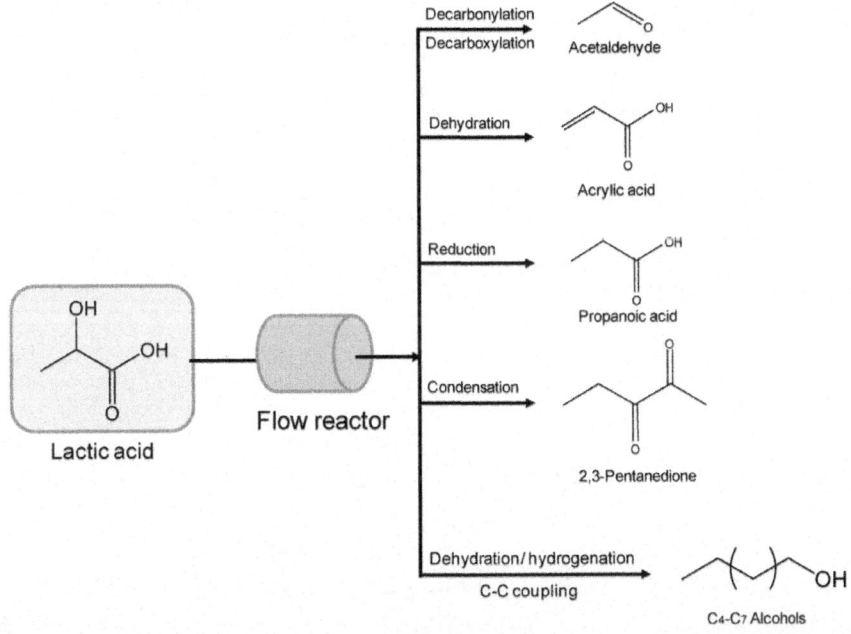

Figure 3. High-added value chemicals and fuels derived from various transformations of lactic acid. Adapted from [2]. Reproduced by permission of the Royal Society of Chemistry.

Gunter and coworkers detailed a novel route to upgrade LA which involves acid dehydration and subsequent condensation of intermediates on phosphate catalysts at moderate temperatures (e.g., 280–350 °C) [28]. The process produces 2,3 pentanedione which is a high valuable fine chemical currently obtained in limited amounts. These authors found that temperatures up to 320 °C and high residence times favored such condensation product as compared to shorter residence times and higher temperatures that lead to acrylic acid and acetaldehyde as main products. Since equimolar amounts of CO_2 and pentadione are generated in the process, a Claysen type condensation mechanism followed by internal molecular restructuring of the adduct is proposed to explain the formation of this interesting product (Scheme 2).

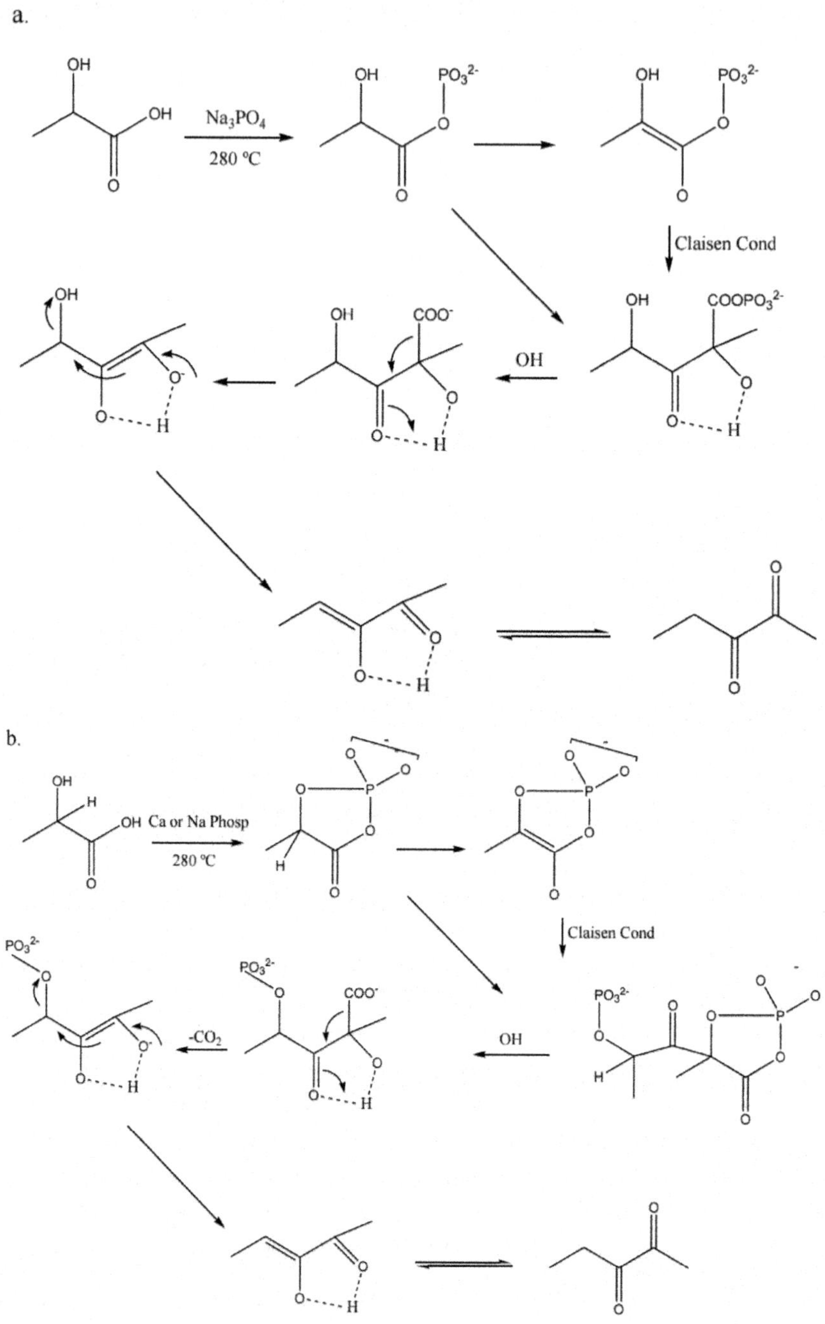

a.

b.

Scheme 2. Proposed mechanisms for 2,3-pentanedione formation from lactic acid using phosphate catalysts. Adapted from [28].

One of the main operating issues of lactic acid lies in its high and sometimes uncontrolled reactivity which typically lead to a mixture of products upon reaction. Recently, a new flow catalytic approach involving oxygen removal (to control the activity of lactic acid) followed by C−C coupling reactions (e.g., aldol-condensation and ketonization) of intermediates has been proposed to upgrade lactic acid to alcohols in the C_4–C_7 range suitable as gasoline-compatible biofuels (Figure 3) [29]. The process makes use of a bi-functional Pt/Nb_2O_5 catalytic bed (with the catalyst containing metal and acidic sites) and moderate temperature and pressures, typically 350 °C and 50 bar. The removal of oxygen from lactic acid is favored by the use of bi-functional catalysts through a cycle of dehydration (on the acid sites of Nb_2O_5) and hydrogenation (catalyzed by platinum), leading to the production of two monofunctional intermediates, namely acetaldehyde and propanoic acid. These monofunctional compounds, along with gaseous CO and CO_2, are the primary products observed when aqueous solutions of lactic acid were converted using Pt/Nb_2O_5. An increase in conversion (by increasing temperature or decreasing space velocity) leads to the onset of an organic layer that spontaneously separates from water. Analysis of the organic phase showed that it was mainly composed of propanoic acid and ketones in the C_4–C_7 range. Interestingly, this organic layer was not observed when a control catalyst supported on inert carbon black (Vulcan) was employed in the reaction. Approximately 50% of the carbon in the feed can be stored in this organic layer when optimum reaction conditions and concentrated solutions of lactic acid are used. The chemistry involved in the formation of the higher ketones is explained on the basis of the condensation of acetaldehyde and propanoic acid by aldolcondensation and ketonization [30] reactions, respectively, both taking place over the niobia support (Scheme 3). As in the case of ethanol steam reforming (Section 2.1), double bed catalytic flow approaches can be combined to improve yields to ketonization products. A second bed of ceriazirconia (following that of Pt/Nb_2O_5) operating at the same conditions of pressure and temperature could achieve a complete ketonization of the remaining unconverted propanoic acid in the first bed [26]. The organic layer obtained in this process can be separated into its components, serving as a source of valuable chemicals. Alternatively, the oil can be treated with Ru/C catalysts to convert the ketones into the corresponding alcohols without any needs of separation. This liquid mixture of alcohols in the C_4–C_7 range can be used as a high-energy density liquid fuel for the transportation sector. Concentrated aqueous solutions of lactic acid (e.g., 60 wt%) can be efficiently processed over Pt/Nb_2O_5 with good stability for more than two days on stream under continuous flow conditions.

As indicated in section 2.2, the treatment of biomass-derived hexoses at moderate temperatures under aqueous acid conditions leads to the production of

HMF. This compound is, however, highly reactive and readily decomposes into an equimolar mixture of formic and levulinic acids (FA, LVA). Technologies such as the Biofine process [31] make the most of this process for the industrial production of LVA from inexpensive lignocellulosic waste including paper mill sludge, urban waste and agricultural residues. LVA could be also potentially obtained from less-recalcitrant hemicellulose (but less abundant and with a more heterogeneous composition) via dehydration of pentoses to furfural and subsequent hydrogenation to furfuryl alcohol (Figure 2).

A number of recent flow catalytic technologies have been developed to upgrade LVA into liquid transportation fuels (Figure 4). The key intermediate in many of these technologies is γ valerolactone (GVL) which is produced via hydrogenation of LVA over non-acidic metal catalysts such as Ru/C achieving near quantitative yields of GVL at mild temperatures (e.g., 150 °C) [32,33]. Interestingly, the formic acid co-produced in the sugar dehydration process can serve as hydrogen source for LVA reduction to GVL, thereby avoiding utilization of external fossil-fuel derived hydrogen [34]. Several flow catalytic routes have been recently developed to upgrade GVL into advanced biofuels, that is, biofuels with high energy densities and 100% compatible with the current transportation infrastructure. For example, GVL can be readily converted to MTHF via hydrogenation with intermediate formation of 1,4-pentanediol (Figure 4) [35]. MTHF is a highly interesting fuel which can be implemented in gasoline engines without the need of drastic changes in vehicles.

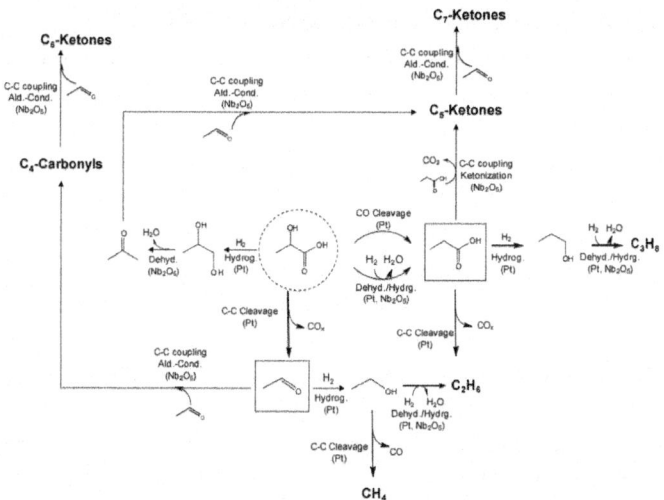

Scheme 3. Proposed reaction scheme for 2hydroxypropanoic acid (LA) catalyzed transformation into C5–C7 ketones using Pt/Nb_2O_5 as catalyst. Copyright Wiley–VCH Verlag GmbH & Co.KGaA. Reproduced with permission from [29].

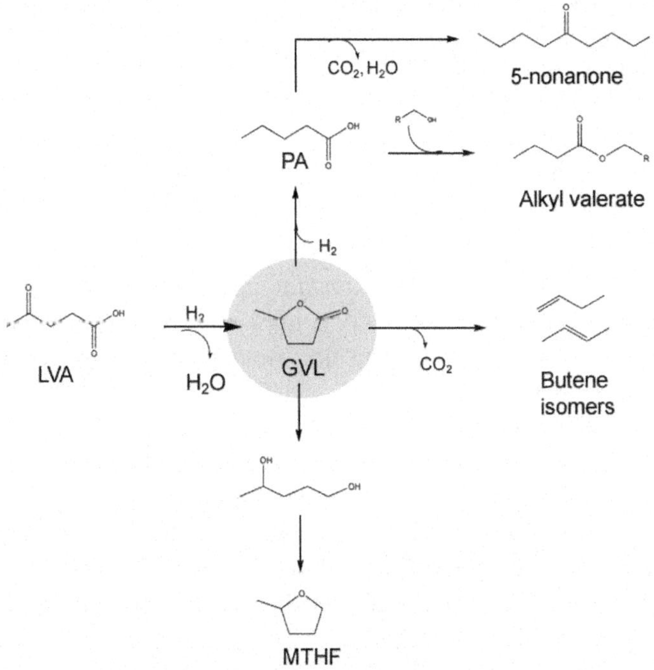

Figure 4. Levulinic acid and γ valerolactone (GVL) as platform molecules for the production of liquid transportation biofuels. Adapted from reference 2. Reproduced by permission of the Royal Society of Chemistry.

An alternative route that is currently gaining momentum involves the conversion of GVL into pentanoic acid (PA) by means of ring-opening (on acid sites) and hydrogenation (on metal sites) reactions employing bifunctional heterogeneous catalysts at moderate temperatures and pressures. Lange and co-workers have recently exploited this route to continuously produce the so-called *valeric biofuels* (*i.e.*, alkyl valerates), a novel family of lignocellulosic biofuels with excellent energy density, polarity and volatility ignition properties to be used in conventional engines without any modifications [36]. Excellent yields of PA (92%) can be achieved from aqueous solutions (e.g., 50 wt%) of LVA over Pd/Nb_2O_5 operating in flow mode [37]. Remarkably, PA was ketonized to 5-nonanone over the niobic support with yields up to 60% when the space velocity in the flow reactor was reduced to the level of WHSV = 0.1 h^{-1}, This yield is however limited by the large number of processes involved which prevent a proper control of intermediate steps (LVA to PA to 5-nonanone). As in the case of lactic acid, one solution to overcome this limitation involves the utilization of double-bed flow reactors in which conditions of each bed are independently adjusted to maximize yields of intermediate steps.

For example, a double-bed configuration of Pd/Nb$_2$O$_5$ (at 275 °C) + Ce$_{0.5}$Zr$_{0.5}$O$_2$ (at 425 °C) allows aqueous solutions of GVL to be transformed into 5-nonanone with yields of 90% [38]. In this configuration, the first bed achieves conversion of PA to GVL which is subsequently converted into 5nonanone in the second bed in excellent final yields. Such C$_9$ ketone can serve as a platform molecule for the production of hydrocarbon fuels with molecular weight and structures adequate for gasoline, diesel and jet fuels applications [39].

Similar dual continuous flow reactor approaches have been designed to upgrade aqueous solutions of GVL into jet fuels through the formation of C$_4$ alkenes (Figure 4) [40]. The first flow reactor is loaded with acidic silica-alumina which achieves decarboxylation of GVL to butenes which are subsequently oligomerized to a distribution of alkenes centered at C$_{12}$ using an acidic Amberlyst catalyst. In this case, the double bed configuration is not feasible since water has to be removed before entering the second reactor to achieve an effective oligomerization of butenes. The flow process is simple and achieves GVL deoxygenation and subsequent C−C coupling in a clean (CO$_2$ can be efficiently sequestrated at the system pressure) and cheap (no external hydrogen required) fashion.

Polyols and Sugars

Glycerol (1,2,3-propanetriol) is perhaps one of the most relevant biomass derivatives whose importance as a renewable platform molecule has dramatically increased in recent years [41]. It is produced in large amounts in biodiesel facilities (100 kg of glycerol per ton of biodiesel) as byproduct of the transesterification of vegetable oils and animal fats with methanol for biodiesel production. Consequently, novel technologies to upgrade this molecule into fuels and chemicals are highly desirable.

Glycerol possesses a rich chemistry derived from the three hydroxyl groups. With regard to flow catalytic routes, the most relevant processes are depicted in Figure 5. As in the case of ethanol, glycerol can serve as a good source of hydrogen by means of steam reforming processes over metal catalysts [42]. Aqueous glycerol solutions can be gasified to CO, H$_2$ and CO$_2$ at mild temperatures (200−280 °C) and with higher hydrogen yields as compared to ethanol reforming. Additionally, as indicated in Figure 5, the process is more versatile and reaction conditions and catalytic materials can be carefully selected to transform concentrated aqueous solutions of glycerol into either syngas (a valuable mixture of CO and H$_2$) for Fischer-Tropsch (F-T) or methanol syntheses [43]. Alternatively, H$_2$-enriched streams can also be obtained by coupling APR with water gas-shift (WGS) processes [44] using double-bed flow reactors.

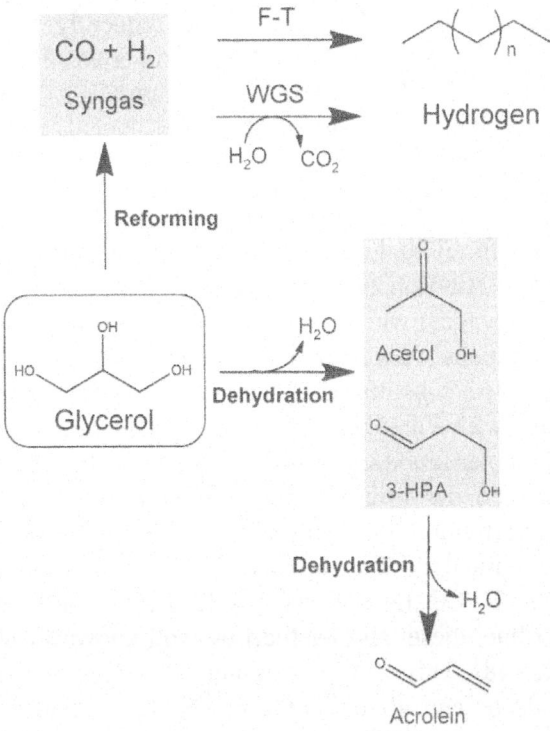

Figure 5. Main routes for the flow catalytic transformation of glycerol into fuels and chemicals. Adapted from [2]. Reproduced by permission of the Royal Society of Chemistry.

Glycerol can also be dehydrated to acrolein using strong solid acids (ZrO_2-WO_3) as catalysts [45]. Acrolein is an important intermediate in the production of acrylic acid which is polymerized to acrylic resins used as superabsorber polymers in diapers and related products. Proper control of temperature and space velocity conditions in the flow reactor allows the production of acrolein in acceptable yields, although unwanted reactions including polymerization and cracking (leading to catalyst deactivation) are difficult to avoid at the high-acidic conditions.

A two-step flow catalytic process has been recently designed to convert biomass sugars and polyols in form of aqueous solutions into liquid hydrocarbon fuels of diverse classes [46]. In a first step, sugars and polyols are partially deoxygenated over a Pt-Re/C catalyst at temperatures near 225 °C to yield a mixture of monofunctional hydrocarbons in the C_4–C_6 range (including acids, alcohols, ketones and heterocycles) which are stored in an organic phase that spontaneously separates from water. The final composition of this organic stream is determined by pressure, temperature and WHSV conditions employed in the flow reactor. Subsequently, the organic stream of monofunctional compounds could be converted into targeted liquid hydrocarbon fuels of different classes through C–C coupling reactions including aldol-condensations and ketonizations.

Synthesis Gas

As in the case of fossil coal and natural gas, biomass can be gasified under controlled oxidant atmosphere to produce a mixture of CO and H_2 denoted as synthesis gas (syngas) which is precursor of fuels and chemicals by well-established continuous flow technologies. The utilization of biomass for syngas generation (in substitution of classical feeds such as coal and natural gas) is associated with a number of processing challenges namely, the control over the H_2/CO molar ratio (since biomass is highly oxygenated as compared to fossil fuels), and the strict cleaning standards required for the biomass derived syngas (extremely high purity is required to downstream processing to fuels and chemicals) [47]. Biomass-derived syngas, once cleaned and with appropriate H_2/CO ratio, serves as feedstock for liquid hydrocarbon fuels (e.g., gasoline, diesel and jet fuel) by well-known Fischer-Tropsch (F-T) technologies [48]. This route is commonly denoted as biomass to liquids (BTL) and could be considered as the renewable version of coal to liquids (CTL) and gas to liquids (GTL) related processes. Commercialization of BTL technologies is being hindered by the elevated cost of the process, although some commercial technologies such as Primus Green Energy [49] and Sundrop fuels™ [50] are currently produced fuels on a large scale. Syngas can be also converted into methanol by methanol synthesis approaches [51], and this methanol subsequently transformed into gasoline by the Mobil´s methanol to gasoline process [52] or improved TIGAS technologies from Topsoe [53]. These alternative processes can help to improve the economics of the BTL route which is, by far, the main limitation of the process.

CONCLUSIONS AND FUTURE PROSPECTS

Continuous flow chemical processes for biomass valorization to fuels and chemicals hold significant potential for future development in our aim to drive our chemistries to more efficient and scalable approaches, while being environmentally sound and sustainable at the same time. The highlighted examples demonstrate the potential of a range of transformations of biomass-derived platform molecules under continuous flow conditions and heterogeneous catalysis. Several high added value chemicals and biofuel precursors can be obtained using different continuous flow chemical methodologies which possess already established markets and developed applications to replace fossil-derived commodities. Many of these and related routes to convert platform molecules into valuable end products under continuous flow conditions offer a significant industrial potential, with some already being developed or under development, taking advantage of the important benefits of continuous flow processing. We envisage a series of topics to be part of future key investigations in the implementation of continuous flow chemical processing of biomass feedstocks:

- • - Design of novel flow processes for an efficient and effective biomass conversion
- • - Design of water-tolerant and stable catalysts able to perform aqueous chemistries in high yields to products controlling the selectivity and reactivity of biomass-derived intermediates
- • - Development of low environmental impact technologies based on multi-step reactors, cheap and readily available transition metal (bifunctional) catalysts, mild reaction conditions, *etc.*

Regardless of the industrial potential benefits of the implementation of continuous flow processes in biomass valorization practices, the environmental advantages of these methodologies in the processing of platform chemicals has to be taken into account. In this regard, continuous flow processing is the future of biomass valorization practices and we hope this manuscript can serve as momentum to both academia and industry to join efforts in order to carry on designing flow processes for biomass processing envisaging their implementation at industrial scale.

ACKNOWLEDGMENTS

RL gratefully acknowledges Ministerio de Ciencia e Innovación, Gobierno de España for the concession of a Ramon y Cajal contract (ref. RYC-2009-04199) and funding under projects CTQ–2010-18126 and CTQ2011-28954-C02-02 (MICINN) as well as Consejeria de Ciencia e Innovación, Junta de Andalucía for funding under project P10-FQM-6711.

REFERENCES

1.   Ragauskas, A.J.; Williams, C.K.; Davison, B.H.; Britovsek, G.; Cairney, J.; Eckert, C.A.; Frederick, W.J.; Hallett, J.P.; Leak, D.J.; Liotta, C.L.; *et al.* The path forward for biofuels and biomaterials. *Science* **2006**, *311*, 484–489.

2. Serrano-Ruiz, J.C.; Luque, R.; Sepúlveda-Escribano, A. Transformations of biomass-derived platform molecules: From high added-value chemicals to fuels via aqueous-phase processing. *Chem. Soc. Rev.* **2011**, *40*, 5266–5281.

3. Bozell, J.J.; Petersen, G.R. Technology development for the production of biobased products from biorefinery carbohydrates—The US Department of Energy's "Top 10" revisited. *Green Chem.* **2010**, *12*, 539–554.

4. Glasnov, T.N.; Kappe, C.O. The microwave-to-flow paradigm: Translating high-temperature batch microwave chemistry to scalable continuous-flow processes. *Chem. Eur. J.* **2011**, *17*, 11956–11968.

5. Ethanol producer magazine, International Ethanol Report 2010. Available online: http://www.ethanolproducer.com/articles/6696/international-ethanol-report-2010 (accessed on 10 May 2012).

6. Haryanto, A.; Fernando, S.; Murali, N.; Adhikari, S. Current status of hydrogen production techniques by steam reforming of ethanol: A review. *Energy Fuels* **2005**, *19*, 2098–2106.

7. Ni, M.; Leung, D.Y.C.; Leung, M.K.H. A review on reforming bio-ethanol for hydrogen production. *Int. J. Hydrog. Energy* **2007**, *32*, 3238–3247.

8. Batista, M.S.; Assaf, E.M.; Assaf, J.M.; Ticianelli, E.A. Double bed reactor for the simultaneous steam reforming of ethanol and water gas shift reactions. *Int. J. Hydrog. Energy* **2006**, *31*, 1204–1209.

9. Bedia, J.; Barrionuevo, R.; Rodríguez-Mirasol, J.; Cordero, T. Ethanol dehydration to ethylene on acid carbon catalysts.*Appl. Catal. B* **2001**, *103*, 302–310.

10. Ethanol producer magazine, Braskem starts up ethanol-to-ethylene plant. Available online: http://www.ethanolproducer.com/articles/7022/braskem-starts-up-ethanol-to-ethylene-plant (accessed on 10 May 2012).

11. Takai, T.; Mochizuki, D.; Umeno, M. Method of producing propylene containing biomass-origin carbon. European Patent Application EP1953129 (A1), 2008.

12. Tu, Y.; Chen, Y. Effects of alkaline-earth oxide additives on silica-supported copper catalysts in ethanol dehydrogenation. *Ind. Eng. Chem. Res.* **1998**, *37*, 2618–2622.

13. Roman-Leshkov, Y.; Barrett, C.J.; Liu, Z.Y.; Dumesic, J.A. Production of dimethylfuran for liquid fuels from biomass-derived carbohydrates. *Nature* **2007**, *447*, 982–986.

14. Lilga, M.A.; Richard, T.H.; Gray, M. Production of Oxidized Derivatives of 5-Hydroxymethylfurfural (HMF). *Top. Catal.* **2010**, *53*, 1264–1269.

15. Surapas, S.; Pham, T.; Prasomsri, T.; Sooknoi, T.; Mallinson, R.G.; Resasco, D.E. Conversion of furfural and 2-methylpentanal on Pd/SiO$_2$ and Pd–Cu/SiO$_2$ catalysts. *J. Catal.* **2011**, *280*, 17–27.

16. Sitthisa, S.; Resasco, D.E. Hydrodeoxygenation of furfural over supported metal catalysts: A comparative study of Cu, Pd and Ni. *Catal. Lett.* **2011**, *141*, 784–791.

17. García-Suárez, E.J.; Balu, A.M.; Tristany, M.; García, A.B.; Philippot, K.; Luque, R. Versatile dual hydrogenation–oxidation nanocatalysts for the aqueous transformation of biomass-derived platform molecules. *Green Chem.* **2012**, *14*, 1434–1439.

18. X-Cube™—Catalysis Made Simple. ThalesNano Nanotechnology Inc: Graphisoft Park, Hungary, 2009. Available online: http://www.thalesnano.com/products/x-cube (accessed on 10 May 2012).

19. Lange, J.P.; van der Heide, E.; van Buijtenen, J.; Price, R. Furfural-A promising platform for lignocellulosic biofuels.*ChemSusChem* **2012**, *5*, 150–166.

20. Werpy, T.; Petersen, G. *Top Value Added Chemicals from Biomass. Volume 1—Results of Screening for Potential Candidates from Sugars and Synthesis Gas*; U.S. Department of Energy: Washington, DC, USA, 2004; pp. 1–67.

21. Fan, Y.; Zhou, C.; Zhu, X. Selective catalysis of lactic acid to produce commodity chemicals. *Catal. Rev.* **2009**, *51*, 293–324.

22. Holm, M.S.; Saravanamurugan, S.; Taarning, E. Conversion of sugars to lactic acid derivatives using heterogeneous zeotype catalysts. *Science* **2010**, *328*, 602–605.

23. Mok, W.S.; Antal, M.J.; Jones, M. Formation of acrylic acid from lactic acid in supercritical water. *J. Org. Chem.* **1989**,*54*, 4596–4602.

24. Gunter, G.C.; Langford, R.H.; Jackson, J.E.; Miller, D.J. Catalysts and supports for conversion of lactic acid to acrylic acid and 2,3-pentanedione. *Ind. Eng. Chem. Res.* **1995**, *34*, 974–980.

25. Sawicki, R.A. Catalyst for Dehydration of Lactic Acid to Acrylic Acid. U.S. Patent 4729978, 1998.

26. Serrano-Ruiz, J.C.; Dumesic, J.A. Catalytic upgrading of lactic acid to fuels and chemicals by dehydration/hydrogenation and C–C coupling reactions. *Green Chem.* **2009**, *11*, 1101–1104.

27. Samuel, U.R.; Kohler, W.; Gamer, A.O.; Keuser, U. Propionic acid and derivatives. In *Ullmann's Encyclopedia of Industrial Chemistry*; VCH Publishers Inc: Weinheim, Germany, 1993.

28. Gunter, G.C.; Miller, D.J.; Jackson, J.E. Formation of 2,3-pentanedione from lactic acid over supported phosphate catalysts. *J. Catal.* **1994**, *148*, 252–260.

29. Serrano-Ruiz, J.C.; Dumesic, J.A. Catalytic processing of lactic acid over Pt/Nb$_2$O$_5$. *ChemSusChem* **2009**, *2*, 581–586.

30. Renz, M. Ketonization of carboxylic acids by decarboxylation: Mechanism and scope. *Eur. J. Org. Chem.* **2005**, *2005*(6), 979–988.

31. Fitzpatrick, S.W. Production of levulinic acid from carbohydrate-containing materials. U.S. Patent 5608105, 1997.

32. Yan, Z.; Lin, L.; Liu, S. Synthesis of γ-Valerolactone by Hydrogenation of Biomass-derived Levulinic Acid over Ru/C Catalyst. *Energy Fuels* **2009**, *23*, 3853–3858.

33. Upare, P.P.; Lee, J.M.; Hwang, D.W.; Halligudi, S.B.; Hwang, Y.K.; Chang, J.S. Selective hydrogenation of levulinic acid to γ-valerolactone over carbon-supported noble metal catalysts. *J. Ind. Eng. Chem.* **2011**, *17*, 287–292.

34. Deng, L.; Li, J.; Lai, D.M.; Fu, Y.; Guo, Q.X. Catalytic conversion of biomass-derived carbohydrates into γ-valerolactone without using an external H$_2$ supply. *Angew. Chem. Int. Ed.* **2009**, *48*, 6529–6532.

35. Bozell, J.J.; Moens, L.; Elliott, D.C.; Wang, Y.; Neuenscwander, G.G.; Fritzpatrick, S.W.; Bilski, R.J.; Jarnefeld, J.L. Production of levulinic acid and use as a platform chemical for derived products. *Res. Conserv. Recycl.* **2000**, *28*, 227–239.

36. Lange, J.P.; Price, R.; Ayoub, P.M.; Louis, J.; Petrus, L.; Clarke, L.; Gosselink, H. Valeric biofuels: A platform of cellulosic transportation fuels. *Angew. Chem. Int. Ed.* **2010**, *49*, 4479–4483.

37. Serrano-Ruiz, J.C.; Wang, D.; Dumesic, J.A. Catalytic upgrading of levulinic acid to 5-nonanone. *Green Chem.* **2010**, *12*, 574–577.

38. Serrano-Ruiz, J.C.; Braden, D.J.; West, R.M.; Dumesic, J.A. Conversion of cellulose to hydrocarbon fuels by progressive removal of oxygen. *Appl. Catal. B* **2010**, *100*, 184–189.

39. Martin-Alonso, D.; Bond, J.Q.; Dumesic, J.A. Catalytic conversion of biomass to biofuels. *Green Chem.* **2010**, *12*, 1493–1513.

40. Bond, J.Q.; Martin-Alonso, D.; Wang, D.; West, R.M.; Dumesic, J.A. Integrated catalytic conversion of γ-valerolactone to liquid alkenes for transportation fuels. *Science* **2010**, *327*, 1110–1114.

41. Pagliaro, M.; Rossi, M. *Future of Glycerol, New Usages for a Versatile Raw Material*; RSC publishing: Cambridge, UK, 2008.

42. Soares, R.R.; Simonetti, D.A.; Dumesic, J.A. Glycerol as a source for fuels and chemicals by low-temperature catalytic processing. *Angew. Chem. Int. Ed.* **2006**, *45*, 3982–3985.

43. Simonetti, D.A.; Rass-Hansen, J.; Kunkes, E.L.; Soares, R.R.; Dumesic, J.A. Coupling of glycerol processing with Fischer-Tropsch synthesis for production of liquid fuels. *Green Chem.* **2007**, *9*, 1073–1083.

44. Kunkes, E.L.; Soares, R.R.; Simonetti, D.A.; Dumesic, J.A. An integrated catalytic approach for the production of hydrogen by glycerol reforming coupled with water-gas shift. *Appl. Catal. Environ.* **2009**, *90*, 693–698.

45. Zhou, C.H.C.; Beltramini, J.N.; Fan, Y.X.; Lu, G.Q.M. Chemoselective catalytic conversion of glycerol as a biorenewable source to valuable commodity chemicals. *Chem. Soc. Rev.* **2007**, *37*, 527–549.

46. Kunkes, E.L.; Simonetti, D.A.; West, R.M.; Serrano-Ruiz, J.C.; Gartner, C.A.; Dumesic, J.A. Catalytic conversion of biomass to monofunctional hydrocarbons and targeted liquid-fuel classes. *Science* **2008**, *322*, 417–421.

47. Serrano-Ruiz, J.C.; Dumesic, J.A. Catalytic routes for the conversion of biomass into liquid hydrocarbon transportation fuels. *Energy Environ. Sci.* **2011**, *4*, 83–99.

48. Dry, M.E. The Fischer-Tropsch process: 1950–2000. *Catal. Today* **2002**, *71*, 227–241.

49. *Biomass to Fuel Conversion*; Primus Green Energy Ltd.: Hillsborough, NJ, USA, 2012. Available online: http://www.primusge.com/how-it-works/biomass-to-fuel-conversion/ (accessed on 15 June 2012).

50. Sundrop Fuels, Inc. Available online: http://www.sundropfuels.com/ (accessed on 15 June 2012).

51. Lange, J.P. Methanol synthesis: A short review of technology improvements. *Catal. Today* **2001**, *64*, 3–8.

52. Exxon Mobil, Research and Engineering, Methanol to gasoline (MTG). Available online: http://www.exxonmobil.com/Apps/RefiningTechnologies/files/sellsheet_09_mtg_brochure.pdf (accessed on 15 June 2012).

53. Haldor Topsøe. Available online: http://www.topsoe.com/business_areas/gasification_based/Processes/Gasoline_TIGAS.aspx (accessed on 15 June 2012).

Chapter 14

MICROWAVE ASSISTED CHEMICAL PRETREATMENT OF MISCANTHUS UNDER DIFFERENT TEMPERATURE REGIMES

Zongyuan Zhu[1], Duncan J. Macquarrie[1], Rachael Simister[2], Leonardo D. Gomez[2] and Simon J. McQueenMason[2]

[1] Department of Chemistry, Green Chemistry Centre of Excellence, University of York, Heslington, York YO10 5DD, UK

[2] Department of Biology, Centre for Novel Agricultural Products, University of York, Heslington, York YO10 5DD, UK.

ABSTRACT

Background

Miscanthus is a major bioenergy crop in Europe and a potential feedstock for second generation biofuels. The most efficient and realistic method to produce fermentable sugars from lignocellulosic biomass is by enzymatic hydrolysis, assisted by thermo-chemical pretreatment. Recently, microwave technology has drawn growing attention, because of its unique effects and performance on biomass.

Result

In this work, microwave energy was applied to facilitate NaOH and H_2SO_4 pretreatment for *Miscanthus* under different temperatures (130–200 °C) for 20 min. The yields of reducing sugars from *Miscanthus* during the pretreatment process increased up to 180 °C and then declined with increasing temperature. Out results here showed a remarkable sugar yield from available carbohydrate (73 %) at the temperature of 180 °C by using 0.2 M H_2SO_4. In comparison with conventional heating pretreatment studied at same temperature with same biomass material, the reducing sugar release in this study was 17 times higher within half the time. It was highlighted that the major sugar component could be tuned by changing pretreatment temperature or pretreatment media.

Optimally, the glucose and xylose yield from available carbohydrate are 47 and 22 % by using 0.2 M H_2SO_4 and NaOH respectively when temperature was 180 °C. The digestibility of pretreated*Miscanthus* was 10 times higher than that of untreated biomass. 68–86 % of the lignin content was removed from biomass by 0.2 M NaOH. Simultaneous saccharification fermentation (SSF) results showed an ethanol production of 143–152 mg/g biomass by using H_2SO_4/NaOH microwave assisted pretreatment, which is 7 times higher than that of untreated *Miscanthus*. Biomass morphology was studied by SEM, showing temperature has a strong influence on lignin removal process, as different lignin deposits were observed. At the temperature of 180 °C, NaOH pretreated biomass presented highly exposed fibres, which is a very important biomass characteristic for improved enzymatic hydrolysis.

Conclusion

Compared to conventional pretreatment, microwave assisted pretreatment is more energy efficient and faster, due to its unique heating mechanism leading to direct interaction between the polar part of biomass and electromagnetic field. The results of this work present promising potential for using microwave to assist biomass thermo-chemical pretreatment.

BACKGROUND

Nowadays, there is a global rise in energy demand and rising concerns about increasing greenhouse gas emissions, hence biofuels derived from lignocellulosic biomass based on the biorefinery philosophy is drawing growing attention [1, 2]. Second generation bioethanol is produced from lignocellulosic biomass following three main processing steps: pretreatment, hydrolysis, and fermentation [3]. Pretreatment is crucial in the conversion of biomass into biofuel via biochemical hydrolysis, so that the recalcitrant structure of biomass can be accessed and sugars released for fermentation [3].

A number of pretreatments have been studied to improve the yields of fermentable sugars from cellulose and hemicellulose, such as mechanical [4, 5], steam explosion [6, 7], ammonia fibre explosion [5, 8], hot water [9], sub/supercritical fluid [10–12], ozone [5], biological [5], ultrasound [13], acid or alkaline pretreatments [14–16], ionic liquid [17, 18] and so forth. It is worth mentioning that novel pretreatment media, such as sub/supercritical fluids and ionic liquid are also drawing attention due to their unique solvent properties. For instance, sub/supercritical pretreatment offers several advantages, such as particle size reduction and low toxicity. Additionally, cellulose accessibility would be improved, because sharply reduced pressure leads to explosive decompression of feedstock [12], Nevertheless, this technology requires equipment capable of withstanding high temperature and pressure [10–12]. Alternatively, ionic liquids have high thermal stability and high solvent power, and can be easily recycled. While ionic liquid toxicity and biodegradability have been controversial issues, [17, 18, 20] recent advances in ionic liquid design have improved this situation. It is also suggested that some of ionic liquid could be as cheap as conventional organic solvent. [19] Therefore, aqueous acid and alkali are more extensively used as pretreatment media during biomass pretreatment [21–30]. In comparison with HCl, HNO_3 and H_3PO_4, H_2SO_4 is cheaper, less corrosive, non-oxidative and stronger, hence it was used in this study. Compared with ionic liquid and sub/supercritical fluids pretreatments, microwave assisted pretreatment offers great advantages because of its unique heating mechanism. In microwave, energy transmission is contributed by dielectric losses, and the magnitude of heating depends on the dielectric properties of the subject [31]. It is more direct, uniform and much faster, due to the direct interaction between the object to be heated and an applied electromagnetic field [31, 32]. When microwave is applied to lignocelluloses, it selectively heats the more polar parts throughout the material (as opposed to conventional heat sources which heat from the outside towards the inside), and creates a 'hot spot' within heterogeneous materials [33]. Hence, it is hypothesized that an 'explosion' effect could occur in the particles, improving the disruption of the recalcitrant structures of lignocellulose. Additionally, it has been claimed that the electromagnetic field used in the microwave might create non-thermal effects that also accelerate the destruction of the crystal structures [31].

Microwaves have been used in the acid or alkali pretreatment of sugar cane bagasse [24], oilseed rape straw [34], switchgrass [28], crystalline cellulose [35], and wheat straw [30]. Ma et al. reported that by using microwave pretreatment of rice straw, the maximal efficiencies of the cellulose, hemicellulose and total saccharification of pretreated biomass were increased by 30.6, 43.3 and 30.3 % respectively. Additionally, microwave pretreatment disrupted the silicified waxy surface on rice straw, broke down the lignin-hemicellulose complex and partially removed silica and lignin [36]. Lu et al. reported that the glucose yield of pretreated rape straw from enzymatic hydrolysis was greatly enhanced (56.2 %) after microwave pretreatment, (11.5 % for untreated rape straw) [34]. These works show that microwave thermo-chemical processes are effective and promising pretreatment methods. Their results focused on pretreated biomass solid fraction, while little data has been reported concerning sugar removal during the pretreatment process. In current work, we monitored the effects of microwave assisted pretreatment in the presence of acid and alkali on *Miscanthus*, at different temperatures (between 130 and 200 °C). In this work, the results were focused on the sugar removal efficiency during pretreatment, rather than on the residual biomass, Moreover, to our knowledge, potential ethanol production from microwave pretreated *Miscanthus* has not been reported yet. [37] In addition, we used the SSF (simultaneous saccharification fermentation) process to investigate the potential ethanol production from microwave pretreated *Miscanthus* solid fraction.Hence, the overall sugar yield in the pretreatment liquid fraction and potential ethanol production from biomass solid fraction are studied here. Biomass morphological characteristics were studied by scanning electron microscopy.

RESULTS AND DISCUSSION

Microwave pretreatment of lignocellulosic material enhances its hydrolysis, and temperature plays a significant role during the pretreatment [38]. A higher temperature typically achieves higher biomass solubility, shortens the pretreatment time, and reduces the biomass recalcitrance more effectively [39]. However, high temperature also leads to the formation of compounds that are harmful to subsequent hydrolysis and fermentation [38]. Hence, different temperatures ranging from 130 to 200 °C were assayed in this work, in order to investigate temperature influence on biomass under microwave irradiation.

Monosaccharides Analysis in the Pretreatment Media

Figure 1 show the total reducing sugars released from *Miscanthus* during pretreatment by using water, NaOH and H_2SO_4 as pretreatment media at various temperatures (130–200 °C). Sugar production increases up to 180 °C, and then decreases at higher temperatures in water, 0.2 M NaOH or 0.2 M H_2SO_4. Compared to water and NaOH, H_2SO_4 gives better sugar yield during pretreatment when temperature is between 130 and 180 °C. The sugar yields in this study are based on the total carbohydrate in biomass. The maximum sugar yield (3 µmol/mg biomass; yield: 73 %) was achieved by using 0.2 M H_2SO_4 pretreatment at 180 °C. The reducing sugar yield of *Miscanthus* by using water and NaOH pretreatment is also remarkably high at the same temperature, 1.3 and 1.76 µmol/mg biomass respectively. The temperature of 180 °C has been recognised as a key temperature in the microwave degradation of cellulose [40]. According to Fan et al., when the temperature is below 180 °C the CH_2OH groups on cellulose are hindered from interacting with microwaves when they are strongly involved in hydrogen bonding within both the amorphous and crystalline regions. When temperatures are above 180 °C, these CH_2OH groups could be involved in a localized rotation under the microwave radiation [41]. Therefore, the rate of cellulose decomposition increases and maximum sugar yield is achieved here at 180 °C. Further increasing pretreatment temperature (200 °C) leads to a significant drop in sugar yield, possibly due to the degradation of sugars under high temperature.

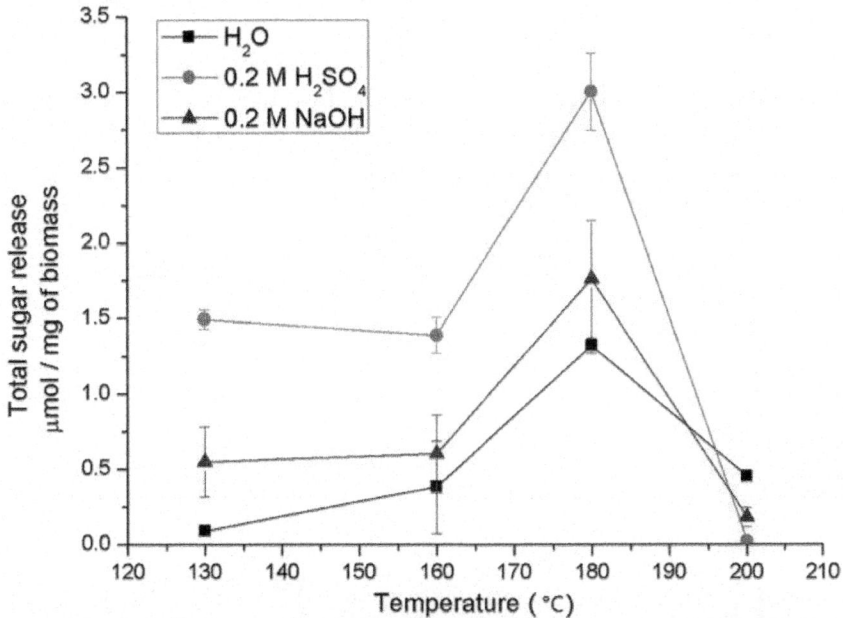

Figure. 1: Total sugar release at different temperatures (130, 160, 180 and 200 °C).

Figure 2 shows the monosaccharide composition in the pretreatment media when temperature ranges from 130 to 200 °C. The monosaccharide composition of the liquor after acid and alkaline microwave pretreatment suggests a breakdown of hemicelluloses from*Miscanthus*, where xylans are the major component. Water, alkali and acid pretreatment have been reported to extract most of the hemicellulose, mainly glucuronoarabinoxylan or 1-arabino-d-xylans [42]. By using a 0.2 M H_2SO_4 pretreatment, we observed a xylose production of 1.01 μmol/mg biomass at 130 °C (Fig. 2a). When the temperature was 160 °C (Fig. 2b), xylose productions were 0.25 μmol/mg biomass and 0.38 μmol/mg biomass with water and NaOH respectively. However, good glucose yield (20 %) was achieved by 0.2 M H_2SO_4, suggesting that cellulose start to be hydrolysed into glucose. The xylose yield decreased significantly under H_2SO_4 condition when temperature increased from 130 to 160 °C, this probably due to its degradation at high temperature and acid conditions into furfural, formaldehyde, formic acid, crotonaldehyde, lactic acid, acetaldehyde, and dihydroxyacetone [43]. Similar results were observed at 180 °C (Fig. 2c). Xylose production increased to 1.01 μmol/mg biomass and 1.05 μmol/mg biomass respectively when water and NaOH are used as pretreatment media, suggesting that higher temperatures can facilitate hemicellulose breakdown. H_2SO_4 gave the maximum glucose production when temperature is 180 °C, which is 1.83 μmol/mg biomass (yield from carbohydrate: 47 %), suggesting the efficient decomposition of cellulose under microwave condition. At 200 °C (Fig. 2d), the sugar amounts are extremely low, which could be explained by sugar degradation. Previous studies suggested that both glucose and xylose can be converted into other chemicals, such as levulinic acid, 5-hydroxymethylfurfural, humins [44, 45], and furfural [46, 47] under high temperature hydrothermal conditions.

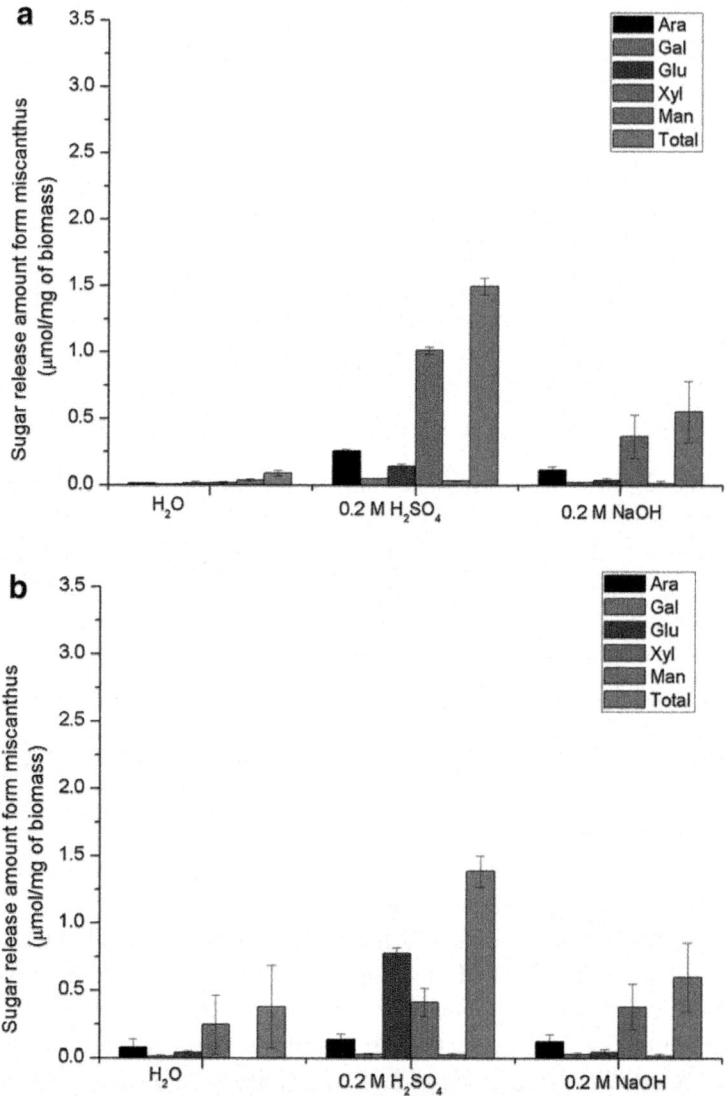

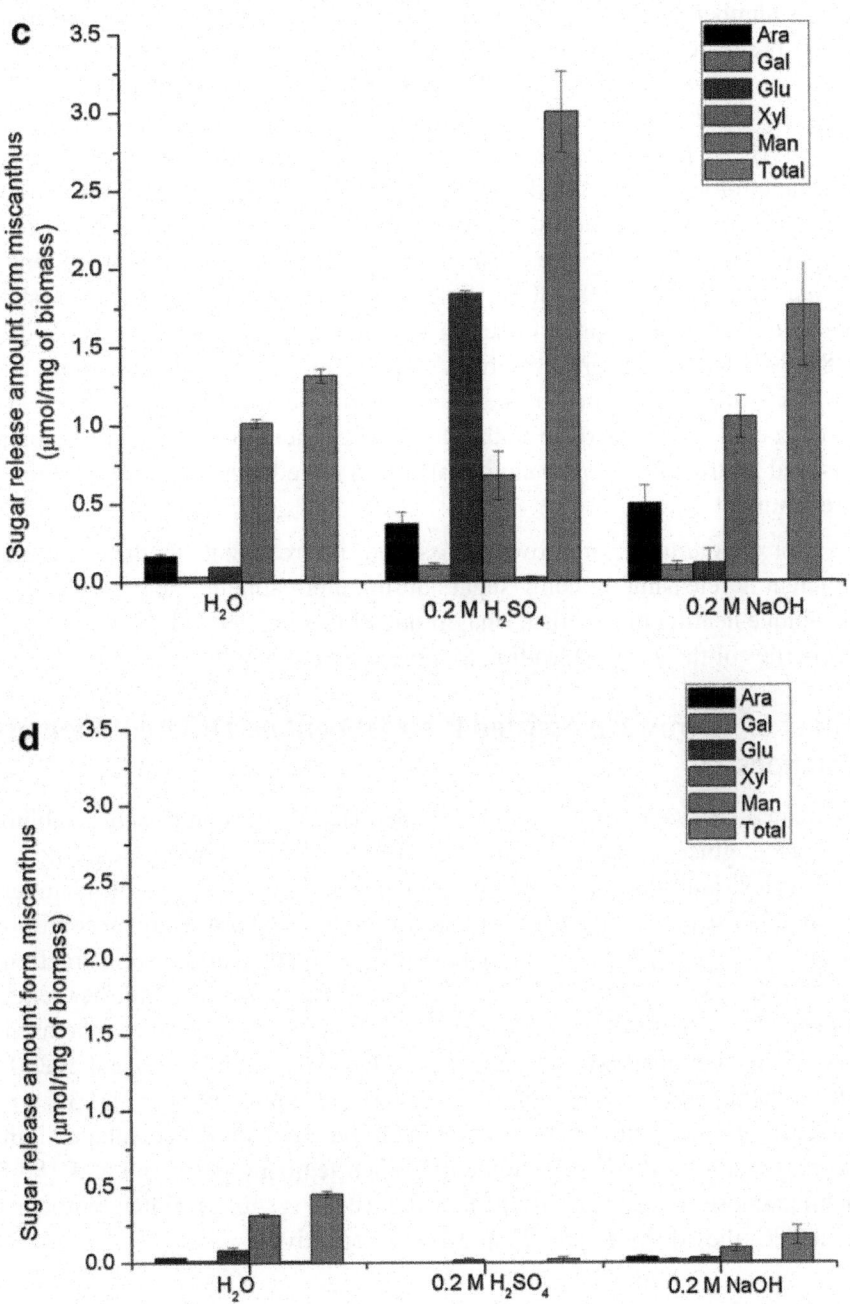

Figure. 2: Monosaccharides released to pretreatment media at various temperatures. **a** 130 °C; **b** 160 °C; **c** 180 °C; **d** 200 °C

A number of pretreatments have been studied on *Miscanthus* before. Yu et al. pretreated *Miscanthus* by using aqueous ammonia/hydrogen peroxide under lower temperature (90–150 °C) with longer holding times (1–4 h), and the results showed lower cellulose removal during the pretreatment (2.4–19.1 %) [48]. Haverty et al. studied peroxide/formic acid assisted pretreament for *Miscanthus* under autothermal conditions, and the results showed 0.3–4.37 % cellulose removal across conditions assayed [49]. One of our co-author, Gomez et al. studied conventional thermo-chemical pretreatment for *Miscanthus* material (20–180 °C, holding time 40 min), and their results shows 6–12 mg reducing sugar release/mg biomass (yield from total carbohydrate: 1.88–3.76 %) when temperature is 180 °C [50]. The reducing sugar yield in this work is 19 times higher within half the time than the result from Gomez et al. In comparison with other pretreatment methods, our microwave assisted pretreatment led to significant yield of reducing sugar release during pretreatment process.

For *Miscanthus*, microwave assisted pretreatment is therefore more efficient in releasing reducing sugars during pretreatment; the reason could be its unique heating mechanism (magnitude of heating depends on the dielectric properties of the subject) leading to more efficient biomass decomposition.

Effect of Microwave Assisted Pretreatment on Different Biomass Fractions

Untreated *Miscanthus* has 42 % of hemicellulose, comprising arabinose, galactose, glucose, xylose, mannose, galacturonic acid and glucuronic acid, with xylose and glucose as major components. Figure 3 shows the changes in hemicellulose percentage in the biomass after different pretreatment conditions. At 130 °C, H_2SO_4 reduced the hemicellulose fraction in the solid fraction to 21 % (Fig. 3a). At 160 °C, the hemicellulose percentage was further reduced to 14.7 % by 0.2 M H_2SO_4. When the temperature was further increased to 180 °C, the hemicellulose fraction was reduced to around 10 % and at 200 °C all the hemicellulose was removed after H_2SO_4 pretreatment (Fig. 3c, d). In the case of water and NaOH pretreatment, hemicellulose percentage slightly decreased after 130 °C pretreatments. When temperature is 160 and 180 °C, hemicellulose percentages are further reduced (Fig. 3b, c). When temperature is 200 °C, they decreased to 25 and 16 % respectively.

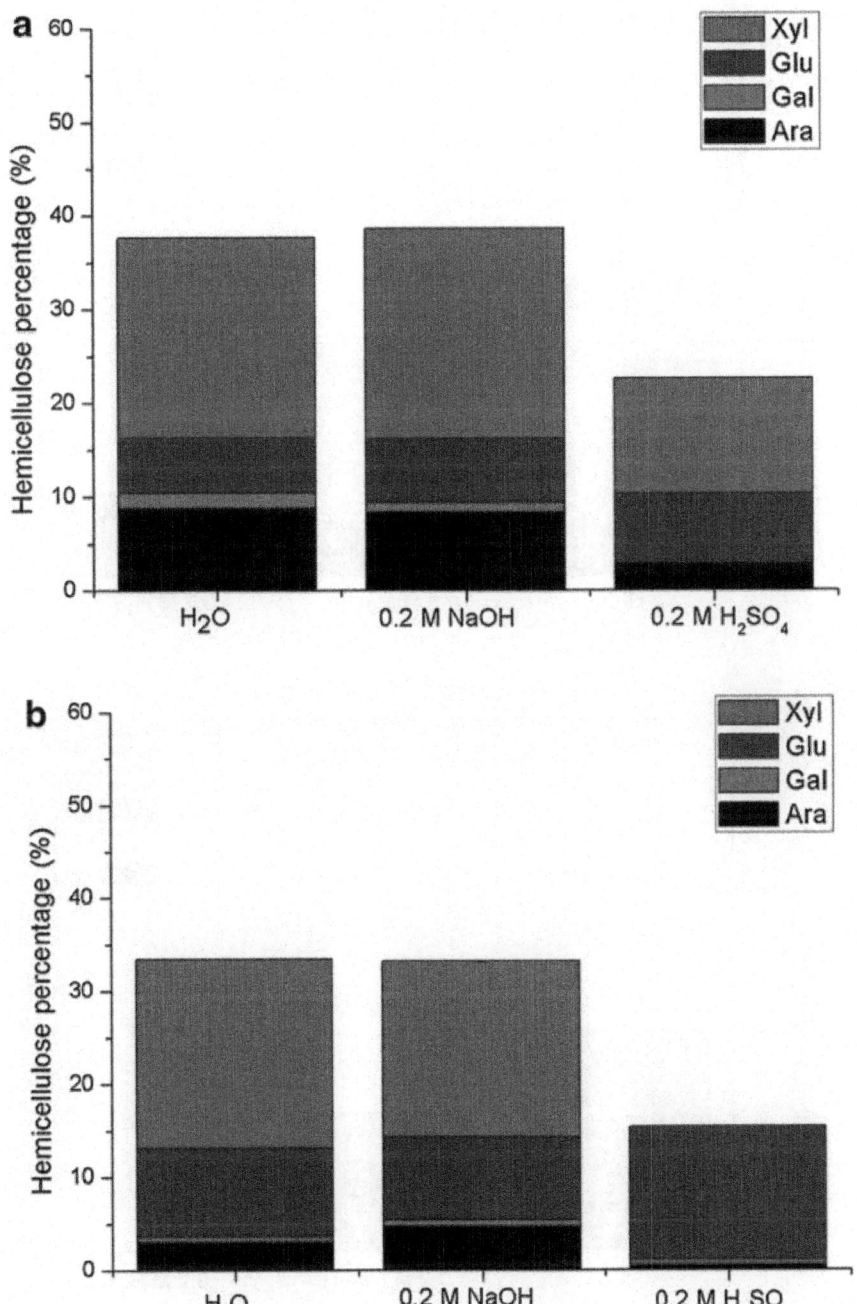

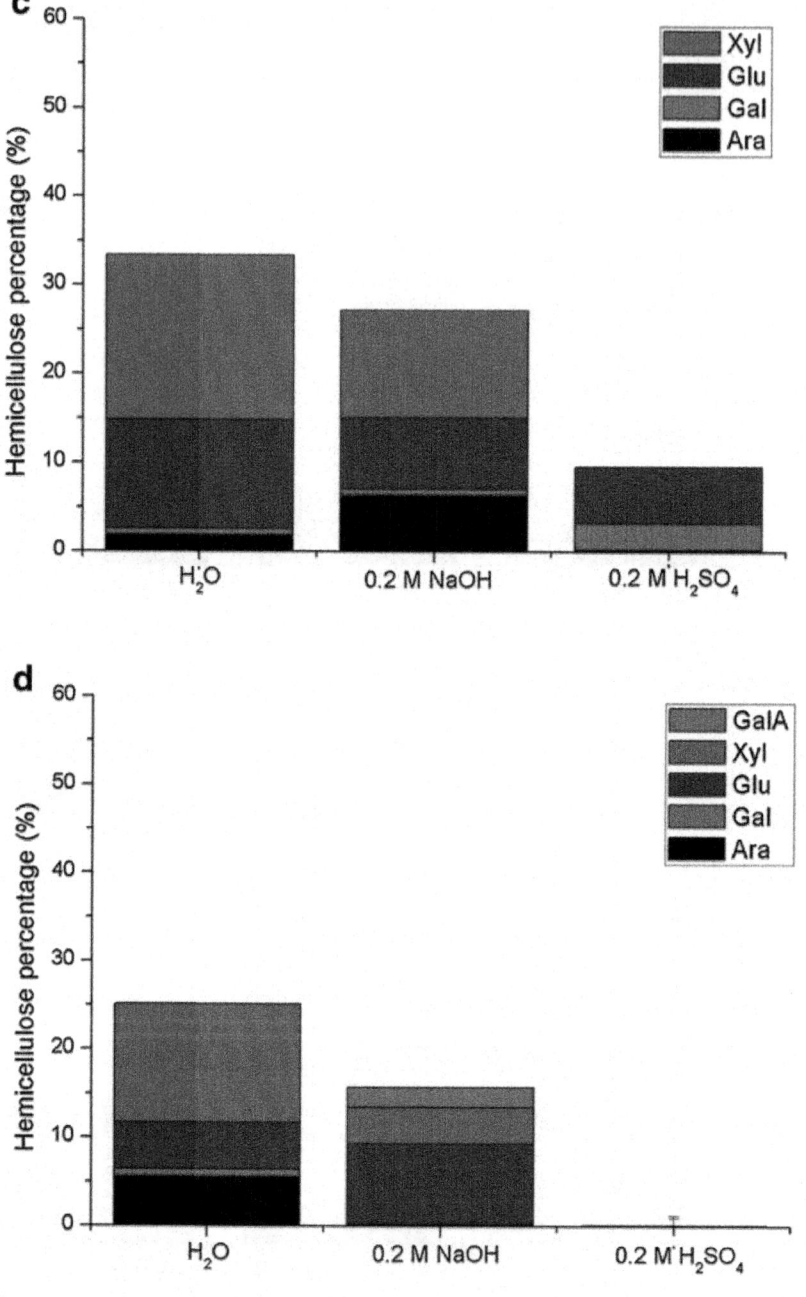

Figure. 3: Hemicellulose percentages after different temperature pretreatments. **a** 130 °C; **b** 160 °C; **c** 180 °C; **d** 200 °C (the standard deviations are presented in Additional file 2: Table S1).

In general, increasing of pretreatment temperature removed increasing proportions of hemicellulose from biomass. Water and NaOH pretreatments have similar effects on hemicellulose degradation, in agreement with previous results on monosaccharides in the pretreatment media (Fig. 2). H_2SO_4 removes hemicellulose more efficiently into pretreatment media, and when temperature was 200 °C, the hemicellulose was completely removed from biomass.

Lignin is composed of phenolic units, although it has multiple potentials for its use as a product feedstock or as a fuel in its own right, it is also considered as a barrier for the efficient enzyme hydrolysis of biomass [51]. Hence, the presence of lignin is considered one of the most important factors limiting the hydrolysis of lignocellulose [22]. Alkaline and oxidative treatments, such as alkaline peroxide and lime and oxygen, have been utilized to remove lignin [27, 52, 53].

In untreated *Miscanthus*, lignin represents 304 mg/g biomass (Fig. 4a). NaOH removed lignin more efficiently than water at lower pretreatment temperatures. All pretreatments conditions remove the same amount of lignin at 200 °C. The lignin removal is up to 221 mg/g of biomass when the temperature was 200 °C in all pretreatments. At 180 °C, the lignin content of biomass pretreated with acid was considerably higher. This could be explained by lignin extraction from the inner regions of the cell wall, and subsequent condensations and re-deposition on the surface as reported for wood samples [54]. In comparison with the results from conventional thermo-chemical pretreatment of *Miscanthus* (180 °C, 40 min) by Gomez et al., and similar amount of lignin is removed by 0.2 M NaOH pretreatment, whereas more lignin (210–240 mg/g biomass) is presented in the biomass after water and 0.2 M H_2SO_4 pretreatments [50]. Under our conditions, the distribution and structure of lignin could be changed under microwave assisted acidic conditions, and the results show a decreasing lignin amount as measured by using acetyl bromide methods. [55] The other possible explanation is that the ester linkages between polysaccharides and lignin were cleaved by microwave effect, leading to the partial solubilisation of lignin.

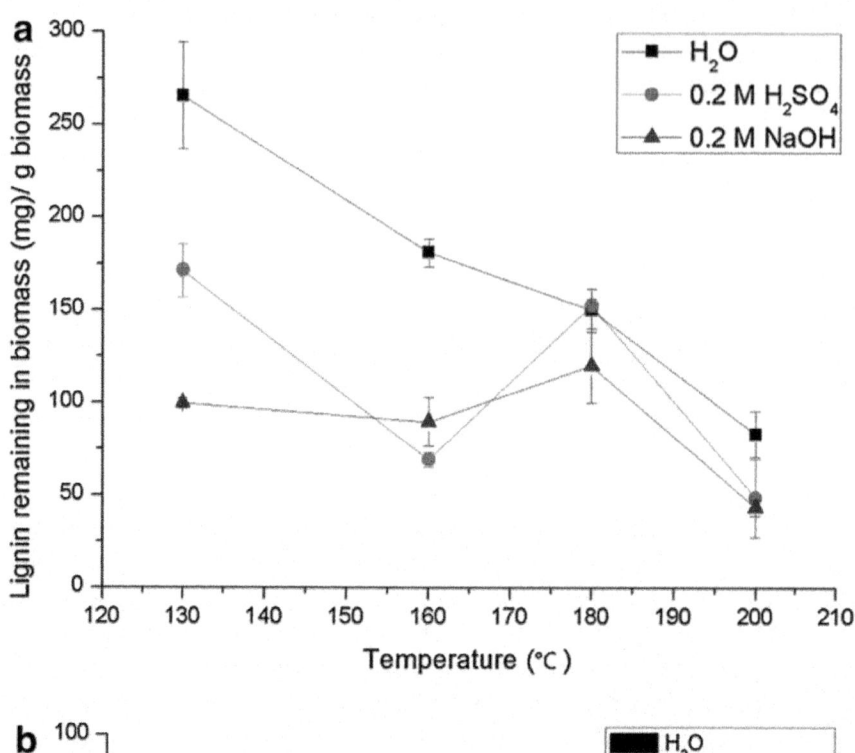

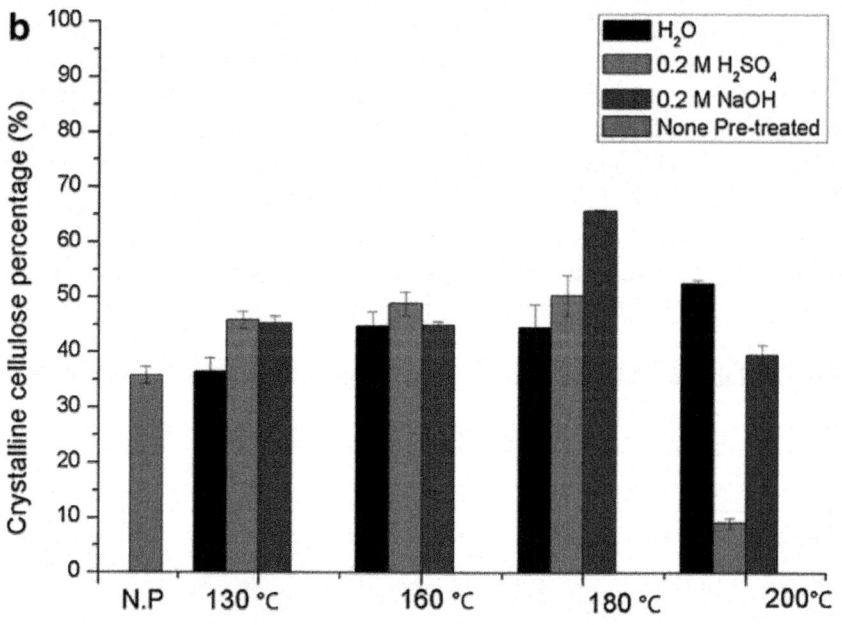

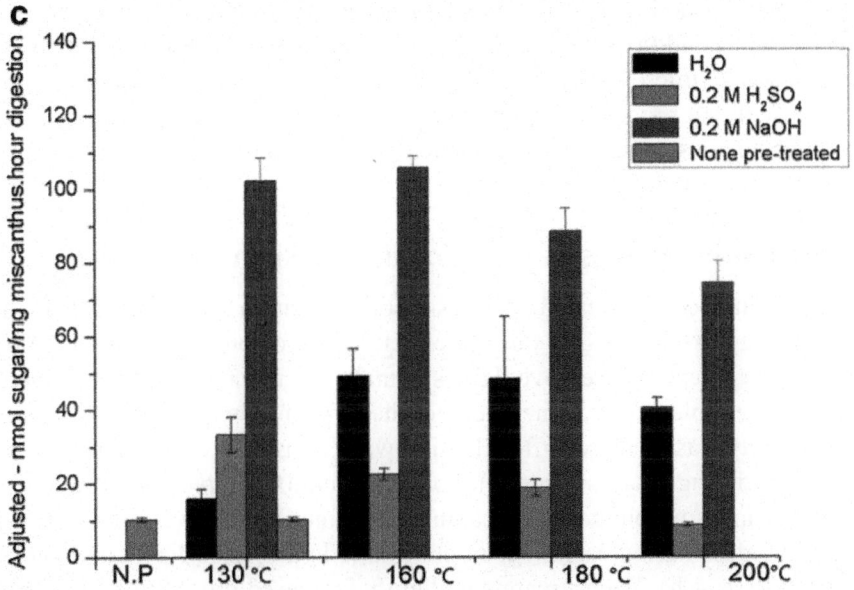

Figure. 4: a Lignin amount remaining in biomass after pretreatment under different temperature; **b** crystalline cellulose percentage after various pretreatment; **c** digestibility analysis of pre-treated biomass. *NP* stands for none pretreatment.

Pretreatment is an important step to make cellulose more accessible to cellulases, enhancing glucose production [26]. The percentage of crystalline cellulose in raw *Miscanthus* is 36 % (Fig. 4b). Microwave assisted water pretreatment has little effect on crystalline cellulose when the temperature was 130 °C. When the temperature was increased to 160, 180, and 200 °C, it increased to 44, 45, and 53 % due to lignin and hemicellulose removal. Under H_2SO_4 pretreatment, crystalline cellulose percentages in solid fraction are similarly enhanced when temperatures were between 130 and 180 °C, but dropped remarkably to 9 % when pretreatment temperature was 200 °C, showing that under more severe acid condition (200 °C), crystalline cellulose was degraded and carbonized (see "SEM"). In the case of NaOH pretreatment, the crystalline cellulose percentage in solid fraction was enhanced to 67 % when the pretreatment temperature was 180 °C, in good agreement with the extensive removal of lignin and hemicellulose observed in these conditions. At 200 °C, the crystalline cellulose percentage was 39 %, showing a reduction of cellulose crystallinity at this temperature [26]. Under microwave irradiation the heat is produced by direct interaction between polar part of biomass and oscillating electromagnetic field. The cellulose fibres could be described as being ionic conducting (crystalline) and non-conducting (amorphous) [40]. A very ordered hydrogen bonded network is contained in the crystalline cellulose which could lead to a proton transport network under an electromagnetic field under right condition [56]. Therefore, the crystalline cellulose is able to act as an active microwave absorber, promoting the biomass decomposition. Along with the process of lignin/hemicellulose removal, crystalline cellulose percentage goes up, enhancing the microwave absorbing effect and promoting biomass degradation.

Digestibility Analysis of Solid Fraction of Biomass

Digestibility of pretreated biomass solid fraction were measured and compared, in order to find out the effectiveness of pretreatment conditions. The pretreatment is to remove lignins to make the remaining biomass fraction more accessible to the enzymes in the enzyme hydrolysis. *Miscanthus* digestibility was increased after all microwave assisted pretreatments, albeit to widely differing extents (Fig. 4c). For untreated *Miscanthus*, the digestibility is 10.25 nmol/mg biomass h, meaning 10.25 nmol glucose is produced from 1 mg biomass during each hour of enzymatic hydrolysis (the total enzymatic hydrolysis is 4 h). Water treatment slightly increased digestibility at 130 °C. It was further enhanced to 40–50 nmol/mg biomass h when the pretreatment temperature was increased from 160 to 200 °C. In the case of H_2SO_4, the digestibility was marginally increased when the holding temperature was 130 °C, thereafter it declines with the temperature. At 200 °C, the digestibility was only 8.7 nmol/mg biomass h. Conversely, NaOH pretreatment remarkably improves *Miscanthus* digestibility. At 130–160 °C, the digestibility of NaOH pre-treated *Miscanthus* was 10 times higher than that of untreated biomass. Because of the delignification effect of NaOH, alkaline pretreated *Miscanthus* with low lignin percentage and higher cellulose percentage generates more sugar in the hydrolysis process. The difference in saccharification after acid or alkali pretreatments can be explained by the fact that the easily hydrolysed sugars are released into the pretreatment liquor, reducing the amount of sugar available for enzymatic digestion.

SEM Analysis of Microwave Pre-Treated *Miscanthus*

Scanning electron microscope was used to study the morphologic characteristics of raw and pre-treated *Miscanthus*. Figure 5 shows micrographs of the surface of raw *Miscanthus* particles, which present flat and smooth surface.

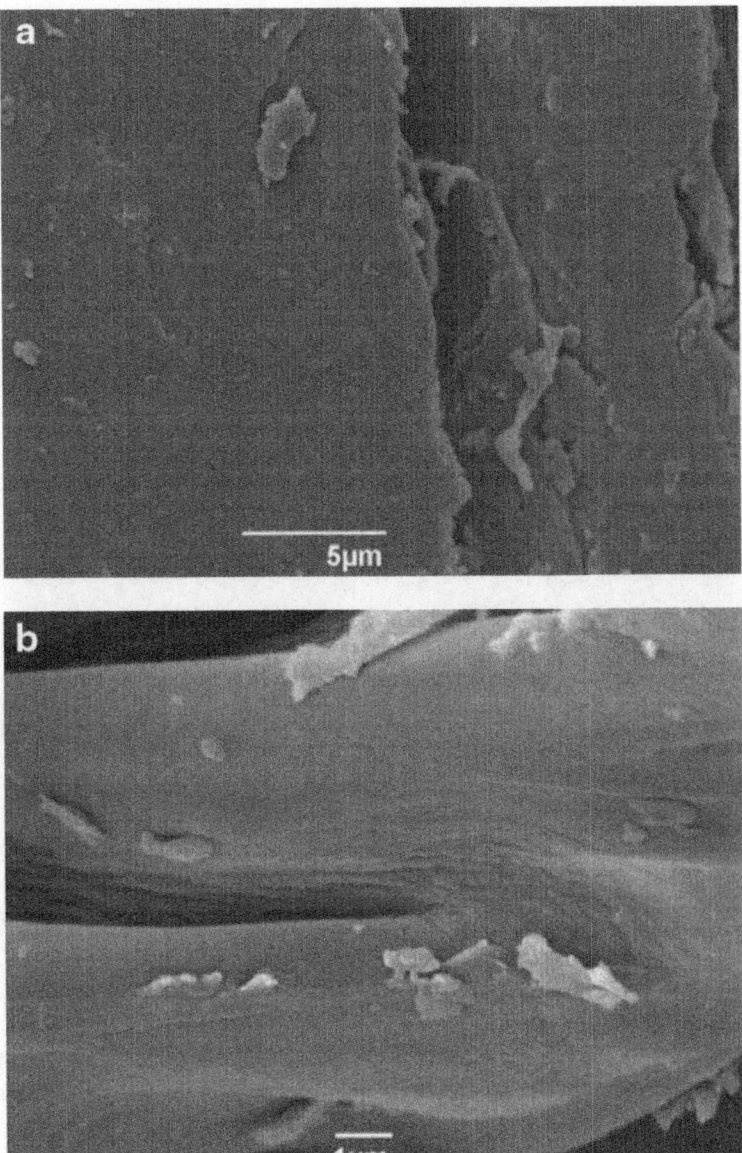

Figure. 5: Surface images of the untreated *Miscanthus* obtained by SEM. Flat surface of a fibre showing, **a** *bar scale* 5 μm; **b** *bar scale* 1 μm.

Images from *Miscanthus* samples pretreated with 0.2 M NaOH under various temperatures are shown in Fig. 6. When pretreatment temperature was 130 °C, parallel strips and small amounts of lignin deposits appear on the biomass surface, indicating that NaOH has a mild influence on the biomass surface (Fig. 6a; Additional file 1: Figure S1). When the temperature was increased to 160 °C, lignin deposits form on the biomass surface. In contrast, at 180 °C, the biomass surface becomes rough, with more exposed cellulose fibres, due to the more complete removal of hemicellulose and lignin (Fig. 6c). When pretreatment temperature was enhanced to 200 °C, a different type of lignin deposit was observed on the biomass surface (Fig. 6d). Despite the fact that similar amounts of lignin in solid fraction of *Miscanthus* are reduced after various 0.2 M NaOH pretreatments (Fig. 6), the impacts of alkali on the biomass surface are remarkably distinctive at different temperatures.

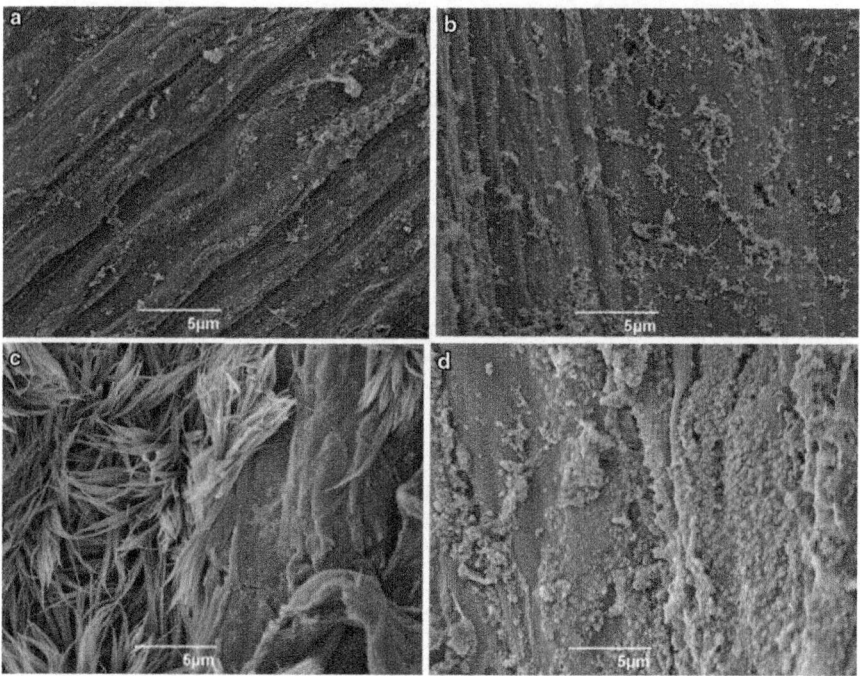

Figure. 6: Surface images obtained by SEM on *Miscanthus* treated with 0.2 M NaOH pretreatment under various temperature; microwave power: 300 W; magnification *scale bar*: 5µm. **a** 130 °C; **b** 160 °C; **c** 180 °C; **d** 200 °C.

Mild water and acid conditions do not change the surface of *Miscanthus* particles (data not shown). Figure 7 shows the surface images of *Miscanthus* after water pretreatment and 0.2 M H_2SO_4 pretreatment at 200 °C. Parallel strips appear on the biomass surface, in addition to the appearance of lignin deposits. Higher magnification shows (Fig. 7c) that the size of these deposits is larger than those observed with NaOH pretreatment at 200 °C (Additional file 1: Figure S1d). Samples treated with H_2SO_4 at 200 °C were carbonized with characteristic spheres appearing on biomass surface [57].

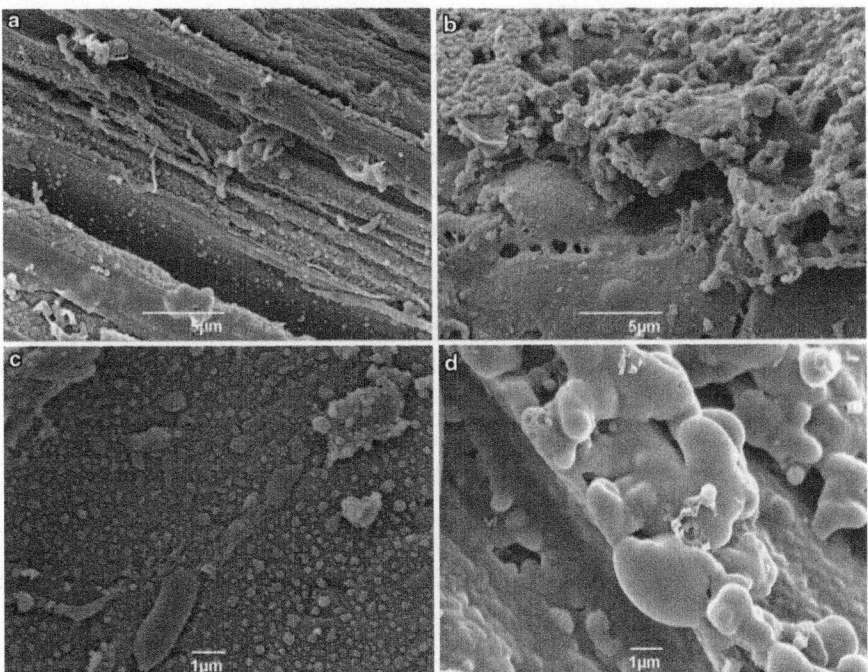

Figure. 7: Surface images obtained by SEM on *Miscanthus* treated with water and 0.2 M H_2SO_4 pretreatment at 200 °C; microwave power: 300 W. **a** Water pretreatment; magnification *bar scale* was 5 μm; **b** 0.2 M H_2SO_4 pretreatment; magnification *bar scale* is 5 μm; **c** water pretreatment; magnification *bar scale* is 1 μm; **d** 0.2 M H_2SO_4 pretreatment; magnification *bar scale* is 1 μm.

Simultaneous Saccharification Fermentation (SSF) of Hydrother-mal-Microwave Pre-Treated Samples

The sugar mixtures resulting from biomasses pretreated using different processing alternatives can be more or less amenable to fermentation into ethanol. We investigated the ethanol production of *Miscanthus* after microwave pretreatment at 180 °C using SSF. We choose this temperature due to the larger effect observed in lignin and hemicellulose removal, and changed the retention times in the microwave to vary the energy applied on the samples. Figure 8 shows the time course of ethanol production in SSF of *Miscanthus* with or without microwave assisted pretreatment. Water pretreated *Miscanthus* gives rise to very low ethanol production, regardless of the increasing pretreatment time and incubation time (Fig. 8a). In the case of H_2SO_4 pretreatment, dramatic differences can be observed when holding time was increased from 5 to 20 min (Fig. 8b). Ethanol production was 143 mg/g biomass when pretreatment holding time was 5 min. Longer pretreatment holding time reduced ethanol production drastically. It could be due to the inhibitors which are produced in the pretreatment process, such as furfural and 5-hydroxymethylfurfural [58]. The other explanation is that the majority of digestible sugars are released during the pretreatment process, meaning the left biomass residue is less digestible. The biomass was washed with ethanol in order to remove possible inhibitors produced in the pretreatment process. Therefore, the first explanation is less likely. The result here is in agreement with the previous results of digestibility study that H_2SO_4 pretreated biomass material is less digestible (Fig. 4c). However, NaOH has a completely different effect. As can be seen from Fig. 8c, when pretreatment time is 5 min, very small amount of ethanol is produced. When the pretreatment time increases from 10 to 20 min, significant amount of ethanol is increasingly produced. An outstanding amount of ethanol production (152 mg/g biomass) is obtained when pretreatment time is 15 min and incubation time is 48 h. The results are in good agreement with previous digestibility study and biomass morphological study that biomass is more digestible due to NaOH performance on biomass structure (Figs. 4c, 6c).

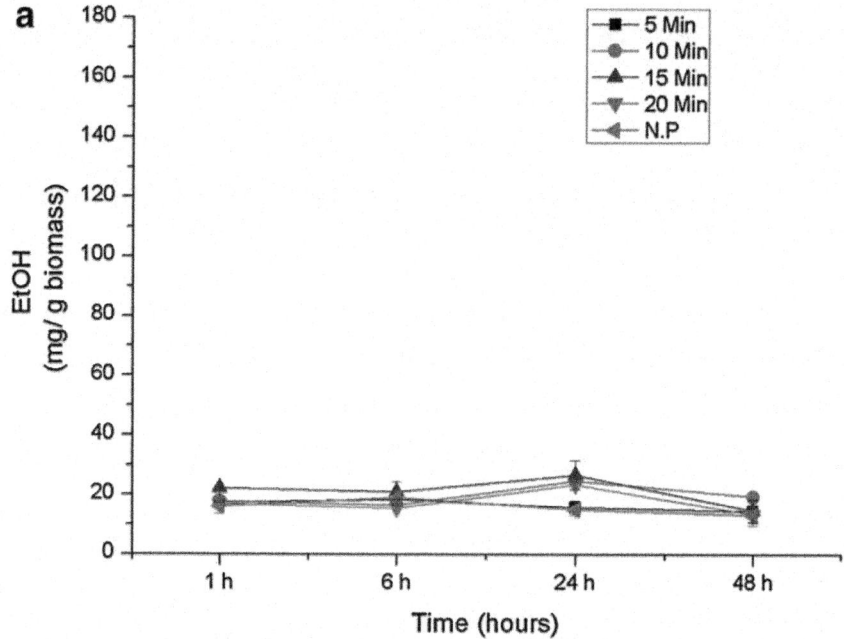

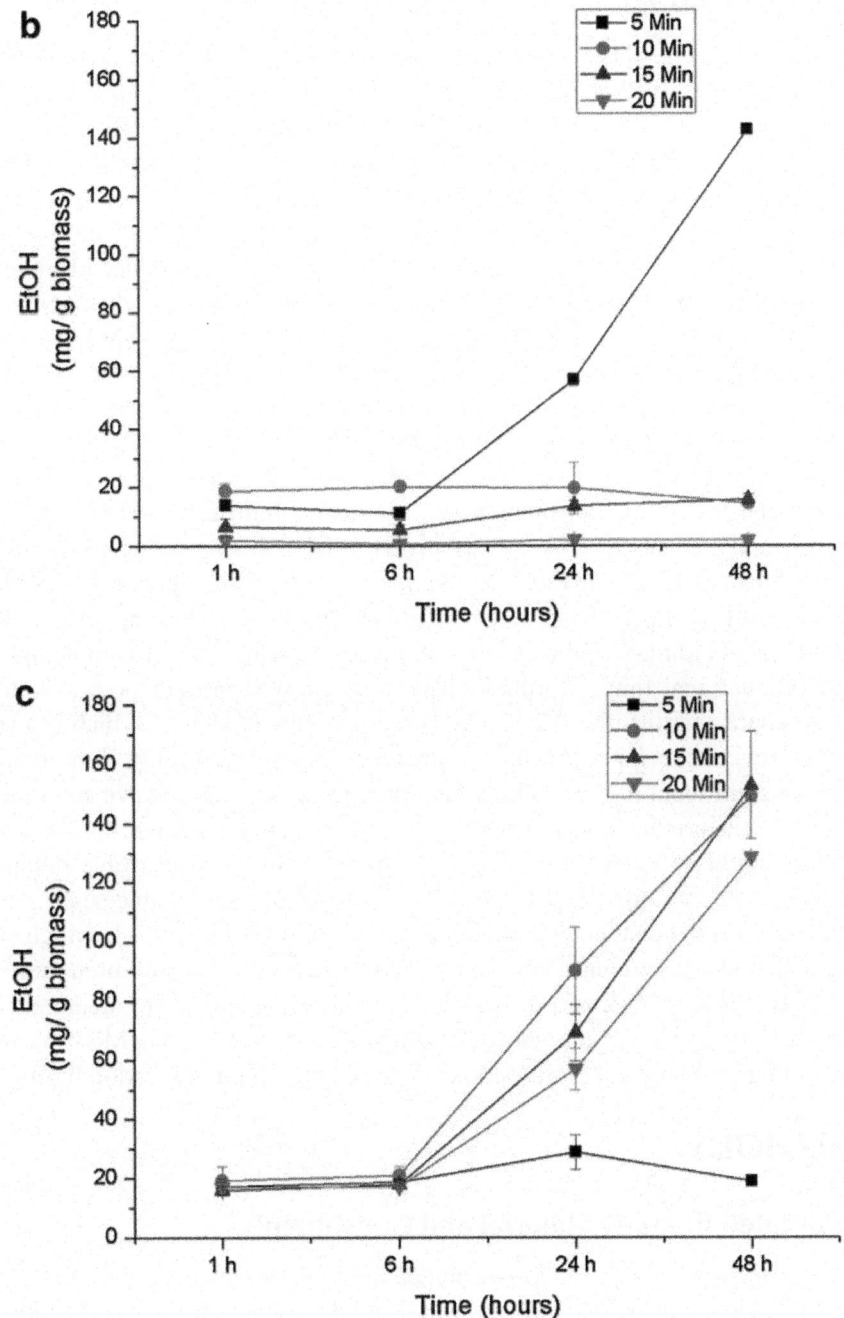

Figure. 8: Ethanol production of untreated and pre-treated *Miscanthus* over 48 h incubation time; **a** water pretreatment; **b** H₂SO₄ pretreatment; **c**NaOH pretreatment.

CONCLUSION

Miscanthus is one of the most promising energy crops in Europe and improving processing alternatives is a priority for second generation biofuel production. In this work, we tested microwave assisted pretreatments in the presence of water, H_2SO_4 and NaOH. Different temperature was assayed here, and the maximum sugar yield (73 %) is obtained by using 0.2 M H_2SO_4 at 180 °C, which is 17 times higher than conventional heating pretreatment within half time less. It was highlighted that xylose and glucose were selectively produced by tuning pretreatment temperature or media, and significant amount of glucose (yield: 47 %) was obtained from available carbohydrate when 0.2 M H_2SO_4 was used for pretreatment under 180 °C. The temperature has a strong influence on the lignin removal process, as different form of lignin deposits are observed from SEM images of biomass surface. Additionally, lignin removal process was improved with microwave assistance, especially in the case of H_2O and H_2SO_4. Due to the effective removals of lignin and hemicellulose, NaOH pretreatment significantly enhances *Miscanthus* digestibility, which was up to 10 times higher than that of untreated *Miscanthus*. It is worth mentioning that the fermentability of pretreated *Miscanthus* is more than 7 times higher than that of untreated biomass, and it can be optimized by changing pretreatment media and pretreatment time. Morphological study showed more exposed biomass fibres characteristic after 0.2 M NaOH pretreatment at 180 °C, which is a very important feature for following enhanced biomass digestibility. Temperature plays a significant role in pretreatment process. Under microwave conditions, 180 °C is a crucial point in the biomass degradation process, as the polar groups could be involved in a localized rotation in the microwave radiation and promote biomass degradation [41]. In our study, remarkable sugar yields and promising bioethanol production were achieved at 180 °C, which was identified as the optimal condition for our microwave assisted pretreatment. Overall, this work extensively studied the microwave assisted pretreatment for *Miscanthus*, and the results showed promising potential of using Microwave to assist thermo-chemical conversion of biomass to second generation biofuels.

METHODS

Untreated Biomass Material and Constituents

Miscanthus giganteus was grown under field conditions near York, UK, and harvested at maturity. The biomass was ground using a hammer mill to produce and average particle size of 100 μm × 57 μm. The biomass compositions of untreated *Miscanthus* are cellulose (34 % ± 2.5 %), hemicellulose (42 ± 2.8 %), lignin (30.4 ± 2 %) and ash (0.83 ± 0.03 %).

Microwave Pretreatment Methods

The pretreatment was conducted in a CEM Discover microwave machine (CEM Discover SP-D, US). The CEM microwave reactor vessel (30 ml) was charged with 0.4 g of *Miscanthus* and 16 ml H_2SO_4 or NaOH solution (0.2, 0.4 and 1). Pretreatment was carried out at various temperatures between 130 and 200 °C for a period of 20 min. After pretreatment, the liquid fraction was separated from biomass solid fraction by filtration. Liquid fraction was neutralized with 150 mM $Ba(OH)_2$ or 1 M HCl. The solid fraction was rinsed with absolute ethanol (3 × 10 ml) and dried at 50 °C overnight.

The CEM MARS 6 (CEM, US) was used for the scale up microwave pretreatment in order to perform Simultaneous saccharification and fermentation (SSF). 3 g of *Miscanthus* and 80 ml H_2SO_4 (0.2 M) or NaOH solution (0.2 M) were added in a 100 ml reaction vessel. The pretreatment was carried out at 180 °C for various holding time (5–20 min). Same procedures (separation, washing and drying) as above were performed in order to conduct SSF.

Analysis of Carbohydrates in Liquid Fraction

The liquid fraction resulting from alkaline and acid pretreatments was neutralized by 1 M HCl or 1 M NaOH solutions, respectively. Then monosaccharide in the liquid fraction was analysed by High Performance Ion Exchange Chromatography (Dionex, ICS-3000PC, Thermal scientific, USA) equipped with electrochemical detector to quantify the sugar content [59].

Hemicellulose Content

Hemicellulose was analysed by using the method developed by Foster et al. [60]. 4 mg of biomass were hydrolysed using 0.5 ml 2 M TFA. After flushing the vial with dry Argon, the vials were heated at 100 °C for 4 h. TFA was removed completely by centrifugal evaporation with fume extraction overnight. Then the biomass was washed with 500 µl of Propan-2-ol twice. The samples were re-suspended in 200 µl of deionised water. After thorough mixing, the supernatant was put into a new tube for analysis using Dionex in order to measure monosaccharides in hemicellulose.

Lignin Quantification

Lignin was quantified as follows: 3.5 mg of un/pretreated biomass was dissolved in 250 µl acetyl bromide solution (25 % v/v acetyl bromide/glacial acetic acid), then 1 ml 2 M NaOH and 175 µl hydroxylamine HCl in a 5 ml volumetric flask were added. The solution was taken to 5 ml with acetic acid and diluted 10 times. The absorbance was read at 280 nm and the percentage of lignin calculated using the following formula [61]:

ABSL%={abs/(coeff×pathlength)}×{(total   volume×100%)/   biomass weight}ABSL%={abs/(coeff×pathlength)}×{(total   volume×100%)/   biomass weight}

Coefficient = 17.75; path length = 1 cm; total volume = 5 ml; biomass = 3.5 mg.

Analysis of Crystalline Cellulose

To determine the percentage of crystalline cellulose in biomass, 10 mg untreated or pre-treated biomass was hydrolysed using 500 μl 2 M TFA (trifluoroacetic acid) at 100 °C for 4 h. The solids were subsequently hydrolysed using Acetic acid:Nitric Acid:Water (8:1:2 v/v) at 100 °C for 30 min. Finally, the resulting residue was hydrolysed into glucose using 175 μl 72 % H_2SO_4 at room temperature for 45 min and then diluted to 3.2 % H_2SO_4 and heated at 120 °C for 2 h. Anthrone assay was used to quantify corresponding glucose [60].

Analysis of Biomass Digestibility

The digestibility of biomass was investigated by using a high throughput saccharification assay which is based on a robotic platform that can carry out the enzymatic digestion and quantification of the released sugars in a 96-well plate format. Enzymatic hydrolysis was carried out using an enzyme cocktail with a 4:1 (v/v) ratio of Celluclast and Novozyme 188 (both Novozymes, Bagsvaerd, Denmark). The enzymes were filtered using a Hi-Trap desalting column (GE Healthcare, Little Chalfont, Buckinghamshire, UK) before use. 0.1 mg biomass was hydrolysed for 8 h with 250 μl enzyme cocktail, in 250 ml of 25 mM sodium acetate buffer at pH 4.5, at 30 °C. Determination of sugars released after hydrolysis was performed using a modification of the method by Anton and Barrett using 3-methyl-2-benzothiazolinonehydrozone (MTBH) [62].

Morphological Studies

Morphological characteristics of the raw materials and pre-treated biomass residue were studied using a scanning electron microscope fitted with tungsten filament cathode (JEOL, JSM-6490LV, Japan). Samples were sputter-coated with 7 nm Au/Pd to facilitate viewing by SEM. Images were obtained under vacuum, using a 5 kV accelerating voltage and a secondary electron detector.

Simultaneous Saccharification and Fermentation (SSF)

The SSF experiments were performed in 100 ml conical flask with 1 g of untreated/pre-treated biomass, 10.75 ml sterile water, 0.250 ml NaOAc buffer, 1 ml enzyme solution (4:1 v/v ratio of Celluclast and Novozyme 188, Novozymes, Bagsvaerd, Denmark), 1.365 ml ATCC medium, and 200 μl yeast extract (the yeast was grown until optical density 5 and added). The flasks were incubated for 48 h in a shaking incubator under at 30 °C and 150 rpm. Samples for ethanol determination were collected after 1, 6, 24, 48 h in GC vials containing 500 μL of 1 M sodium chloride and 0.04 % 1-propanol. Ethanol concentrations at different time points were measured by using a standard curve of ethanol.

AUTHORS' CONTRIBUTIONS

DJM and LDG planned the pre-treatments. ZZ carried out the pretreatments and the determination of monosaccharides, chemical compositions, scanning electron microscopy study and SSF, as well as the analysis of the results. RH measured biomass digestibility. ZZ, DJM and LDG prepared the manuscripts. SMM coordinated the overall study. All authors suggested modifications to the draft. All authors read and approved the final manuscript.

ACKNOWLEDGEMENTS

The present work was funded by the European Community's Seventh Framework Programme SUNLIBB (FP7/2007-2013) under the grant agreement no. 251132. The authors gratefully acknowledge Dr Andrew Hunt for advising and support.

ADDITIONAL FILES

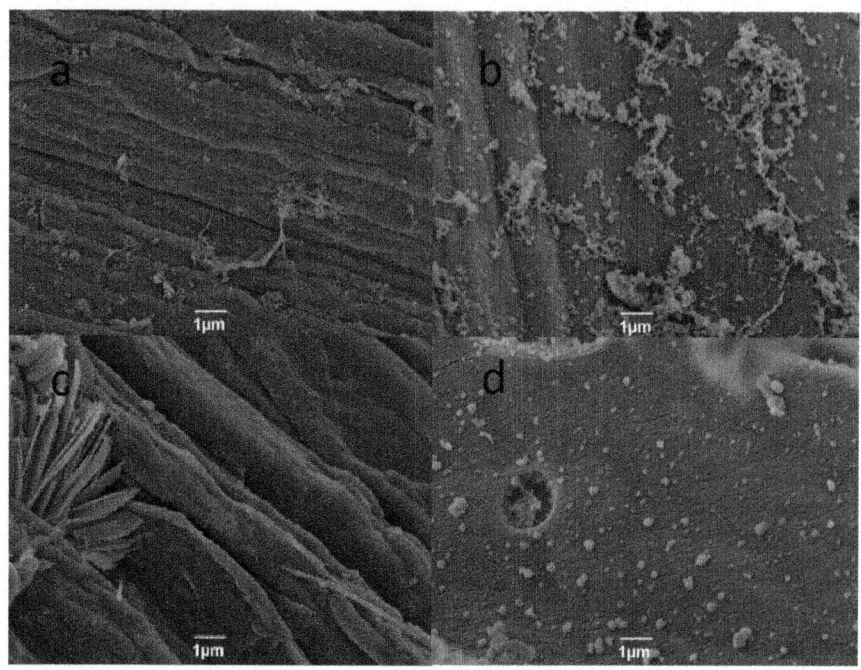

Figure S1. Surface images obtained by SEM on *Miscanthus* treated with 0.2 M NaOH pretreatment under various temperature; microwave power: 300 W; magnification scale bar: 1 μm. **a** 130 °C; **b** 160 °C; **c** 180 °C; **d** 200 °C.

Table S1. Standard deviation table for hemicellulose percentage in biomass

130 °C	Ara	Gal	Glu	Xyl	
H_2O	0.00104	1.13E-04	9.69E-04	0.04993	
NaOH	0.00272	2.80E-04	6.80E-04	0.01523	
H_2SO_4	0.00137	6.21E-06	0.00361	0.02608	
160 °C	Ara	Gal	Glu	Xyl	
H_2O	5.38E-04	6.30E-06	0.0034	0.01294	
NaOH	0.00488	4.90E-05	0.01154	0.0481	
H_2SO_4	3.41E-05	7.39E-06	0.00365	-	
180 °C	Ara	Gal	Glu	Xyl	

H2O	1.14E-04	1.63E-05	0.00754	0.00622	
NaOH	0.00292	6.71E-05	3.02E-04	0.00681	
H$_2$SO$_4$	3.36E-05	0.00416	0.01079	-	
200 °C	Ara	Gal	Glu	Xyl	GalA
H$_2$O	0.00161	5.96E-05	0.00108	0.00341	-
NaOH	4.75E-06	-	0.00116	6.06E-04	6.96E-06
H$_2$SO$_4$	-	-	0.92622	1.15678	-

REFERENCES

1. Ragauskas AJ, Williams CK, Davison BH, Britovsek G, Cairney J, Eckert CA, Frederick WJ, Hallett JP, Leak DJ, Liotta CL et al (2006) The path forward for biofuels and biomaterials. Science 311(5760):484–489

2. Arthur J, Ragauskas MN, Kim DH, Eckert Charles A, Hallett Jason P, Liotta CL (2006) From wood to fuels: integrating biofuels and pulp production. Ind Biotechnol 2(1):55–65

3. Bensah EC, Mensah M (2013) Chemical pretreatment methods for the production of cellulosic ethanol: technologies and innovations. Int J Chem Eng 2013:21

4. Rivers DB, Emert GH (1987) Lignocellulose pretreatment—a comparison of wet and dry ball attrition. Biotechnol Lett 9(5):365–368

5. Sun Y, Cheng JY (2002) Hydrolysis of lignocellulosic materials for ethanol production: a review. Bioresour Technol 83(1):1–11

6. Kovacs K, Macrelli S, Szakacs G, Zacchi G (2009) Enzymatic hydrolysis of steam-pretreated lignocellulosic materials with Trichoderma atroviride enzymes produced in-house. Biotechnol Biofuels 2:14.doi:10.1186/1754-6834-2-14

7. Balat M, Balat H, Oz C (2008) Progress in bioethanol processing. Prog Energ Combust 34(5):551–573

8. Alizadeh H, Teymouri F, Gilbert TI, Dale BE (2005) Pretreatment of switchgrass by ammonia fiber explosion (AFEX). Appl Biochem Biotech 121:1133–1141

9. Wang W, Yuan TQ, Wang K, Cui BK, Dai YC (2012) Combination of biological pretreatment with liquid hot water pretreatment to enhance enzymatic hydrolysis of Populus tomentosa. Bioresour Technol 107:282–286

10. Kim KH, Hong J (2001) Supercritical CO_2 pretreatment of lignocellulose enhances enzymatic cellulose hydrolysis. Bioresour Technol 77(2):139–144

11. Schacht C, Zetzl C, Brunner G (2008) From plant materials to ethanol by means of supercritical fluid technology. J Supercrit Fluids 46(3):299–321

12. Morais ARC, da Costa Lopes AM, Bogel-Łukasik R (2015) Carbon dioxide in biomass processing: contributions to the green biorefinery concept. Chem Rev 115(1):3–27

13. Nikolic S, Mojovic L, Rakin M, Pejin D, Pejin J (2011) Utilization of microwave and ultrasound pretreatments in the production of bioethanol from corn. Clean Techn Environ Policy 13(4):587–594

14. Xu N, Zhang W, Ren SF, Liu F, Zhao CQ, Liao HF, Xu ZD, Huang JF, Li Q, Tu YY et al (2012) Hemicelluloses negatively affect lignocellulose crystallinity for high biomass digestibility under NaOH and H_2SO_4 pretreatments in Miscanthus. Biotechnol Biofuels 5:58

15. Canilha L, Santos VTO, Rocha GJM, Silva JBAE, Giulietti M, Silva SS, Felipe MGA, Ferraz A, Milagres AMF, Carvalho W (2011) A study on the pretreatment of a sugarcane bagasse sample with dilute sulfuric acid. J Ind Microbiol Biot 38(9):1467–1475

16. Kaar WE, Holtzapple MT (2000) Using lime pretreatment to facilitate the enzymic hydrolysis of corn stover. Biomass Bioenergy 18(3):189–199

17. da Costa Lopes AM, Bogel-Łukasik R (2015) Acidic ionic liquids as sustainable approach of cellulose and lignocellulosic biomass conversion without additional catalysts. Chemsuschem 8(6):947–965

18. da Costa Lopes A, Joao K, Morais AR, Bogel-Lukasik E, Bogel-Lukasik R (2013) Ionic liquids as a tool for lignocellulosic biomass fractionation. Sustain Chem Process 1(1):3

19. Chen L, Sharifzadeh M, Mac Dowell N, Welton T, Shah N, Hallett JP (2014) Inexpensive ionic liquids: [HSO4]–based solvent production at bulk scale. Green Chem 16(6):3098–3106

20. Agnieszka Brandt JG, Hallett JP, Welton T (2012) Deconstruction of lignocellulosic biomass with ionic liquids. Green Chem 15:550–583

21. Ju Y-H, Huynh L-H, Kasim NS, Guo T-J, Wang J-H, Fazary AE (2011) Analysis of soluble and insoluble fractions of alkali and subcritical water treated sugarcane bagasse. Carbohydr Polym 83(2):591–599

22. Rezende CA, de Lima MA, Maziero P, deAzevedo ER, Garcia W, Polikarpov I (2011) Chemical and morphological characterization of

sugarcane bagasse submitted to a delignification process for enhanced enzymatic digestibility. Biotechnol Biofuels 4:1–18

23. Hong B, Xue GX, Weng LQ, Guo X (2012) Pretreatment of moso bamboo with dilute phosphoric acid. Bioresources 7(4):4902–4913

24. Chen W-H, Tu Y-J, Sheen H-K (2011) Disruption of sugarcane bagasse lignocellulosic structure by means of dilute sulfuric acid pretreatment with microwave-assisted heating. Appl Energy 88(8):2726–2734

25. Jensen JR, Morinelly JE, Gossen KR, Brodeur-Campbell MJ, Shonnard DR (2010) Effects of dilute acid pretreatment conditions on enzymatic hydrolysis monomer and oligomer sugar yields for aspen, balsam, and switchgrass. Bioresour Technol 101(7):2317–2325

26. Mittal A, Katahira R, Himmel ME, Johnson DK (2011) Effects of alkaline or liquid-ammonia treatment on crystalline cellulose: changes in crystalline structure and effects on enzymatic digestibility. Biotechnol Biofuels 4:41. doi:10.1186/1754-6834-4-41

27. Banerjee G, Car S, Scott-Craig JS, Hodge DB, Walton JD (2011) Alkaline peroxide pretreatment of corn stover: effects of biomass, peroxide, and enzyme loading and composition on yields of glucose and xylose. Biotechnol Biofuels 4:16. doi:10.1186/1754-6834-4-16

28. Keshwani DR, Cheng JJ (2010) Microwave-based alkali pretreatment of switchgrass and coastal bermudagrass for bioethanol production. Biotechnol Progr 26(3):644–652

29. Gupta R, Lee YY (2010) Investigation of biomass degradation mechanism in pretreatment of switchgrass by aqueous ammonia and sodium hydroxide. Bioresour Technol 101(21):8185–8191

30. Zhu S, Wu Y, Yu Z, Chen Q, Wu G, Yu F, Wang C, Jin S (2006) Microwave-assisted alkali pre-treatment of wheat straw and its enzymatic hydrolysis. Biosyst Eng 94(3):437–442

31. de la Hoz A, Diaz-Ortiz A, Moreno A (2005) Microwaves in organic synthesis. Thermal and non-thermal microwave effects. Chem Soc Rev 34(2):164–178

32. Lancaster M (2002) Green chemistry: an introductory text. R Soc Chem, Cambridge

33. Macquarrie DJ, Clark JH, Fitzpatrick E (2012) The microwave pyrolysis of biomass. Biofuels Bioprod Biorefin 6(5):549–560

34. Lu X, Xi B, Zhang Y, Angelidaki I (2011) Microwave pretreatment of rape straw for bioethanol production: focus on energy efficiency. Bioresour Technol 102(17):7937–7940

35. Wu YY, Fu ZH, Yin DL, Xu Q, Liu FL, Lu CL, Mao LQ (2010) Microwave-assisted hydrolysis of crystalline cellulose catalyzed by biomass char sulfonic acids. Green Chem 12(4):696–700

36. Ma H, Liu WW, Chen X, Wu YJ, Yu ZL (2009) Enhanced enzymatic saccharification of rice straw by microwave pretreatment. Bioresour Technol 100(3):1279–1284

37. Lee W-C, Kuan W-C (2015) Miscanthus as cellulosic biomass for bioethanol production. Biotechnol J 10(6):840–854

38. Zhu SD, Wu YX, Yu ZN, Zhang X, Li H, Gao M (2006) The effect of microwave irradiation on enzymatic hydrolysis of rice straw. Bioresour Technol 97(15):1964–1968

39. Luo J, Cai M, Gu T (2013) Pretreatment of lignocellulosic biomass using green ionic liquids. In: Gu T (ed) Green biomass pretreatment for biofuels production. Springer, Netherlands, pp 127–153

40. Budarin VL, Clark JH, Lanigan BA, Shuttleworth P, Macquarrie DJ (2010) Microwave assisted decomposition of cellulose: a new thermochemical route for biomass exploitation. Bioresour Technol 101(10):3776–3779

41. Fan JJ, De Bruyn M, Budarin VL, Gronnow MJ, Shuttleworth PS, Breeden S, Macquarrie DJ, Clark JH (2013) Direct microwave-assisted hydrothermal depolymerization of cellulose. J Am Chem Soc 135(32):11728–11731

42. Ju YH, Huynh LH, Kasim NS, Guo TJ, Wang JH, Fazary AE (2011) Analysis of soluble and insoluble fractions of alkali and subcritical water treated sugarcane bagasse. Carbohydr Polym 83(2):591–599

43. Lee YY, Iyer P, Torget RW (1999) Dilute-acid hydrolysis of lignocellulosic biomass. In: Tsao GT, Brainard AP, Bungay HR, Cao NJ, Cen P, Chen Z, Du J, Foody B, Gong CS, Hall P (eds) Recent progress in bioconversion of lignocellulosics, vol 65. Springer, Berlin, pp 93–115

44. Jing Q, Lu XY (2008) Kinetics of non-catalyzed decomposition of glucose in high-temperature liquid water. Chin J Chem Eng 16(6):890–894

45. Girisuta B, Janssen LPBM, Heeres HJ (2006) Green chemicals: a kinetic study on the conversion of glucose to levulinic acid. Chem Eng Res Des 84(5):339–349

46. Moller M, Schroder U (2013) Hydrothermal production of furfural from xylose and xylan as model compounds for hemicelluloses. Rsc Adv 3(44):22253–22260

47. Jing Q, Lü X (2007) Kinetics of non-catalyzed decomposition of d-xylose in high temperature liquid water*. Chin J Chem Eng 15(5):666–669

48. Yu G, Afzal W, Yang F, Padmanabhan S, Liu Z, Xie H, Shafy MA, Bell AT, Prausnitz JM (2014) Pretreatment of Miscanthus × giganteus using aqueous ammonia with hydrogen peroxide to increase enzymatic hydrolysis to sugars. J Chem Technol Biotechnol 89(5):698–706

49. Haverty D, Dussan K, Piterina AV, Leahy JJ, Hayes MHB (2012) Autothermal, single-stage, performic acid pretreatment of Miscanthus x giganteus for the rapid fractionation of its biomass components into a lignin/hemicellulose-rich liquor and a cellulase-digestible pulp. Bioresour Technol 109:173–177

50. Gómez L, Vanholme R, Bird S, Goeminne G, Trindade L, Polikarpov I, Simister R, Morreel K, Boerjan W, McQueen-Mason S (2014) Side by side comparison of chemical compounds generated by aqueous pretreatments of maize stover, Miscanthus and sugarcane bagasse. Bioenerg Res 7(4):1466–1480

51. Donohoe BS, Decker SR, Tucker MP, Himmel ME, Vinzant TB (2008) Visualizing lignin coalescence and migration through maize cell walls following thermochemical pretreatment. Biotechnol Bioeng 101(5):913–925

52. Chang VS, Nagwani M, Kim CH, Holtzapple MT (2001) Oxidative lime pretreatment of high-lignin biomass—poplar wood and newspaper. Appl Biochem Biotech 94(1):1–28

53. Mosier N, Wyman C, Dale B, Elander R, Lee YY, Holtzapple M, Ladisch M (2005) Features of promising technologies for pretreatment of lignocellulosic biomass. Bioresour Technol 96(6):673–686

54. Li JB, Henriksson G, Gellerstedt G (2007) Lignin depolymerization/repolymerization and its critical role for delignification of aspen wood by steam explosion. Bioresour Technol 98(16):3061–3068

55. Selig MJ, Viamajala S, Decker SR, Tucker MP, Himmel ME, Vinzant TB (2007) Deposition of lignin droplets produced during dilute acid pretreatment of maize stems retards enzymatic hydrolysis of cellulose. Biotechnol Progr 23(6):1333–1339

56. Fan J, Debruyn M, Zhu Z, Budarin V, Gronnow M, Gomez LD, Macquarrie D, Clark J (2013) Microwave-enhanced formation of glucose from cellulosic waste. Chem Eng Process Process Intensif 71:37–42

57. Titirici M-M, Antonietti M, Baccile N (2008) Hydrothermal carbon from biomass: a comparison of the local structure from poly- to monosaccharides and pentoses/hexoses. Green Chem 10(11):1204–1212

58. Larsson S, Palmqvist E, Hahn-Hägerdal B, Tengborg C, Stenberg K, Zacchi G, Nilvebrant N-O (1999) The generation of fermentation

inhibitors during dilute acid hydrolysis of softwood. Enzyme Microb Tech 24(3–4):151–159

59. Jones L, Milne JL, Ashford D, McQueen-Mason SJ (2003) Cell wall arabinan is essential for guard cell function. Proc Natl Acad Sci USA 100(20):11783–11788

60. Foster CE, Martin TM, Pauly M (2010) Comprehensive compositional analysis of plant cell walls (lignocellulosic biomass) Part II: carbohydrates. J Vis Exp 37:e1837. doi:10.3791/837

61. Foster CE, Martin TM, Pauly M (2010) Comprehensive compositional analysis of plant cell walls (lignocellulosic biomass) Part I: lignin. J Vis Exp 37:e1745. doi:10.3791/1745

62. Gomez LD, Whitehead C, Barakate A, Halpin C, McQueen-Mason SJ (2010) Automated saccharification assay for determination of digestibility in plant materials. Biotechnol Biofuels 3(1):23

CITATION

CHAPTER 1

Shu-Kun Lin, "The Nature of the Chemical Process. 1. Symmetry Evolution –Revised Information Theory Similarity Principle Ugly Symmetry," Int. J. Mol. Sci. 2001, 2(1), 10-39; doi:10.3390/i2010010.

CHAPTER 2

M. Opgenorth, W. McDermott, P. Laz and C. Lengsfeld, "A Combined Probabilistic and Optimization Approach for Improved Chemical Mixing Systems Design," *Engineering*, Vol. 3 No. 6, 2011, pp. 643-652. doi:10.4236/eng.2011.36077.

CHAPTER 3

M. Zimmer, K. Krieg and J. Rentsch, "Online Process Control of Alkaline Texturing Baths: Determination of the Chemical Concentrations," *American Journal of Analytical Chemistry*, Vol. 5 No. 3, 2014, pp. 205-210. doi:10.4236/ajac.2014.53025.

CHAPTER 4

László Dobos, András Király, and János Abonyi, "Economic-Oriented Stochastic Optimization in Advanced Process Control of Chemical Processes," The Scientific World Journal, vol. 2012, Article ID 801602, 10 pages, 2012. doi:10.1100/2012/801602.

CHAPTER 5

Mohammad Gias, "Study on the color levelness of silk fabric dyed with vegetable dyes," Sustainable Chemical Processes20153:10, DOI: 10.1186/s40508-015-0038-1.

CHAPTER 6

Robert Wojcieszak, Francesco Santarelli, Sébastien Paul, Franck Dumeignil, Fabrizio Cavani and Renato V Gonçalves, "Recent developments in maleic acid synthesis from bio-based chemicals," Sustainable Chemical Processes20153:9, DOI: 10.1186/s40508-015-0034-5.

CHAPTER 7

Christoph R Müller, Andreas Rosen and Pablo Domínguez de María, "Multi-step enzyme-organocatalyst C–C bond forming reactions in deep-eutectic-solvents: towards improved performances by organocatalyst design," Sustainable Chemical Processes20153:12, DOI: 10.1186/s40508-015-0039-0.

CHAPTER 8

B A Fachri, R M Abdilla, C B Rasrendra and H J Heeres, "Experimental and modelling studies on the uncatalysed thermal conversion of inulin to 5-hydroxymethylfurfural and levulinic acid," Sustainable Chemical Processes20153:8, DOI: 10.1186/s40508-015-0035-4.

CHAPTER 9

Joyeeta Mukherjee and Munishwar Nath Gupta, "Biocatalysis for biomass valorization," Sustainable Chemical Processes20153:7, DOI: 10.1186/s40508-015-0037-2.

CHAPTER 10

Yong Huang, Esteban E. Ureña-Benavides, Afrah J. Boigny, Zachary S. Campbell, Fiaz S. Mohammed, Jason S. Fisk, Bruce Holden, Charles A. Eckert, Pamela Pollet and Charles L. Liotta, "Butadiene sulfone as 'volatile', recyclable dipolar, aprotic solvent for conducting substitution and cycloaddition reactions," Sustainable Chemical Processes20153:13, DOI: 10.1186/s40508-015-0040-7.

CHAPTER 11

Felipe K Sutili, Halliny S Ruela, Daniel De O Nogueira, Ivana CR Leal, Leandro SM Miranda and Rodrigo OMA De Souza, "Enhanced production of fructose ester by biocatalyzed continuous flow process," Sustainable Chemical Processes20153:6, DOI: 10.1186/s40508-015-0031-8.

CHAPTER 12

Sandeep S Nair, JY Zhu, Yulin Deng and Arthur J Ragauskas, "High performance green barriers based on nanocellulose," Sustainable Chemical Processes20142:23, DOI: 10.1186/s40508-014-0023-0.

CHAPTER 13

Juan Carlos Serrano-Ruiz, Rafael Luque, Juan Manual Campelo and Antonio A. Romero, "Continuous-Flow Processes in Heterogeneously Catalyzed Transformations of Biomass Derivatives into Fuels and Chemicals," Challenges 2012, 3, 114-132; doi:10.3390/challe3020114.

CHAPTER 14

Zongyuan Zhu, Duncan J. Macquarrie, Rachael Simister, Leonardo D. Gomez and Simon J. McQueen-Mason, "Microwave assisted chemical pretreatment of *Miscanthus* under different temperature regimes," Sustainable Chemical Processes2015 3:15, DOI: 10.1186/s40508-015-0041-6.

INDEX